Creating a Physical Biology

MATHEMATISCH-PHYSIKALISCHE KLASSE · FACHGRUPPE VI
Biologie
Neue Folge · Band 1 · Nr. 13

NACHRICHTEN

VON DER

GESELLSCHAFT DER WISSENSCHAFTEN

ZU

GÖTTINGEN

Über die Natur der Genmutation und der Genstruktur

Von

N. W. Timoféeff-Ressovsky, K. G. Zimmer und M. Delbrück

1935
WEIDMANNSCHE BUCHHANDLUNG / BERLIN SW 68

Einzelpreis RM. 3.—

Cover of Max Delbrück's personal copy of the reprint of the Three-Man Paper, Delbrück Papers, California Institute of Technology, box 36, folder 12. Courtesy of the Archives, California Institute of Technology.

Creating a Physical Biology

The Three-Man Paper and
Early Molecular Biology

EDITED BY PHILLIP R. SLOAN
AND BRANDON FOGEL

THE UNIVERSITY OF CHICAGO PRESS   CHICAGO AND LONDON

QH
506
.C73
2011

PHILLIP R. SLOAN is professor emeritus in the Program of Liberal Studies and the Program in History and Philosophy of Science at the University of Notre Dame. BRANDON FOGEL is the Collegiate Assistant Professor in the Division of Humanities at the University of Chicago.

The University of Chicago Press, Chicago 60637
The University of Chicago Press, Ltd., London
© 2011 by The University of Chicago
All rights reserved. Published 2011.
Printed in the United States of America
20 19 18 17 16 15 14 13 12 11   1 2 3 4 5

ISBN-13: 978-0-226-76782-6 (cloth)
ISBN-13: 978-0-226-76783-3 (paper)
ISBN-10: 0-226-76782-5 (cloth)
ISBN-10: 0-226-76783-3 (paper)

Library of Congress Cataloging-in-Publication Data

Creating a physical biology : the Three-Man Paper and early molecular biology / edited by Phillip R. Sloan and Brandon Fogel.
     p. cm.
  Includes bibliographical references and index.
  ISBN-13: 978-0-226-76782-6 (cloth : alk. paper)
  ISBN-10: 0-226-76782-5 (cloth : alk. paper)
  ISBN-13: 978-0-226-76783-3 (pbk. : alk. paper)
  ISBN-10: 0-226-76783-3 (pbk. : alk. paper) 1. Timofeev-Resovskii, N. V. (Nikolai Vladimirovich), 1900–1981. Über die Natur der Genmutation und der Genstruktur 2. Delbrück, Max. 3. Zimmer, Karl Günter, 1911– 4. Molecular biology—History—20th century. 5. Genetics—History—20th century. I. Zimmer, Karl Günter, 1911– II. Delbrück, Max. III. Sloan, Phillip R. IV. Fogel, D. Brandon. V. Timofeev-Resovskii, N. V. (Nikolai Vladimirovich), 1900–1981. Über die Natur der Genmutation und der Genstruktur. English.
  QH506.C73 2011
  572.809—dc23
                                                           2011023775

♾ This paper meets the requirements of ANSI/NISO Z39.48-1992 (Permanence of Paper).

Contents

Preface

The origins of this volume lie in a term project by one of the editors, Brandon Fogel, for a graduate seminar taught by the other editor, Phillip Sloan. The task was to produce a translation and commentary on a portion of the collaborative paper by Nikolai Timoféeff-Ressovsky, Karl Zimmer, and Max Delbrück, "On the Nature of Gene Mutation and Gene Structure" of 1935. The scarcity of this paper in the original—a text more often mentioned than read—and its role as the background for Erwin Schrödinger's influential essay *What Is Life?* (1944), suggested that a full translation of the text was needed. Summer support by the University of Notre Dame Graduate School allowed Fogel to devote substantial effort on the translation over two summers. A symposium at the History of Science Society annual meeting in 2005 allowed the authors of the accompanying essays to develop their views on the historical context and significance of this paper, and ultimately to develop a collaborative strategy for this volume as a whole. National Science Research grant 0646732 and a research leave from the University of Notre Dame gave support for Sloan in carrying out the research for his chapter as a portion of a larger project on the history of the conception of life in modern molecular biology.

As the project deepened, it became increasingly necessary to understand the significance of this work in its historical context, without the filter provided by Schrödinger's interpretation and a presentist history of molecular biology. This required situating the paper in the complex world of the Kaiser Wilhelm Society Institutes in the turbulent period of the final days of the Weimar Republic. It also required understanding the paper outside of the context of the discovery of the structure of DNA. Read in its historical context, the paper's intellectual origins were revealed to be as closely tied to molecular research on photosynthesis as they were to genetics.

The result was a change of our own perspectives on this paper, and as developed in the chapters that follow, the genesis and subsequent fate of this paper are not quite what has usually been assumed in the literature.

Completion of this work was possible only with the assistance of several individuals and institutions. Through the receipt of a Maurice A. Biot Archives Fund award from the California Institute of Technology, one of the authors, Daniel McKaughan, was able to make an initial visit to the Max Delbrück archives in Pasadena, and his research has benefited two of the other authors of this volume (Phillip Sloan and Nils Roll-Hansen). We wish to thank Judith Goodstein, Bonnie Ludt, and Loma Karklins of the California Institute of Technology library for additional assistance, including supplying the photograph of the only copy of the "Green Pamphlet" reprint—Max Delbrück's own copy—we have been able to locate. We also express our appreciation for assistance from the University of Chicago rare-book room in obtaining materials from the James Franck papers, and from the University of Delaware Library for important information on Kurt Wohl. We are indebted to Antje Kreienbring of Humboldt University in Berlin for valuable assistance in obtaining information about dissertation titles for Hans Gaffron, Karl Walter Kofink, and Karl Zimmer. Considerable assistance from Wulf Wulfhekel and Klaus Nippert of the Karlsruhe Institute of Technology was deeply appreciated in unraveling a mystery about the identity of "Werner" Kofink, about which more will be said later.

We are also indebted to correspondence and discussions with Ernst Peter Fischer of the University of Constanz and Peter Starlinger of the Max Delbrück Institute for Genetics in Cologne, who shared important information related to Delbrück. Peter Herrlich of the Leibniz Institute for Age Research in Jena was of considerable assistance in obtaining additional information and photographic materials on Karl Zimmer, with further assistance rendered by Dietrich Harder, Horst Jung, and Hermann Dertinger and the staff of the Max Planck Society archives. Diane Paul of the Museum of Comparative Zoology at Harvard has been a constant source of advice and assistance on all matters pertaining to Timoféeff-Ressovsky, including the sharing of printed materials. Mary Jo Nye of Oregon State University shared in unpublished form her paper on Michael Polanyi and scientific social groups in Berlin in the 1930s. Pnina Abir-Am of Brandeis University, Matthew Meselsohn of Harvard University, and Gino Segrè of the University of Pennsylvania have each offered encouragement and assistance at several points. Barbara Gaffron very kindly supplied unpublished materials on her father-in-law, Hans Gaffron. Elinor "Elly" Welt of Seattle, Washington, supplied important information on the historical sources of her novel *Berlin Wild*, centered at the Subunit for Genetic Research within the Kaiser Wilhelm Institute for Brain Research during the war years, which was based on extensive interviews with her husband, Peter Welt. Substantial assistance was rendered in developing the discussions of photosynthesis by comments from Kärin Nickelsen of the University of Berne, to whom we are also deeply indebted for an extraordinarily astute vetting of the translation.

We also acknowledge the excellent services of Tamara Bakerinwood of the University of Notre Dame, who rendered the entire German text of the Three-Man Paper in electronic form. We wish to thank Karen Merikangas Darling of the University of Chicago Press, who gave us considerable encouragement for the project at many points along the way, and her assistant Abby Collier for technical help in bringing this complex project to press. Helpful comments from Erik Peterson and James Barham on drafts and chapters is acknowledged. James Barham also completed the onerous task of preparing the bibliography for the translation of the paper itself from its cryptic original references, including verification of each citation. This has been a very substantial contribution to the project.

Phillip R. Sloan
Brandon Fogel

CHAPTER ONE

Introduction

Phillip R. Sloan and Brandon Fogel

The 1935 paper "On the Nature of Gene Mutation and Gene Structure" by Russian *Drosophila* geneticist Nikolai Timoféeff-Ressovsky (1900–1981) and two Germans, radiation physicist Karl Günther Zimmer (1911–88) and theoretical physicist Max Delbrück (1906–81), is a landmark in the history of the development of biophysics.[1] The paper was commonly known as the "Three-Man Work" or "Three-Man Paper" (*Dreimännerwerk*, *Dreimännerarbeit*), from which derives the acronym (3MP) that will be used throughout this volume. Though its particular scientific conclusions would not, at least in specific detail, stand the test of time, the 3MP was an important stimulus in the development of the new molecular biology, both methodologically and conceptually. Through a novel mixture of experimental genetics and chemistry with the most current theoretical physics, used to model the gene as a specifically molecular structure, and utilizing X-ray technology and contemporary "target" theory, the paper demonstrated the potential of such cross-disciplinary collaboration to provide detailed, physicalistic explanations of biological phenomena. Though its historical importance has never been in doubt, the paper has been largely inaccessible, due to its manner of publication. In presenting the first English translation, accompanied by analyses of its historical and philosophical significance, the current volume aims to provide access to this celebrated work and to contextualize its origins.

The 3MP was published in the *Reports* of the biology section of the Göttingen Academy of Sciences in 1935, a journal that folded after only three issues.[2] Delbrück later commented that to publish it in the *Nachrichten* at that time was to give it a "funeral first class,"[3] although, as Richard

Beyler details in his chapter below, this oft-quoted remark is an inaccurate account both of its impact and of its dissemination. In fact, the paper survived its initial publication quite well, through restatements of its arguments in subsequent publications, and physically extending its influence through the circulation of a large number of reprints among a select group of individuals. These reprints were enclosed in a green cover and soon became known as the "Green Pamphlet" (a reproduction of which is the frontispiece of this volume). In 1942, a copy of the Green Pamphlet was lent to Erwin Schrödinger by Paul Ewald, a physicist interested in crystallography who, like Schrödinger, was a refugee from Nazi Europe. Schrödinger drew heavily on the paper for his public lectures of February 1943, published in 1944 in his popular treatise on the nature of life and heredity, *What Is Life?* It is through that text, often credited with supplying an intellectual impetus for the migration of many physical scientists into biology after World War II, that much of the English-speaking world came to know of the 3MP and its contents.[4] However, as will be developed in detail below, Schrödinger misconstrued the paper's main theoretical conclusions, due to his own theoretical agenda in biophysics. He ignored clearly expressed reservations in the 3MP concerning the complex relationship between developmental and structural genetics in favor of a thoroughgoing reductionism, a position that many would later mistakenly assume was drawn from the 3MP.[5]

The purpose of this introductory discussion is to situate the 3MP within the broader context of issues that surrounded the debates over the relation of physics and biology in the mid-1930s, thereby allowing the reader to appreciate the paper's significance. It will also address some of the enduring philosophical questions that surround the application of physics to biology.

Following some historiographical reflections and a brief biographical sketch of the three authors, a discussion of the concept of the gene around 1930 will provide a background against which the paper can be read. These perspectives will supply a framework for the more detailed studies of the paper's historical and conceptual contexts in the chapters that follow.

Writing the History of Molecular Biology

Some attention to historiographical concerns is required in discussing the historical importance of the 3MP. A particular pitfall is the "presentism" of viewing phage and DNA research as the apotheosis of molecular biology; in such a view, the beginnings of modern biophysics are mere pre-history to the discovery of the structure of DNA in 1953.[6] But the early visions of the new molecular approach to biology were not limited to investigations of genetic phenomena, and as developed in detail in this volume, the paper emerged from a rich context of radiation research, medical physics, and other explorations of biophysics in the 1920s and 1930s. The first appearance of the term "molecular biology," in a report to the Rockefeller Foundation in 1938 by Warren Weaver, the director of the Natural Sciences Division of the Rockefeller Foundation from 1932 to 1955, refers to "the studies . . . in a relatively new field . . . in which delicate modern techniques are being used to investigate ever more minute details of certain life processes."[7] To encourage this, the Rockefeller Foundation during Weaver's tenure supplied much of the primary research funding for biophysical research both in the United States and abroad in the pre–World War II period.[8]

In the earliest intellectual histories of molecular biology, published in the 1960s, the 3MP was included unproblematically as part of the "origins" of modern molecular biology, as we see one of its authors, Karl Zimmer do in his retrospective discussion of the target theory. Likewise, Donald Fleming gives it a similar importance in a well-known analysis of the impact of the cross-disciplinary migration of physicists into biology.[9] Nonetheless, recent detailed scholarly work on the history of molecular biology has restricted the designator "molecular" to the post–World War II period, with its origins closely related to a complex interaction of institutional developments, technological inventions, and new experimental practices arising in the 1950s and 1960s.[10] Two recent examples of this new historiography illustrate this trend: Soraya de Chadarevian's study of the transformation of the Medical Research Council Unit for the Study of the Molecular Structure of Biological Systems at Cambridge University, which changed its name to the MRC Unit for Molecular Biology only in 1957, has highlighted the institutional factors in the definition of "molecular" biology.[11] A similar study is that of Bruno Strasser on the postwar development of biophysics at the University of Geneva, which details the transformation in 1963 of the Laboratoire de biophysique, previously devoted to electron microscopy, into a new Institut de biologie moléculaire.[12] These transformations are also located within the broader transnational "molecularization" of biology after World War II, with the first journal bearing this name, the *Journal of Molecular Biology*, commencing publication in 1959.[13] The shift in this recent historiography away from history of scientific ideas toward local microhistory, with emphasis on the material history of instruments (e.g., the electron microscope), on new techniques of analysis, and on experimental practices within laboratory groups, in interaction with funding agencies and local political contexts, has required new ways of approaching the 3MP.

Another historiographical issue concerns the attempt to identify the "origins" of a discipline by identification with certain key figures or even with key papers. As exemplified particularly by the now-famous Watson and Crick papers of 1953, much of the identification of such significance is retrospective, coming some time after the events themselves.[14] In the case of the 3MP, it can be argued that it was the retrospective reading of Schrödinger's later popularized presentation of important aspects of the paper in 1944, rather than the paper itself, that has played this role in "origins" stories. As detailed in Richard Beyler's chapter, the central theory of the 3MP, the target theory, did not, in fact, survive intact. But aspects of this theory were elaborated upon and institutionalized in a series of developments that preceded Schrödinger's exposition.

The 3MP intersects with several issues that have been of concern in the historiographical analysis of contemporary biophysics. First, it is the product of an "intellectual migration" from physics into biology by two of its authors, Max Delbrück and Karl Zimmer, the first moving from high-level theoretical physics, and the other from experimental radiology. These migrations were not those of refugee scientists driven from Europe by the rise of totalitarianism and moving into biology, the familiar story told in the better-known "origins" story of the history of molecular biology.[15] Rather, their discipline jumping occurred within the general confines of a single complex institution, the Kaiser Wilhelm Society (Kaiser-Wilhelm Gesellschaft) in Berlin.

The paper is also a result of major technological developments of the 1920s and 1930s, specifically the technology for producing artificial mutations in fruit flies through techniques of irradiation with X-rays and beta and gamma radiation. This required developments of precise techniques for targeting flies in varying stages of development, for administering specific quantities of radiation, and for measuring the outcomes by way of interbreeding experiments. The difficulties in solving these technical questions are easily overlooked. The development of special containers was required, as well as means of conducting precise radiation experiments on a small volume of biological material. The collaboration of Timoféeff-Ressovsky and Zimmer began in the early 1930s with techniques developed by others, particularly Hermann Muller, yet they attained a new

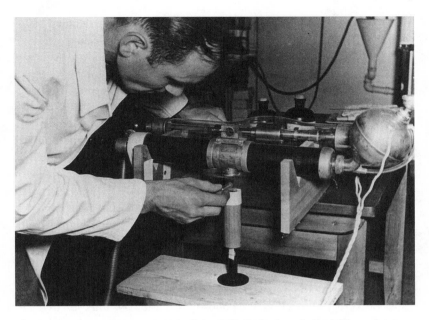

Assistant of the Genetics Division of the Kaiser Wilhelm Institute for Brain Research preparing an X-radiation experiment on *Drosophila* in 1938. Archives of the Max Planck Society, Berlin-Dahlem (Kaiser Wilhelm Institute for Brain Research, no. III/14).

level of quantitative precision through development of specific means of managing the flies in radiation experiments and administering different wavelengths of radiation.

The 3MP's origins also involve specific institutional issues associated with the interplay of distinct institutes within the Kaiser Wilhelm Society in the early 1930s; these are outlined in the Sloan chapter in this volume. The research for the paper was funded by the Emergency Fund for German Science, established through the initiative of Max Planck, Fritz Haber, and Ernst von Harnack in 1920, which served as a major agency for interwar German scientific research. Finally, the paper's origins and character cannot be fully understood without attention to the political conditions created by the rapid nazification of Germany after January 1933, which displaced numerous German academics for either ethnic or political reasons and resulted in the creation of the ephemeral interdisciplinary discussion seminar organized by Max Delbrück that ran from 1934 to 1937. It was from this seminar that the 3MP emerged. Different aspects of these issues will be explored in several chapters of this book.

If we must acknowledge important microhistorical and detailed histo-riographical concerns that surround the concept of "molecular" biology as a discipline, there are also "big picture" issues connected with origins of modern biophysics, whether the designator "molecular" biology is applied or not, that bear on the 3MP. The intellectual project of ontological reduc-tionism—the claim that life can be fully explained through the categories of the physical sciences—has major implications for the understanding of human identity, the nature of "life," and the many social and political issues surrounding the relations of organismic biology to analytic biology. It eventually involves the status of consciousness itself. These issues, with contemporary as well as historical interest, inevitably lead us back to an examination of the history of efforts to connect the biological and the physical sciences. The interaction of these domains, though centuries old, developed novel features and built on new assumptions in the 1920s and 1930s, when new forms of interaction between physics and biology devel-oped, spurred by the efforts of a group of individuals, primarily German-speaking, to bring together the new quantum mechanics with traditional issues of biological function, the philosophy of biology, and even psychol-ogy. For some, the resulting theoretical advances promised to provide the conceptual resolution of the long controversy over mechanism and vitalism that was particularly heated in the period after the inaugura-tion of neovitalism by embryologist-turned-philosopher Hans Driesch (1867–1941). Although such hopes were not quite fulfilled, examination of this dynamic theoretical discourse can help clarify the motivations of key players who managed to achieve various institutional realignments. It is within the larger intellectual nexus of the 1930s that one major measure of significance of the 3MP lies, and in this book we illuminate some of this little-known history, centering on this key early moment in what would become a vast cross-disciplinary conversation.

Dramatis Personae

The biographies of Timoféeff-Ressovsky, Zimmer, and Delbrück are enmeshed in the tumultuous political and social events of the 1930s and 1940s. Nikolai Vladimirovich Timoféeff-Ressovsky, often rendered in German publications as Nikolai W., the largest contributor to the 3MP in terms of length, was the best-known scientist of the three in 1935.[16] He was originally a population geneticist who was one of the early introducers

Nikolai Timoféeff-Ressovsky around 1940. Courtesy of Dr. Elinor Welt (personal collection).

of the neosynthetic theory into Germany in the late 1920s. By the mid-1930s he had also become known as a leading experimental *Drosophila* geneticist whose important papers, appearing in Russian, German, and English, had already introduced some of the fundamental concepts that are discussed in the first section of the 3MP.

Aerial photograph of the Berlin-Buch Kaiser Wilhelm Institute for Brain Research and surrounding environs in the 1930s looking to the south. The large building in the center is the main institute for brain research. The building to the left is the main clinic, and the building to the right of the main institute housed the coworkers. The smaller building above the main institute is the home that the Timoféeff-Ressovsky and Vogt families shared. The large churchlike building is the mausoleum chapel, no longer in existence, for the planned cemetery. Archives of the Max Planck Society, Berlin-Dahlem (Kaiser Wilhelm Institute for Brain Research, no. I/3). We are deeply indebted to Dr. Dietrich Harder, formerly of the Institute of Physics and Biophysics at Göttingen University, and now at the University of Oldenburg, for identification of the buildings.

Timoféeff-Ressovsky studied biology as an undergraduate at Moscow University under the population geneticist Sergei Chetverikov (1880–1959), although the Russian civil war prevented him from actually completing the degree. After the revolution, despite having no advanced degrees, he secured a position at the research institute of Nikolai Kol'tsov (1872–1940). In 1925 he was invited to join the Kaiser Wilhelm Institute (KWI) for Brain Research (*Hirnforschung*), located in the far northeast Berlin suburb of Buch, following extended contact with its director, Oskar Vogt (1870–1959), who had come to Moscow to study the brain of the recently deceased Vladimir Lenin.

Timoféeff-Ressovsky was to assist Vogt in developing a theory of the genetic basis of hereditary mental disorder; the active research group he assembled in Buch, which included several political refugees, survived through the end of World War II.[17]

Although ordered to return to the Soviet Union in 1937, Timoféeff-Ressovsky made a fateful decision to remain in Nazi Germany, due largely to the rise of Lysenkoism under Stalin. Despite the outbreak of war between Germany and the Soviet Union in June of 1941, he continued his work at the KWI without significant interruption. His activity during the war period has been the subject of conflicting scholarship. One group has regarded him as a firm anti-Nazi who worked to shelter several people in his unit and whose son, Dmitri, died in the Mauthausen concentration camp following arrest for anti-Nazi activities.[18] Others have pointed out the several compromises he evidently made with the Nazi regime that could indicate sympathy with its eugenicist policies.[19] After the war he was arrested by the Soviet authorities, charged with being an enemy collaborator, and sentenced to ten years of hard labor in the Gulag prison camps; he spent time there with Aleksandr Solzhenitsyn and appears in the *Gulag Archipelago* leading informal scientific seminars.[20]

In 1947 when the Soviets realized he might be useful to their atomic-bomb project, Timoféeff-Ressovsky was transferred to what became known as Laboratory B of the uranium-production-plant group at Sungul' near Sverdlovsk in the Urals, headed by fellow captive and one-time coauthor Nikolaus Riehl (1901–90).[21] Timoféeff-Ressovsky became head of a biophysics research unit working on radiation and biology, where he was ultimately reunited with the radiochemist Hans-Joachim Born, the physician and radiation biologist Alexander Catsch, and his long-time collaborator Karl Zimmer, all of whom had worked with him at the Buch genetics unit during the war. After the death of Stalin in 1953, Timoféeff-Ressovsky was granted amnesty in 1955 and allowed to move to Sverdlovsk, where he organized a biophysics laboratory with the support of the Russian Academy of Sciences. Following the resignation of Nikita Khrushchev as Russian prime minister in 1964, Trofim Lysenko was finally removed from power, and the new prime minister, Leonid Brezhnev, allowed some liberalization of state control of biological science, particularly inquiry in traditional Mendelian genetics. Timoféeff-Ressovsky then moved to a new Institute of Medical Radiology at Obninsk, south of Moscow, where he organized a department of genetics and radiobiology. His reputation was still not fully rehabilitated, however, and he was subjected to further persecution by Lysenko's successor, N. P. Dubinin, and finally forced out of active science in 1971. Efforts by Max Delbrück and other Western scientists, first to prevent his imprisonment in 1946, then to gain his release, and finally to ensure proper recognition in the Soviet Union for

Photograph of the staff of the Kaiser Wilhelm Institute for Brain Research in the late 1930s in front of the mausoleum chapel. In the center is the director, Oskar Vogt. At far right is Timoféeff-Ressovsky, and on the same row six places toward the center is probably young Karl Zimmer. Archives of the Max Planck Society, Berlin-Dahlem (Kaiser Wilhelm Institute for Brain Research, no. III/19).

Karl Günther Zimmer in 1971 in his office at the Institute for Radiation Biology at the Karlsruhe Technological Institute. Courtesy of the Archives of the Forschungszentrum, Karlsruhe.

his scientific achievements, were all unsuccessful. In the post-glasnost era, the publication of popular biographies praising his opposition to Stalinism has made Timoféeff-Ressovsky one of the celebrated cases of maltreated Soviet-era scientists.[22]

Karl Günther Zimmer, the author of the second section of the 3MP, was Timoféeff-Ressovsky's primary collaborator in the period between late 1933 and 1945.[23] Zimmer was born in Breslau, now Wroclaw, Poland, in July 1911, as the son of a local bailiff. After early education in the St. Elisabeth Gymnasium in Breslau and the Helmholtz-Realgymnasium in Berlin, he received his PhD in July 1934 in photobiophysics from the University of Berlin under the direction of the physical chemist Max Bodenstein (1871–1942) with a dissertation on the action of ultraviolet light on biochemical reactions. Before completing his doctorate, he was a lecturer and tutor in matter theory and general science in 1931–32 in the Technische Hochschule in Berlin-Charlottenburg. In late 1933 he became Timoféeff-Ressovsky's assistant, and after receiving his diploma in 1934, he became affiliated as technical physicist with the radiology unit of the Cecilienhaus Hospital, also in Berlin-Charlottenburg, and located near the main Kaiser Wilhelm campus in the southwest suburb of Berlin-Dahlem. His employment in Timoféeff-Ressovsky's experimental genetics unit at the KWI for Brain Research in the far-east Berlin suburb of Buch was finalized in 1939, and he would remain there until 1945.[24] Together he and Timoféeff-Ressovsky developed an intensive program of radiation research, funded by the Emergency Association for German Science and its successor, the German Research Association (Deutsche Forschungsgemeinschaft). Zimmer was also employed by the Auer Society in Berlin, which specialized in the production of radionucleotides for research and which later became involved with the German atomic-bomb project. Following the fall of Berlin in April 1945, he was detained by the Soviets as a scientist relevant to atomic-bomb research and transported to the prisoner-of-war camp in Krasnogorsk. Through the intervention of his friend and collaborator Nikolaus Riehl, who had also engineered Timoféeff-Ressovsky's transfer from the Gulag, Zimmer was moved to the uranium-production plant at Laboratory B at Sungul', where he contributed to the work on radiation biology mentioned above. After his release in 1955, he took a temporary position at the Swedish Forest Research Institute at the Royal University in Stockholm (1955–56), and in 1957 he was appointed to direct the Institute for Radiation Biology of the Center for Nuclear Research at Karlsruhe, where he conducted radiation research, particularly on bacteriophage and the structure of DNA. His final years were spent as an emeritus professor at the Ruprecht Karls University in Heidelberg.

The third and best-known figure in this triumvirate of authors is Max Delbrück, the charismatic young physicist-turned-molecular biologist, who would go on to win the Nobel Prize for Physiology and Medicine in 1969 for his work in molecular genetics. Since he has been the subject of substantial biographical treatment, only the highlights need to be mentioned here.[25]

Born into an academic family in the southwest Berlin suburb of Grünewald near Dahlem, Delbrück grew up in a rarified intellectual world; his father was a professor of history at the University of Berlin, and

Max Delbrück in the early
1940s during his residence
in Nashville, Tennesee.
Courtesy of the Eskind
Biomedical Library, Van-
derbilt University.

close neighbors and family friends included the von Harnacks, Plancks,
and Bonhoeffers. His first university-level training was in astronomy and
astrophysics at the University of Tübingen, succeeded by studies at the
University of Berlin and the University of Bonn. At the University of Göttingen, where he completed his first degree, he was drawn into work in
the new quantum mechanics, with doctoral research under Walter Heitler
(1904–81) on the quantum-mechanical theory of the homopolar chemical
bond in lithium. From September 1929 to March 1931 he was in England
studying in the physics department at the University of Bristol at the H. H.
Wills physics laboratory at the invitation of John E. Lennard-Jones.[26]

Following his Bristol stay, a Rockefeller Foundation fellowship allowed
him to move to Copenhagen in March 1931 to study physics for six months
at Niels Bohr's famous Institute for Theoretical Physics in Copenhagen.
It was in this period that Niels Bohr became actively engaged in theoreti-

cal discussions on the relations of biology and physics with the brilliant young physicist and coauthor of another famous "three-man paper" (cited below), Ernst Pascual Jordan (1902–80).[27] Although Delbrück was in residence in Copenhagen at this initiation of the Bohr-Jordan conversations on theoretical biology, it is not evident that he had yet developed interests in biological questions. From September 1931 to March 1932, he was with Wolfgang Pauli in Zurich, continuing his study of theoretical physics. This was followed by a return to Bristol to work again with Lennard-Jones in March 1932, where he remained, with a visit to the theoretical physics seminar in Copenhagen in late March, until September, when he left Bristol permanently to return to Berlin. Correspondence from Bristol indicates that he had by that date developed strong, if unspecified, interests in biology, especially in botany, and resolved then to return to Berlin as a research assistant to the nuclear physicist Lise Meitner (1878–1968), whom he had met at the 1932 Bohr seminar and who was stationed at the KWI for Chemistry. This was instead of accepting a second research fellowship to work with Pauli.[28] Moving to Berlin allowed him to work in the "neighborhood of the very fine Kaiser Wilhelm Institute for Biology, to which I am entertaining friendly relations."[29] That summer he returned once again to Copenhagen for Bohr's famous "Light and Life" lecture; it was a lecture that had a life-transforming effect on him.

With the return to Berlin, he first came into contact with Timoféeff-Ressovsky in the Genetics Unit of the KWI for Brain Research in Berlin-Buch, and with two other researchers at the KWI for Chemistry in Dahlem, the physical chemist Kurt Wohl (1896–1962) and the biochemist Hans Gaffron (1902–79). These three scientists would figure prominently in Delbrück's later Grünewald seminars. It was at this same time that Hermann Muller (1890–1967), a future Nobel laureate, came to Berlin-Buch on a Guggenheim fellowship to work with Timoféeff-Ressovsky in his genetics unit. However, Muller and Delbrück appear not to have made contact until Bohr's theoretical physics seminar in Copenhagen in April 1933, which Muller attended. This led to further conversations between them in Berlin.[30]

It is only with the publication of the 3MP that Delbrück became professionally associated with genetics and biology; as of 1936, he was still "among the physicists" at Bohr's annual meeting in Copenhagen, displaced this year to June 17–20, if buried in the back row.

After the publication of the 3MP in April 1935, he attended a conference organized by Bohr in Copenhagen from 27 to 29 September 1936,

The June 1936 meeting of theoretical physicists at the Institute for Theoretical Physics in Copenhagen. The front row from right to left: James Franck, George Hevesy, Lise Meitner, Max Born, Werner Heisenberg, Pascual Jordan, and Paul Ehrenfest. Delbrück is third from the left in row five. Courtesy of the Archives of the University of Chicago and used with permission of the Niels Bohr Archives, Copenhagen.

specifically devoted to the question of radiation and genetics.[31] Although increasingly interested in biological questions, Delbrück remained attached to Meitner's unit at the KWI until 1937, when a Rockefeller fellowship allowed him to travel to the California Institute of Technology to study genetics and natural selection theory in *Drosophila*, presumably under T. H. Morgan's general guidance. There he came across the bacteriophage work of Emory Ellis; he saw in the bacteriophage an extremely simple system—in the words of one commentator, the "hydrogen atom" of biology—much more amenable to biophysical analysis than fruit flies.[32] It is his work as a phage geneticist that would eventually earn him the Nobel Prize.

The collaboration of Timoféeff-Ressovsky, Zimmer, and Delbrück that produced the 3MP featured several novelties that will be developed in detail in later chapters. The particular combination of physical and biological theory found in the paper was new in the 1930s and was recognized as such by contemporaries. Though common today, collaborative papers were not standard at the time, and such a cross-disciplinary effort was unusual, though not without precedent.[33] Collaborative research was becoming common in physics, as evidenced by two other famous "three-man papers" from the period, the landmark 1925 paper on quantum mechanics by Max Born, Werner Heisenberg, and Pascual Jordan,[34] and the 1935 critique of quantum mechanics by Albert Einstein, Boris Podolsky, and Nathan Rosen.[35] But this was not the case in biology, especially not in combination with theoretical physics. It is primarily for bringing together physical and biological theory and experimental methods, not the paper's actual scientific conclusions, that the 3MP continues to be acknowledged as foundational in the literature of molecular biology. To the extent that the paper's scientific conclusions were tied to the ill-fated "target theory" of the gene, they did not stand the test of time, a matter discussed in detail in the chapter by Richard Beyler.

The 3MP is also distinguished by its deliberate synthetic and theoretical character. It draws heavily upon an extensive prior body of experimental work on radiation and mutation, and is a valuable resource for this reason alone. But it goes beyond this goal to develop a concrete theory of the gene and gene mutation. In this it displays both connections to, and important differences from, the reflections of that other pioneering figure in theoretical genetics of the period, Hermann Muller. These theoretical ambitions led to the incorporation of contemporary quantum theory into experimental genetics. The paper also lays out a set of research questions and draws conclusions on the nature of the gene, both of which helped focus further research on more precise issues of gene structure. Although a comprehensive history of the gene concept before 1930 would take us beyond the confines of this volume,[36] a brief summary of the state of the gene concept around 1930 is required.

Materializing the Gene

By 1930, gene theory had been most prominently codified in the anglophone world by Thomas Hunt Morgan (1866–1945) and his students. This conceptual development became the basis of the chromosomal genetics that emerged from the famous "Fly Room" at Columbia University. It was systematized initially in the paradigm-creating textbook of

1915, *The Mechanism of Mendelian Heredity*, coauthored by the main Fly Room workers—Morgan, Alfred Sturtevant (1891–1970), Calvin Bridges (1889–1938), and Hermann Muller—in which the theory and techniques of "factor" mapping, crossover theory, and other methods associated with the Morgan school were developed.[37] These reconciled the observable chromosomal mechanics and the empirical results of breeding experiments with the underlying assumptions of Mendelian genetics, particularly the laws of recessivity, dominance, and independent assortment. By 1930 the influence of this approach had spread widely, if differentially, across the international community of researchers.[38] As an original member of the Morgan group, Hermann Muller serves as a personal bridge between the Morgan school and the 3MP, both through his original research work and also by his personal presence in Timoféeff-Ressovsky's research unit in Berlin from November 1932 through September 1933.

The concept of the gene within Morgan's own work was, however, an issue of some uncertainty. In his 1925 Silliman lectures, Morgan adopted Wilhelm Johannsen's term "genes" to denote what had previously been designated in his work only as Mendelian "factors." Morgan was emphatic on three points: first, there was no simplistic "one gene–one character" relationship between genes and phenotype.[39] Second, genes were theoretical entities whose existence was inferred from empirical data, and this did not require genes to be unitary material entities.[40] Morgan's flexible theoretical stance toward the gene concept was maintained in his Nobel Prize lecture, delivered only two years before the publication of the 3MP: "At the level at which the genetic experiments lie, it does not make the slightest difference whether the gene is a hypothetical unit, or whether the gene is a material particle."[41] Third, he was sensitive to the complex relationship between genes and their developmental expression, a sensitivity that was not shared by his students.

Morgan's tentativeness about the ontology of genes was in the foreground in the period in which the 3MP was conceptualized, and it was a particularly divisive issue within discussions in German genetic circles. Jonathan Harwood's important study has described a theoretical divide that existed between the more elite academic geneticists located at the Kaiser Wilhelm Institutes and the major German universities on one side, and the applied animal and plant geneticists at the technical schools on the other. The first group, including such individuals as Alfred Kühn (1885–1968), Richard Goldschmidt (1878–1958), Karl Henke (1895–1956), and Fritz von Wettstein (1895–1945), were more interested in issues of developmental biology and cytoplasmic inheritance, and were disinclined to accept the notion of unitary and atomistic bearers of heredity. Geneticists at such institutions as the Berlin Agricultural College, and subsequently at the KWI for Breeding Research in Müncheberg, who were carrying out the program of Erwin Bauer (1875–1933) and his students, focused on practical plant and animal breeding. This group quickly supported Morgan's transmission genetics. This theoretical division between German genetics traditions provides one of the more general features of the German genetics landscape in the 1920s and 1930s within which the 3MP is located.[42]

Morgan's tentativeness about the materiality of the gene was not shared by his students, and especially not by Hermann Muller, who held generally materialist views about biological processes and was not shy about expressing them. As an undergraduate at Columbia, Muller was inspired to study biology by the charismatic Edmund B. Wilson, and he remained at the university for doctoral studies under Morgan, becoming an original member of the "Fly Room" group and collaborated in the *Mechanism of Mendelian Heredity* textbook.

Personal frictions and intellectual differences led Muller to break with the group, and he subsequently pursued a different path from Morgan, Sturtevant, and Bridges.[43] Rather than investigating gene mapping and chromosome mechanics, he examined spontaneous mutation, "position" effects, and the conception of the gene as a material entity. Appointed to the biology faculty at the University of Texas in 1920, Muller formed an alliance with the microscopist and cell biologist Theophilus Painter (1889–1969), whose chromosome-staining techniques enabled specific areas of the chromosome to be identified visually and tracked across generations. It was at Texas that Muller also pursued the experimental study of the artificial production of mutations by X-rays.

When Muller began his mutation studies in the mid-1920s, mutations were thought to be the result of chemical changes in the nuclear material. It had also been known for some time that X-rays and the radiation produced by radium could affect living tissues.[44] But until Muller's work, this relation was primarily concerned with tissue damage from radiation. He changed this focus by looking at ways to induce mutations artificially and controllably, ultimately settling on X-radiation.

Muller began intensive work on this research program in the fall of 1926, with a series of experiments irradiating *Drosophila* with X-rays. Crossing experiments between irradiated and nonirradiated fruit flies allowed the effects of the X-rays on the specific traits to be determined. Muller found that the artificial mutations were produced in linear relation to the radiation dosage in roentgens (r), and that the artificial mutations were identical, both in phenotypic expression and in transmission, to those occurring naturally. These results were first presented to the public in a short paper in July 1927, followed by a longer version delivered to the Fifth International Congress of Genetics in Berlin the same summer. A longer paper followed in the *Proceedings of the National Academy of Sciences* a year later, developing in some detail the ramifications of his experimental work for the conception of the gene itself.[45]

The ability to produce mutations artificially by manipulating external conditions—temperature, chemicals, and especially radiation—reinforced Muller's preexisting belief that the gene must be a material entity with a definite structure of a fairly simple kind. In the *PNAS* paper he argued that the evidence for the stability of the gene, and the phenomena of reproducible "back" and "forward" mutations, was indicative of a simple structure: "We may conclude that the gene ordinarily consists of not more than one molecule, or at least not of several molecules of the same kind."[46]

Furthermore, X-radiation did not result in a "crazy-quilt pattern" of inheritance that would suggest some more general reorganization of the genome under radiation. Rather, mutation occurred only as single character changes, suggesting that the "transmuting action of the x-rays is thus spatially narrowly circumscribed, being confined to one gene even when there are two identical genes close together."[47] He understood this to support a theory of a material and localized conception of the gene, and he saw confirmation of this in Painter's work on tracking visual banding patterns on the chromosomes through meiosis: "It should be noted that such cytological verification not only establishes the claims made with respect to the individual translocations involved, but also serves to prove directly that our genetic methods of reasoning are sound: that is, that the genes really do lie in the chromosome in linear arrangement, in the physical order in which we have theoretically mapped them—a cardinal principle which not all those who believe in the chromosome theory in a more general way have hitherto admitted."[48]

Prior to the initiation of the radiation work, Muller also presented a major theoretical paper on the gene concept to a symposium on the topic at the International Congress of Plant Sciences in Ithaca in August 1926, though publication of the paper was delayed until 1929, after the appearance of the early radiation papers. In this paper, "The Gene as the Basis of Life," Muller developed a strongly essentialist conception of the material gene that neatly paired with the radiation studies that were to follow. Muller's way of conceiving the gene can be contrasted with the theory put forth at the same symposium by Harvard experimental plant geneticist Edward M. East (1879–1938). Like Morgan, East emphasized in his paper the tentative and hypothetical nature of the gene, and the error of thinking of it as more than this.[49] By contrast, Muller set out a theory of genes as the essential ingredients of all living substance, the primordial entities at the origins of life, and the governors of all life processes, including the transmission of life between generations. In his view, the gene accounts for "all specific, generic, and phyletic differences, of every order, between the highest and the lowest organisms."[50] Life itself, for Muller, is connected to these genes, conceived as autocatalytic and self-reproducing substances. They are the basis of the "whole mechanism of biological evolution." Genes "arose coincidentally with growth and 'life' itself."[51]

The foregoing makes clear that the gene concept was in flux in the early 1930s when the 3MP was created, and the paper demonstrates some sensitivity to the nuances of this debate. While the 3MP resists Muller's strong essentialism, the conclusions of the paper are closer to Muller's view of the gene than that of Morgan and East; the stated goal of the paper is, after all, to produce a physical model of the gene.

But if Muller's important papers on radiation genetics from 1927 to 1934 are all cited in the bibliography to the paper, and his impact is in evidence in several places, the 3MP is also far from a set of footnotes to Muller's work. Timoféeff-Ressovsky's and Zimmer's sections include original contributions to radiation genetics, and Delbrück's section includes an atomic model of gene mutation, one that only a theoretical physicist could have devised. Indeed, as we shall see below, the fourth section of the 3MP argues explicitly against a Mullerian-style reductionism.

Creating the 3MP

The specific circumstances that brought Timoféeff-Ressovsky, Zimmer, and Delbrück into collaboration on the 3MP are the subject of the Sloan chapter and will be only summarized here. The work originated from meetings between the three at Timoféeff-Ressovsky's apartment on the campus of the KWI Berlin-Buch and from informal seminars involving a wide range of physicists and biologists organized by Delbrück at his family home in Grünewald. These commenced in the fall of 1934. Since the fall of 1933, Timoféeff-Ressovsky had drawn Zimmer into his radiation-genetics research, and their joint work, certainly indebted to Muller's pioneering efforts and residency in Timoféeff-Ressovsky's unit in 1932–33, involved detailed experiments with X-rays and gamma rays in the production of mutations. Zimmer, in particular, was expert in experimental techniques associated with the "target theory" of the gene, a concept explored in detail in the chapters by Summers and Beyler in this volume.

The intense political turmoil within academic and scientific institutions created by the rapid nazification of Germany also played a significant role in the genesis of the 3MP. Following Hitler's assumption of power in January 1933 and the passage of the new Civil Service Laws in March, many prominent scientists, along with a wide array of other academics, artists, and intellectuals, were forced to resign their positions, for being either "non-Aryan" or married to such. A few others, such as Nobel physicist James Franck, Jewish but initially exempted from forced resignation for valorous service in World War I, resigned voluntarily in protest. The Kaiser Wilhelm Institutes, although insulated from the initial purging of the universities, nonetheless soon became prime targets. In March 1933 Vogt's brain research unit at the KWI Buch was raided and trashed by storm troopers because Vogt was unwilling, as Muller reported, to "fire or denounce any of the Jews, Social Democrats, or communists" employed at the institute.[52] Lise Meitner, Delbrück's research director, was Jewish and lost her faculty position at the University of Berlin, although she was able, as an Austrian citizen, to maintain her appointment at the KWI for Chemistry for several more years, until the Anschluss in 1938. In the life sciences, it has been estimated that a total of thirty major biologists of Jewish descent or married to Jews were dismissed, imprisoned, or even executed under the Nazis, and another fifteen non-Jewish biologists were dismissed or emigrated for political reasons.[53] As a result of these academic purges, by the fall of 1934 there were many scientists in the Berlin area who had been forced out of academic positions and who were trying to find alternative means of support, many looking for opportunities abroad.

It was thus partly in response to this academic turmoil that Delbrück organized his seminars in the fall of 1934 at his family home in the nearby Berlin suburb of Grünewald, rather than on the premises of the KWI at Dahlem. From this collaboration, created by both scientific interests and political circumstances, the 3MP emerged. It should also be noted, to avoid an exclusive emphasis on genetics, that another product of this collaboration was an important theoretical paper on photosynthesis. This is discussed in more detail in the Sloan chapter.

The publication of the 3MP, a long—fifty-three pages of printed text—
and complex paper, and its circulation among a select group of physicists
and biologists via an extensive reprint mailing, allowed the dissemination
of its claims to an important network of researchers in Europe and to at
least some authorities in the United States.[54] Those who read and understood it were able to discern a new model for how genetics and physics
could collaborate at a high theoretical and experimental level. They also
learned in detail about the new research in the target theory relating radiation to mutation. The paper displayed the way in which the complex
domain of theoretical physics could provide the basis for claims about the
nature and even the size of the gene.

Moving behind *What Is Life?*

The main vehicle for popularization of the arguments of the 3MP in the
anglophone world was Schrödinger's *What Is Life?*, a 1944 publication of
a series of lectures delivered at Trinity College, Dublin, in February 1943.
Schrödinger's intention was to use quantum mechanics to explain certain
remarkable features of biological systems, most importantly, the stability
of their genetic information over long periods of time. Two of the book's
seven chapters are spent explicitly discussing the experimental results of
the 3MP and Delbrück's model of the gene, without any noticeable critical
comment. However, Schrödinger's invocation of the 3MP is misleading in
at least one major respect; one of the main themes of his book is the claim
that a complete reduction of biological to physical systems is made possible by quantum mechanics, a claim that is not found in, nor supported
by, the 3MP. Given the historical role of *What Is Life?* in the origins of
postwar molecular biology and its explicit connection to the 3MP, it is important to make clear exactly how Schrödinger departed from, and even
misread, the paper.

   At the heart of Schrödinger's argument is the claim that quantum mechanics resolves the apparent discord between the statistical nature of
physical-chemical laws and the seeming determinateness of the influence
of genes over development. While the sense of this claim is consonant with
the 3MP's attempt to explain the stability of genetic information by way of
quantum-mechanical descriptions of molecular structures, Schrödinger's
thermodynamical discussion is much more elaborate and reaches far more
dramatic conclusions. Statistical mechanics describes the appearance of

"order from disorder"; for example, gases of randomly colliding molecules can exhibit regular, lawlike behavior as long as the scale of action is large enough. But genes, understood by Schrödinger as single molecules, exhibit regular, lawlike behavior on their own at very small sizes; the order they generate is grounded directly in their own structure, not on the disordered aggregate action of a large number of them. This "order from order" principle is "the real clue to the understanding of life," according to Schrödinger.[55] In one of the best-known conclusions of the book, he argues that organisms maintain well-ordered internal structures by consuming "negative entropy" from their surroundings. Despite his claims to be merely elaborating Delbrück's model of the gene, nothing of this sort is to be found in the 3MP.

Schrödinger's conclusion on the relevance of quantum mechanics for biology was that the new physics could supply a "bottom up," physicalistic explanation of organic life. This is simply superimposed upon the claims of the 3MP, which can be seen by reading Schrödinger's arguments in *What Is Life?* and those of the concluding section of the 3MP. The complex relation of structural to developmental genetics, left as an open problem by the conclusions of the 3MP, is simply ignored by Schrödinger.[56] Instead, he endorses an "essentialist" theory of the material gene close to that advocated in the 1929 paper by Muller, which Schrödinger had read carefully in 1932, but which is not referenced in *What Is Life?*[57] The gene is a "law-code and executive power—or, to use another simile, they are architect's plan and builder's craft—in one."[58]

The most important differences between *What Is Life?* and the 3MP center on the role of genes in development and on the related question of whether biology can be fully reduced to chemistry and physics. Schrödinger characterizes the gene, famously, as an "aperiodic crystal," determined by the target theory of the 3MP to be perhaps only 1,000 cubic atomic distances in size.[59] He characterizes the power and function of the gene as follows:

> In biology we are faced with an entirely different situation [than in physics]. A single group of atoms existing only in one copy produces orderly events marvelously tuned in with each other and with the environment according to the most subtle laws . . . . Every cell harbours just one of [each gene] (or two, if we bear in mind diploidy). Since we know the power this tiny central office has in the isolated cell, do they not resemble stations of local government dispersed through the body, communicating with each other with great ease, thanks to the code that is common to all of them?[60]

The introduction of a cascade of fertile metaphors related to gene action—"code," "government," "central office"—along with the vision of a deterministic control of the organism by its genes, is the best-known feature of the biophysical vision put forth in Schrödinger's short text.[61]

If this genic determinism was enabled by laws that were somehow outside of physics, then Schrödinger's picture would not be fully reductionist. Some passages in *What Is Life?* seem to support this reading: "We must be prepared to find [living matter] working in a manner that cannot be reduced to the ordinary laws of physics."[62] But this is misleading. What he intends to maintain is the claim that the structure of living matter is simply different from that for which the standard laws of classical physics were formulated. The task of deciphering biological systems is thus more of an engineering challenge than a theoretical one. He states plainly that "the new principle is not alien to physics . . . . It is, in my opinion, nothing else than the principle of quantum theory over again."[63] Schrödinger's vision is thoroughly reductionist: genes determine the organism by means of the laws of physics.[64]

The fourth section of the 3MP, cowritten by all three authors but reflecting Delbrück's hand more than the others, makes clear that no such reductionism is intended there. In one paragraph, they outline the reductionist picture, likely inspired by Muller's 1929 paper, which Delbrück cites in his section of the 3MP: "The genes are thus conceived as the immediate 'starting points' of the chains of reactions constituting the developmental processes . . . . The cell, thus far proving itself so magnificently as the unit of life, dissolves into the 'ultimate units of life,' the genes."[65] The phrasing of the passage is such that a casual reading might mistake this viewpoint for the authors'. But the next passage is unambiguous: "Our ideas about the gene challenge this picture . . . . Such genes are likely incapable of directly forming the morphogenic substances . . . . Therefore, we need not dissolve the cell into genes, and the 'starting points' of the developmental sequences are not attributed to individual genes, but rather to operations of the cell, or even to intercellular processes (which are all eventually controlled by the genome)."[66] Despite the parenthetical comment at the end, the presented viewpoint clearly acknowledges that genes are merely components of complicated systems with many operative parts; this is very different from Schrödinger's "master architect" picture. On the other hand, the 3MP does acknowledge a more specific reductionism: "Genes are physical-chemical units; perhaps the whole chromosome (to be sure the part containing genes) consists of such a unit . . . with many individual, largely autonomous subgroups."[67] But despite this materiality, the conclusion of the 3MP insists on thinking of genes as parts of systems, not themselves explanatory of the whole organism. The 3MP simply does not support the claim that all of biology—or even its characteristic features—is reducible to physics and chemistry.

Some influential early theoreticians of molecular biology nonetheless endorsed what they took to be Schrödinger's reductionist interpretation of the relations of physics to biology. Francis Crick thus viewed the revolution in physics of the 1920s to have supplied the foundation for explaining "*all* biology in terms of physics and chemistry."[68] As he interpreted Schrödinger's claim that quantum principles resolved the issue of genetic stability and transmission, this conveyed for Crick "in an exciting way the idea that, in biology, molecular explanations would not only be extremely important but also that they were just around the corner. This had been said before, but Schrödinger's book was very timely and attracted people who might otherwise not have entered biology at all."[69]

Schrödinger's popular lectures were, however, weak biology, and although they popularized some of the conclusions of the 3MP, they also misread them. As the Nobel laureate Max Perutz (1914–2002) comments in a 1987 retrospective: "To my disappointment, a close study of the book and of the related literature has shown me that what was true in his book was not original, and most of what was original was known not to be true even when it was written . . . . In retrospect, the chief merit of *What is Life?* is its popularization of the Timoféeff-Ressovsky, Zimmer and Delbrück paper that would otherwise have remained unknown outside the circles of geneticists and radiation biologists."[70]

Reading the full 3MP reveals more clearly how one strand of the "revolution in physics" did in fact affect biology. This historic paper was to suffer because its publication history was complex, its theoretical reliance upon the target theory was later undermined in important ways, and its broader philosophical conclusions were misinterpreted by its most famous expositor, Schrödinger. Nonetheless, it offered a sophisticated and subtle model of how physicists and biologists could collaborate on a common problem.

Chapter Summaries

The chapters that follow explore the scientific, historical, and philosophical issues of the 3MP from several perspectives. In chapter 2, William

Summers surveys the historical developments in physics, and in particular the developments in radiation physics, that commenced with Einstein's conception of radiation as particulate, illuminating its importance for the target theory. Placing this theory in its wider context, we can follow Zimmer's application to genetics in the second part of the 3MP.

The chapter by Phillip Sloan describes in detail the social network from which the 3MP originated, a series of informal seminars organized by Max Delbrück in the fall of 1934. These discussions, orchestrated and interconnected by Delbrück, represent his first efforts to transfer the "Copenhagen spirit," which he had absorbed as a young physicist attending the theoretical physics seminars held annually by Niels Bohr, into a context that directly sought to relate biology and physics.

The third historical chapter, by Richard Beyler, takes the historical scenario beyond the confines of the 3MP and explores the reception of this paper and the fate of the target theory after 1935. This chapter extends and further articulates the importance of this theory, and the theoretical difficulties it encountered. This chapter also explores the little-known reception of the 3MP and the way in which its claims were modified in the face of criticisms.

The analysis then shifts to interpretive issues in the philosophy of biology, with chapters by Nils Roll-Hansen and Daniel McKaughan, focusing on the relation between Bohr's and Delbrück's views on the question of reductionism in biology. The degree to which the 3MP does or does not support a reductionist reading, and the meaning of reductionism in relation to Bohr's concept of "complementarity," has spawned a substantial literature. Presenting these two chapters together allows the reader to assess the historiographical and interpretive questions involved.

Roll-Hansen's chapter reaches back to the foundations of Niels Bohr's often misunderstood reflections on the importance of complementarity for biology. Contextualizing these reflections against the background of Harald Høffding's version of neo-Kantianism, Roll-Hansen illuminates the "dual aspect" approach to the relation of the biological and physical that resolved for Bohr the issues of mechanism and vitalism. This is then extended into a discussion of the features of Bohr's philosophical positions that were of relevance to Max Delbrück's interpretations of these relationships, both in the 3MP and beyond.

Further analysis of Delbrück's subsequent reflections on "biological complementarity" forms the topic of McKaughan's chapter. He makes the case for Delbrück's "empirical" antireductionism through analysis of his discussions, lectures, and statements through the 1970s. The slightly different readings of these texts in the Roll-Hansen and McKaughan chapters illuminate some of the difficulties in the exegesis of the 3MP itself, and in the interpretation of Delbrück's thought.

Following these interpretive discussions, the complete text of the paper is presented, translated into English by Brandon Fogel. This is preceded by an interpretive preface that includes a summary of each section of the 3MP. Bringing together the reflections of a geneticist and two different kinds of physicist, with a joint focus on the concept of the gene and the nature of mutation, the 3MP shows us the transition from the transmission genetics of the Morgan school to the biophysical genetics that succeeded it, in which the nature of the gene itself became a focus of investigation through the methods of physics.

Notes

1. N. V. Timoféeff-Ressovsky, K. G. Zimmer, and M. Delbrück, "Über die Natur der Genmutation und der Genstruktur," *Nachrichten von der Gesellschaft der Wissenschaften zu Göttingen, mathematisch-physikalische Klasse, Fachgruppe VI: Biologie* 1 (1935): 189–245. References to specific pages in this paper will be given by page number as found in the translation in this volume, followed by the pagination in the original paper in parentheses. For example 3MP 263 (235) refers to the specific passage on page 263 of this volume, and to page 235 of the German original.

2. The *Nachrichten* series ceased publication under this name in 1940. WorldCat lists fifty-two libraries with this journal, almost all in the United States.

3. Max Delbrück, oral interview by Carolyn Harding, July 14–September 11, 1978, Oral History Project, California Institute of Technology Archives, 50. Retrieved March 15, 2005, from http://oralhistories.library.caltech.edu/16/01/OH_Delbruck_M.pdf. Hereafter cited as "Harding interview." All quotations from this document in this volume are with permission of the Archives of the California Institute of Technology.

4. E. Schrödinger, *What Is Life? The Physical Aspect of the Living Cell with Mind and Matter and Autobiographical Sketches* (Cambridge: Cambridge University Press, 2000; first published Cambridge, 1944). All further references are to this edition. For extensive discussion on the history, publication, and impact of Schrödinger's essay, see R. C. Olby, "Schrödinger's Problem: What Is Life?," *Journal of the History of Biology* 4 (1971): 119–48; E. J. Yoxen, "Where Does Schrödinger's 'What Is Life?' Belong in the History of Molecular Biology?," *History of Science* 17 (1979): 17–52; and Yoxen's more extended discussions in his PhD dissertation, "The Social Impact of Molecular Biology" (Department of History

and Philosophy of Science, Cambridge University, 1978), chap. 5; see also M. Perutz, "Erwin Schrödinger's *What Is Life?* and Molecular Biology," in *Schrödinger: Centenary Celebration of a Polymath*, ed. C. W. Kilmister (Cambridge: Cambridge University Press, 1987), 234–51; K. R. Dronamraju, "Erwin Schrödinger and the Origins of Molecular Biology," *Genetics* 153 (1999): 1071–76. A deconstructive history is developed by P. Abir-Am, "Themes, Genres and Orders of Legitimation in the Consolidation of New Scientific Disciplines: Deconstructing the Historiography of Molecular Biology," *History of Science* 23 (1985): 73–117. For comments on the immediate circumstances of Schrödinger's acquaintance with the paper, see Walter Moore, *Schrödinger: Life and Thought* (Cambridge: Cambridge University Press, 1989), 395. It would seem unlikely that he did not know of this paper in the 1930s, but there is no documentary support for this.

5. For an examination of this reading see Daniel J. McKaughan, "The Influence of Niels Bohr on Max Delbrück: Revisiting the Hopes Inspired by 'Light and Life,'" *Isis* 96 (2005): 507–29.

6. We will use the more inclusive term "biophysics" to characterize various efforts to relate biology and physics in the period under consideration. In view of the restrictions above, we use the descriptor "molecular" with caution.

7. W. Weaver, "Molecular Biology: Origins of the Term," *Science* 170 (1970): 582.

8. On the role of the Rockefeller Foundation in developing molecular biology and biophysics as part of its "vital processes" project under the leadership of Max Mason and Warren Weaver, see Lily E. Kay, *The Molecular Vision of Life: Caltech, the Rockefeller Foundation, and the Rise of the New Biology* (Oxford: Oxford University Press, 1993), chap. 1; Robert Kohler, "Warren Weaver and the Rockefeller Foundation Program in Molecular Biology: A Case Study in the Management of Science," in *The Sciences in the American Context: New Perspectives*, ed. N. Reingold (Washington, DC: Smithsonian Institution, 1979), 249–93.

9. K. G. Zimmer, "The Target Theory," in *Phage and the Origins of Molecular Biology*, ed. J. Cairns, G. Stent, and J. D. Watson (Cold Spring Harbor, NY: Cold Spring Harbor Press, 1966), 33–42; D. Fleming, "Émigré Physicists and the Biological Revolution," *Perspectives in American History* 2 (1968): 176–213, reprinted in *The Intellectual Migration: Europe and America, 1930–1960*, ed. D. Fleming and B. Bailyn (Cambridge, MA: Harvard University Press, 1969), 152–89.

10. See, for example, N. Rasmussen, *Picture Control: The Electron Microscope and the Transformation of Biology in America, 1940–1960* (Stanford, CA: Stanford University Press, 1997).

11. S. de Chadarevian, *Designs for Life: Molecular Biology after World War II* (Cambridge: Cambridge University Press, 2002), chap. 1; Pnina Abir-Am, by contrast, has been willing to extend these origins into the 1930s, relating the first developments of molecular biology, as distinct from biochemistry, to work in the 1930s on protein structure. See P. Abir-Am, "The Molecular Transformation of

Twentieth-Century Biology," in *Science in the Twentieth Century*, ed. J. Krige and D. Pestre (Amsterdam: Harwood, 1997), 495–524; and Abir-Am, "The Biotheoretical Gathering, Trans-disciplinary Authority and the Incipient Legitimation of Molecular Biology in the 1930s: New Perspective on the Historical Sociology of Science," *History of Science* 25 (1987): 1–70.

12. B. J. Strasser, *La fabrique d'une nouvelle science: La biologie moléculaire à l'âge atomique (1945–1964)* (Florence: Olschki, 2006), esp. "Introduction."

13. See, for example, S. de Chadarevian and B. J. Strasser, "Molecular Biology in Postwar Europe: Towards a 'Glocal' Picture," *Studies in History and Philosophy of Biological and Biomedical Sciences* 33 (2002): 361–65. This volume contains papers by several scholars and is devoted to the general issue of the creation of molecular biology in postwar Europe and America.

14. See Abir-Am, "Themes, Genres and Orders." On the "quiet" initial reception of the Watson-Crick papers, see B. J. Strasser, "Who Cares about the Double Helix?," *Nature* 422 (2003): 803–4; and R. C. Olby, "Quiet Debut for the Double Helix," *Nature* 421 (2003): 402–6.

15. See Fleming, "Émigré Physicists." A new look at these issues with particular relevance to the 3MP and its context is Ute Deichmann, "Emigration, Isolation and the Slow Start of Molecular Biology in Germany," *Studies in History and Philosophy of Biological and Biomedical Sciences* 33 (2002): 449–71.

16. Biographical details have been taken from Diane B. Paul and Costas B. Krimbas, "Nikolai V. Timoféeff-Ressovsky," *Scientific American* 266 (February 1992): 86–92; K. G. Zimmer, "N. W. Timoféeff-Ressovsky, 1900–1981," *Mutation Research* 106 (1982): 191–93; V. Korogodin, G. Polikarpov, and V. Velkov, "The Blazing Life of N. V. Timoféeff-Ressovsky," *Journal of Bioscience* 25 (2000): 125–31; and R. L. Berg, "In Defense of Timoféeff-Ressovsky," *Quarterly Review of Biology* 65 (1990); 457–79. Additional details have been derived from B. Gausemeier, *Natürliche Ordnungen und politische Allianzen: Biologische und biochemische Forschung an Kaiser-Wilhelm-Instituten, 1933–1945* (Göttingen: Wallstein, 2005), chap. 3. Use has also been made of the Delbrück-Timoféeff-Ressovsky correspondence (Max Delbrück Papers, California Institute of Technology, box 22, folders 2–3; hereafter cited like this: DP 22.2–3). We acknowledge the valuable assistance of Diane Paul on several matters pertaining to Timoféeff-Ressovsky's biography.

17. On Timoféeff-Ressovsky's genetics work and his location within the German genetics community, see Jonathan Harwood, *Styles of Scientific Thought: The German Genetics Community, 1900–1933* (Chicago: University of Chicago Press, 1993), 55–56 and 111–13; and Gausemeier, *Natürliche Ordnungen*, chap. 3. For the importance of the Russian school of population genetics in which he was trained, see Mark Adams, "Sergei Chetverikov, the Kol'tsov Institute, and the Evolutionary Synthesis," in *The Evolutionary Synthesis: Perspectives on the Unification of Biology*, ed. E. Mayr and W. Provine (Cambridge, MA: Harvard University Press, 1980), 242–78. Several valuable insights into the history of Timoféeff-Ressovsky's

unit during the war years in Berlin can be obtained in fictionalized form from Elly Welt's Holocaust novel, *Berlin Wild* (New York: Viking, 1986), which is based on extensive interviews with her late husband, Peter Welt, who worked in Timoféeff-Ressovsky's laboratory from 1943 to 1945 and who is the model for the young half-Jewish laboratory assistant Josef Bernhardt. At least one publication emerged from this collaboration (A. Catsch, A. Kanellis, G. Radu, and P. Welt, "Über die Auslösung von Chromosomenmutationen bei *Drosophila melanogaster* mit Röntgenstrahlen verschiedener Wellenlänge," *Die Naturwissenschaften* 32 [1944]: 228). In Welt's novel, Timoféeff-Ressovsky is the model for the fictional laboratory director Nikolai Alexandrovich Avilov, and Zimmer for Professor Maximilian Kreutzer. We are indebted for this information to an oral interview on May 19, 2008, with Dr. Elinor "Elly" Welt.

18. Zimmer, "Timoféeff-Ressovsky, 1900–1981"; Berg, "Defense." See also D. Granin, *The Bison: A Novel about a Scientist Who Defied Stalin*, trans. A. W. Bouis (New York: Doubleday, 1990), for a hagiographical account.

19. B. Müller-Hill, "Heroes and Villains," *Nature* 336 (1988): 721–22; Gausemeier (*Natürliche Ordnungen*, chap. 3) has given a nuanced and generally favorable assessment of Timoféeff-Ressovsky's work during the war years, but downplayed the image of the subunit for genetics that he directed as the center of resistance depicted in the Welt novel.

20. A. I. Solzhenitsyn, *The Gulag Archipelago, 1918–1956*, trans. T. P. Whitney (New York: Harper and Row, 1974), 597–600.

21. N. Riehl, N. V. Timoféeff-Ressovsky, and K. G. Zimmer, "Mechanismus der Wirkung ionisierender Strahlen auf biologische Elementareinheiten," *Die Naturwissenschaften* 29 (1941): 625–29. For a brief discussion of the Soviet project and Riehl's role in this, see Pavel V. Oleynikov, "German Scientists in the Soviet Atomic Project," *Non-Proliferation Review* (summer 2000): 1–30.

22. Granin, *Bison*.

23. Details on Zimmer's early background have been obtained from a brief biographical sketch that accompanied his inaugural dissertation, "Der Reaktionsmechanismus der photochemischen Umwandlung von o-Nitrobenzaldehyd zu o-Nitrosobenzoesäure im ultravioletten Lichte," Philosophical Faculty, University of Berlin, received on July 27, 1934. Subsequent information has been obtained from P. Herrlich, "In Memoriam: Karl Günther Zimmer (1911–1988)," *Radiation Research* 116 (1988): 178–80; the obituary by U. Hagen and J. T. Lett in *Radiation and Environmental Biophysics* 27 (1988): 245–46; Oleynikov, "German Scientists in the Soviet Atomic Project"; and the sixtieth-birthday sketch by his associate during the war years, Alexander Catsch, *Strahlentherapie*, 142 (1971): 124–25; See also Gausemeier, *Natürliche Ordnungen*, 171–73. We also acknowledge personal assistance from Peter Herrlich, Dietrich Harder, Horst Jung, Hermann Dertinger, Alexander von Schwerin, Barbara Batchler, Antje Kreienbring, and Susan Uebele in obtaining additional details on Zimmer's career.

24. Although Gausemeier implies (*Natürliche Ordnungen*, 172) that Zimmer was not employed by Timoféeff-Ressovsky's unit until 1939, he had been given some appointment to Timoféeff-Ressovsky's unit by October 1933 with funding supplied by the Emergency Fund for German Science, with the permanent appointment in 1939 to the KWI. The first collaborative paper with Timoféeff-Ressovsky dates from 1935 ("Strahlengenetische Zeitfaktorversuche an *Drosophila melanogaster*," *Strahlentherapie* 53 (1935): 134–38).

25. Biographical details have been drawn primarily from Delbrück's own autobiographical accounts in the Harding interview; the incomplete autobiography dated March 3, 1981 (DP 40.4); a detailed CV prepared for his induction into the Royal Society of London (DP 40.2); and more general biographical discussions in Ernst Peter Fischer and Carole Lipson, *Thinking about Science: Max Delbrück and the Origins of Molecular Biology* (New York: Norton, 1988). We have also utilized Simone Wenkel and Ute Deichmann, eds., *Max Delbrück and Cologne: An Early Chapter of German Molecular Biology* (Hackensack, NJ: World Scientific, 2007); and the William Hayes biography in *Biographical Memoirs of the Royal Society* 28 (1982): 59–90, reprinted in *Biographical Memoirs of the National Academy of Sciences* 62 (1993): 67–117. Some additional insights have been drawn from Ernst Peter Fischer, "Max Delbrück," *Genetics* 177 (2007): 673–76.

26. From this emerged his first published paper: M. Delbrück, "Possible Existence of Multiply Charged Particles of Mass One," *Nature* 130 (1932): 626–27. Other papers from this period include Delbrück and G. Gamow, "Uebergangswahrscheinlichkeiten von angeregten Kernen," *Zeitschrift für Physik* 72 (1931): 492-99; and his short speculative "Note Added in Proof" (Zusatz bei der Korrektur) to L. Meitner and H. Kösters, "Ueber die Streuung kurzwelliger Y-strahlen," *Zeitschrift für Physik* 84 (1933): 137–44, in which he offered the hypothesis that the effect of short-wave radiation on different elements was due to a photoeffect.

27. Finn Aaserud, *Redirecting Science: Niels Bohr, Philanthropy, and the Rise of Nuclear Physics* (Cambridge: Cambridge University Press, 1990), 82–90.

28. The 1932 Copenhagen meeting, which Delbrück attended, was specifically dedicated to the structure of the nucleus, an issue of interest to Delbrück in his Bristol research. Meitner, along with Fritz Strassman and Otto Hahn, developed in its wake the concept of nuclear fission. Delbrück had drawn the attention of the group by his production of a parody on *Faust* at this meeting. See Gino Segrè, *Faust in Copenhagen: A Struggle for the Soul of Physics* (New York: Viking, 2007), esp. chap. 12.

29. Delbrück to Bohr, June 28 and July 1, 1932, Niels Bohr Scientific Correspondence 1930–1945, Niels Bohr Library and Archives, Microfilm BSC 28, American Institute of Physics. Quoted with permission of the Niels Bohr Archives, Copenhagen.

30. Muller was in Berlin from November 9, 1932, to September 16, 1933. On Muller's attendance at the Copenhagen meeting, see E. A. Carlson, *Genes, Radia-*

tion, and Society: The Life and Work of H. J. Muller (Ithaca, NY: Cornell University Press, 1981), 188. The relations between Muller and Delbrück seem distant in this period, and quite unlike those between Muller and Timoféeff-Ressovsky. The second documented contact between Muller and Delbrück was during their travel together from Berlin to Copenhagen to attend the September 1936 meeting organized by Bohr to examine the relations of radiation and genetics in the wake of the 3MP. Muller had returned to Berlin from the Soviet Union to make this journey.

31. According to the registration book of Bohr's institute, the meeting was attended by Delbrück, Timoféeff-Ressovsky, and Muller, and the Norwegian scientists Otto L. Mohr (genetics), Kirstine Bonnevie (biology), and Vilhelm Bjerknes (geophysics), along with the director of the Physiology Department of the Carlsberg Laboratory in Denmark, the yeast geneticist Øjvind Winge. This resulted in a "consensus document" authored by Timoféeff-Ressovsky and Delbrück (DP 29.11). We thank Felicity Pors and Finn Aaserud of the Niels Bohr Archive for important information on this conference.

32. Fischer, "Delbrück," 674.

33. See Summers chapter, this volume.

34. M. Born, W. Heisenberg, and P. Jordan, "Zur Quantenmechanik," *Zeitschrift für Physik* 34 (1925): 858–88; 35 (1925): 557–615.

35. A. Einstein, B. Podolsky, and N. Rosen, "Can Quantum-Mechanical Description of Reality Be Considered Complete?," *Physical Review* 47 (1935): 777–80.

36. See James Schwartz, *In Pursuit of the Gene: From Darwin to DNA* (Cambridge, MA: Harvard University Press, 2008); T. Everson, *The Gene: A Historical Perspective* (Westport, CT: Greenwood, 2007); J. Beurton, R. Falk and H-J. Rheinberger, eds., *The Concept of the Gene in Development and Evolution* (Cambridge: Cambridge University Press, 2000); P. Portin, "The Concept of the Gene: Short History and Present Status," *Quarterly Review of Biology* 68 (1993): 173–223; Lenny Moss, *What Genes Can't Do* (Cambridge, MA: MIT Press, 2003); E. A. Carlson, *The Gene: A Critical History* (Philadelphia: Saunders, 1966).

37. T. H. Morgan, A. H. Sturtevant, H. J. Muller, and C. B. Bridges, *The Mechanism of Mendelian Heredity*, first edition (New York: Holt, 1915). This textbook still used the language of "factors" rather than "genes" in the second edition of 1922.

38. See R. E. Kohler, *Lords of the Fly:* Drosophila *Genetics and the Experimental Life* (Chicago: University of Chicago Press, 1994); E. A. Carlson, *Mendel's Legacy: The Origin of Classical Genetics* (Cold Spring Harbor, NY: Cold Spring Harbor Press, 2004). See also Harwood, *Styles of Scientific Thought*, chap. 4.

39. The denial of this point was also maintained in the 1915 textbook. See J. Schwartz, "The Differential Concept of the Gene: Past and Present," in Beurton, Falk, and Rheinberger, *Concept of the Gene*, 26–39.

40. T. H. Morgan, *The Theory of the Gene* (New Haven: Yale University Press, 1926), 25.

41. T. H. Morgan, "The Relation of Genetics to Physiology and Medicine," *Scientific Monthly* 41 (1935), 7–8, reprinted in *Physiology or Medicine* (Nobel Foundation; Amsterdam: Elsevier, 1967), 1:315–16.

42. Harwood, *Styles of Scientific Thought*. For a critique of Harwood's "stylistic" analysis, see Deichmann, "Emigration, Isolation," 457–60.

43. See Schwartz, *Pursuit of the Gene*, chaps. 12–13. For a comprehensive study of radiation studies in Germany in the period leading up to the 1930s, see Alexander von Schwerin, *Experimentalisierung des Menschen: Der Genetiker Hans Nachtsheim und die vergleichende Erbpathologie, 1920–1945* (Göttingen: Wallstein, 2004).

44. See Summers chapter, this volume.

45. H. J. Muller, "Artificial Transmutation of the Gene," *Science* 66 (1927): 84–87; Muller, "The Problem of Genic Modification," in *Verhandlungen des V. Internationalen Kongresses für Vererbungswissenschaft, Berlin 1927* (Berlin: Borntraeger, 1928): 234–60; Muller, "The Production of Mutations by X-Rays," *Proceedings of the National Academy of Sciences* 14 (1928): 714–26.

46. Muller, "Production of Mutations," 717.

47. Ibid.

48. Ibid., 721.

49. E. M. East, "The Concept of the Gene," in *Proceedings of the International Congress of Plant Sciences, Ithaca, New York, August 16–24, 1926*, vol. 1, ed. B. M. Duggar (Menasha, WI: George Banta, 1929), 889–95.

50. H. J. Muller, "The Gene as the Basis of Life," in *Proceedings of the International Congress of Plant Sciences, Ithaca, New York, August 16–24, 1926*, vol. 1, ed. B. M. Duggar (Menasha, WI: George Banta, 1929), 897–921.

51. Ibid., 916–17.

52. Carlson, *Genes, Radiation and Society*, 190.

53. The major monographic series, edited by Reinhard Rürup and Wolfgang Schieder, *Geschichte der Kaiser-Wilhelm-Gesellschaft im Nationalsozialismus*, is examining many aspects of this history in detail and is now in its eighteenth volume. See also Ute Deichmann, *Biologists under Hitler*, trans. Thomas Dunlap (Cambridge, MA: Harvard University Press, 1996), chap. 1. The physics community is discussed in less detail in Alan Beyerchen, *Scientists under Hitler: Politics and the Physics Community in the Third Reich* (New Haven: Yale University Press, 1977).

54. Delbrück reports that the paper was familiar to some of his contacts when he visited Caltech in 1937 (Harding interview, 50). Details on the dissemination of this paper are given in the Beyler chapter, this volume.

55. Schrödinger, *What Is Life?*, 82.

56. Harwood, *Styles of Scientific Thought*, 51–61.

57. Muller is not cited in the *What Is Life?* text, but Schrödinger knew of his paper as early as 1932. This is confirmed by his reading notes in his notebook labeled "Warum" preparatory for his early paper on biophysical topics, "Warum sind die

Atome so klein?," *Forschungen und Fortschritte* 9 (1933): 126. We are indebted for this reference to Yoxen, "Social Impact," chap. 5.

58. Schrödinger, *What Is Life?*, 22.

59. Ibid., 44. Although attributed to the 3MP, this size estimate is not specifically stated in the paper itself.

60. Ibid., 79.

61. On Schrödinger's gene concept, see Moss, *What Genes Can't Do*, 62–64. For a critical analysis of the historiographical tradition that made Schrödinger the originator of the genetic code concept, see Lily E. Kay, *Who Wrote the Book of Life? A History of the Genetic Code* (Stanford, CA: Stanford University Press, 2000), esp. 59–66.

62. Schrödinger, *What Is Life?*, 76.

63. Ibid., 81.

64. The opposite conclusion—that Schrödinger was advocating an antireductionist position based on the existence of "new laws" of biology—has been a common interpretation. We take this conclusion to be based on a misreading of his final conclusions of chapter 7 of *What Is Life?* See on this issue McKaughan, "Influence of Niels Bohr."

65. 3MP, 270 (240).

66. Ibid., 270 (240).

67. Ibid.

68. F. Crick, *Of Molecules and Men* (Seattle: University of Washington Press, 1966), 10–11.

69. F. Crick, "Recent Research in Molecular Biology: An Introduction," *British Medical Bulletin* 21 (1965): 184.

70. M. Perutz, "Physics and the Riddle of Life," *Nature* 326 (1987): 555–58.

PART I

Historical Origins of the Three-Man Paper

CHAPTER TWO

Physics and Genes: From Einstein to Delbrück

William C. Summers

In canonical accounts of the origin of molecular biology, for example, those by Stent[1] or Fischer and Lipson,[2] in April 1935 a paper with the title "Über die Natur der Genmutation und der Genstruktur" [On the nature of gene mutation and gene structure] was presented to the Göttingen Academy of Sciences,[3] and thus was born Molecular Biology. This paper outlined a theory of the gene about which Erwin Schrödinger (1887–1961) said: "If the Delbrück picture should fail, we would have to give up further attempts."[4]

What was important about this paper and why has it become a "landmark"? Discounting the post hoc canonization of early work by the historical "winners," there are two significant aspects of this paper that have appealed to the imagination of scientist and historian alike. First, it was a bold attempt to tackle a biological problem with a new set of tools, those of quantum physics. Second, the paper represented a conscious collaboration between a geneticist, a biophysicist, and an atomic physicist. Such a collaboration was seen by the three participants as giving them the opportunity to bring different disciplinary perspectives to bear on a problem of common interest. One of the authors, Karl Günther Zimmer (1911–88), later recalled: "Its friends and critics used to refer to it as the 'Green Pamphlet' [because of the bright green cover on the offprints] or, somewhat more deprecatingly, as the 'Dreimännerwerk' ('Three-men-paper'): team work was not very usual in Germany thirty years ago, and inter-disciplinary team work appeared rather strange to some scientists."[5]

Nikolai Timoféeff-Ressovsky (1900–1981), the geneticist, outlined the biological problem: what is the nature of the mutation process and what does it say about the nature of the gene? Karl Zimmer, the biophysicist, outlined an approach to the structure of the gene using physical experiments based on the recently discovered action of X-rays in causing mutations. Max Delbrück (1906–81) used the evidence and concepts from the target-theory experiments to construct a model of mutation and then a theory of gene mutation and structure, in Gunther Stent's words, "a 'quantum mechanical' model of the gene."[6]

This chapter focuses on the developments in physics leading up to this important paper and shows that one of the crucial aspects of this paper involved the application and appreciation of physical concepts growing out of the early work on radioactivity and the physical nature of X-rays. The tools of the "new physics" were important elements in this dawning of molecular biology.

X-rays and Genes

There were several reasonable candidates for agents that might cause mutations or inactivate genes: heat, various chemicals, and radiation. In the early work on fruit-fly genetics all these potentially harmful treatments were employed.[7] The interesting and relevant historical question immediately presents itself: Why radiation? That is, how did it come about that the use of radiation, both ionizing and ultraviolet, became the standard methodological approach applied to the physical study of genes between the 1930s and the 1960s? Universities established departments devoted to this subject;[8] there were courses developed and journals founded, all grounded in the premise that radiation biology was going to provide the keys to unlocking the secrets of the gene. There is, of course, no one simple account of these developments. Certainly the discovery of X-ray mutagenesis by Hermann Muller[9] and Lewis J. Stadler (1896–1954)[10] in the 1920s suggested that radiation was a useful tool with which to induce genetic change, which up to then seemed to be mystifyingly spontaneous and fickle. The development of atomic energy during and after World War II certainly contributed to this impetus in many and complex ways. The more limited question addressed in this paper is: what intellectual and personal antecedents to the fundamental theoretical framework made radiation biology so appealing to scientists in the 1920s and 1930s? A key concept in this nascent field was the idea that radiation is particulate—quantized, in the jargon of the physicist—and that it could be used to explore animate matter in the same way it was being used to explore inanimate matter. The conceptual basis for this approach came to be known as the "target theory" and was the subject of a substantial portion of the famous "Three-Man Paper" (3MP) in 1935. The subsequent development and efflorescence of this approach is the subject of the chapter by Richard Beyler in this volume.

The "target theory" is the name now given to a theoretical model (more precisely, several related models) and a set of experimental approaches to understanding the way radiation interacts with cells.[11] The two basic features of this model are (1) radiation is considered to be random projectiles, and (2) the components of the cell are considered as the targets to be bombarded (and inactivated or otherwise affected) by these projectiles. From the nature of the dose-response relationship for a specific type of radiation, and inactivation (or alteration) of a specific biological function, the number and size of the subcellular target for that particular function can be calculated. This approach has been used to estimate the size and shape of enzymes, viruses, and as noted above, genes.[12] It has also been applied to study the subcellular apparatus that synthesizes proteins and DNA, as well as such global physiologic processes as respiration and ion transport. From the mid-1930s to the mid-1950s, the target theory and its applications were major preoccupations of the field that became known as biophysics. Its full exposition reached its culmination in the early 1940s as described in Douglas E. Lea's *Actions of Radiations on Living Cells*, first published in 1946.[13]

The target theory depended on two fundamental principles that were emerging from the new physics at the beginning of the twentieth century: the *particulate nature of alpha and beta rays* and the *quantum theory of light*. The former suggested that a projectile-target approach might be useful, and the latter concept, involving photons, allowed these ideas to be extended to X-rays and ultraviolet radiation. The key corollary to these notions was the belief that these quanta interact with molecules to produce localized ionizations that are the ultimate cause of the observed biological changes produced by radiation.

Later research in the 1960s would show that these two assumptions about the nature of radiation and its interactions with cells were overly simplistic. Cellular repair reactions, controlled by genes that were discovered in the 1960s, could modify radiation damage after it occurred, thus reversing a hit or its full effect. That is, hits were not all-or-none as envisioned by the target theorists but rather became substrates for cellular metabolism that varied from cell to cell and depended on the particular state of the cell and its environment. The beautiful simplicity of the target theory of the physicists would succumb to the complexity of biochemistry.

The New Physics

A brief summary of the relevant physical studies will provide the context of this investigation. Around the turn of the century Joseph John (J. J.) Thomson (1856–1940) and his colleagues in England, and Henri Becquerel (1852–1908) and his colleagues in France, were interested in the nature of the newly discovered radiation emanating from uranium. One outcome of these investigations was the discovery that there were two kinds of radiation given off by uranium, distinguished on the basis of their penetrating power; alpha rays were absorbed by very thin layers of material, while beta rays were more penetrating. The initial characterization of these emanations was as "rays," and the similarity of these alpha rays and beta rays to Roentgen's X-rays was debated. When it was found that beta rays were deviated by a magnetic field, they achieved the new status of particles. As Heilbron noted: "Magnetic deviability . . . was the touchstone for a stream of charged particles."[14] Alpha rays did not exhibit the expected attenuation of other kinds of rays, that is, light rays and X-rays, and Marie Curie (1867–1934) suggested that alpha rays, "being larger than electrons and possessing at the same time a smaller velocity, . . . have more difficulty in traversing obstacles and form rays that are less penetrating."[15] It was one of Thomson's students, Robert J. Strutt (1875–1947; son of the famous physicist Lord Rayleigh), who provided an experimental test of this hypothesis and managed to convince both William Crookes (1832–1919) and Ernest Rutherford (1871–1937) of the particulate nature of alpha rays.[16]

These discoveries of the particulate nature of alpha and beta rays led, in turn, to an interest in understanding the absorption process itself, that is, the ways that the radiations interact with the absorbing material. Not only were these radiations absorbed, but it was found that they were scattered. A detailed description of the work of Thomson and Rutherford on scattering of alpha and beta radiation has been presented by Heilbron,[17] and his paper is a good background for the arguments presented here. In 1906 Thomson, professor of physics at Cambridge, published what Heilbron calls "one of the most important papers on atomic structure ever written."[18] This paper described his quantitative attempts to determine the order of magnitude of the number of electrons contained in an atom of atomic weight A. Thomson's approach to study of the internal structure of the atom was to employ the scattering of X-rays, the absorption of beta rays, and dispersion of light by gases. His object was to test a model advanced by William H. Bragg (1862–1942) and Rutherford that the atom consisted of a large swarm of moving electrons (the beehive model).

In contrast to the progress in understanding alpha and beta radiation, the nature of X-rays remained quite mysterious. As Abraham Pais has noted: "At least until 1910 most physicists' conception about the absorption of radiation by matter was incorrect."[19] In 1905 Albert Einstein (1879–1955) published his now famous paper on the photoelectric effect, in which he advanced the quantum theory of light. It was this concept of light that would provide the eventual explanation of the nature of X-rays. Einstein's paper, which introduced what has been termed the "reckless hypothesis" of light quanta, had the rather uninformative title "Über einen die Erzeugung und Verwandlung des Lichtes betreffenden heuristischen Gesichtspunkt" [On a heuristic point of view concerning the production and transformation of light].[20] In this paper he formulated the hypothesis that "when a light ray spreads out from a point source, the energy is not distributed continuously over an ever-increasing volume but consists of a finite number of energy quanta that are localized at points in space, move without dividing, and can be absorbed or generated only as complete units."[21] Einstein showed how this radical hypothesis provided an explanation for three observed phenomena: Stoke's rule for fluorescence, the ionization of gases by ultraviolet light, and the photoelectric effect. From 1905 until the early 1920s, in spite of Einstein's growing reputation, most of the physics community was strongly opposed to the light-quanta hypothesis. Indeed, in 1913, in proposing Einstein for membership in the Prussian Academy, Planck, Nernst, Rubens, and Warburg wrote warmly about his eminent status in physics, but went on to note "that he may sometimes have missed the target in his speculations, as, for example, in his hypothesis of light-quanta, cannot really be held too much against him, for it is not possible to introduce really new ideas even in the most exact sciences without sometimes taking a risk."[22]

The turning point, at least for most in the physics community, if not for Einstein, came in the early 1920s when the experiments of Arthur H. Compton (1892–1962) on the frequency shifts for X-rays scattered from matter, modeled after a collision between a light quantum (christened a "photon" by G. N. Lewis in 1926)[23] and a nearly free electron, provided confirmation of Einstein's light quanta and put to rest the theoretical objections of Niels Bohr (1885–1962), John C. Slater and Hendrik "Hans" Kramers, to mention a few of his eminent opponents.[24] Einstein summed up the state of affairs in a popular article in 1924: "The positive results of the Compton experiment proves that radiation behaves as if it consisted of discrete energy projectiles, not only in regard to energy transfer but also in regard to momentum transfer."[25]

During these same two decades at the inception of the twentieth century, two other results were widely heralded in physics, each of which contributed to the notion of projectiles and targets. In 1912 Charles T. R. Wilson (1869–1959) published his remarkable photographs of visible tracks of "ionizing particles" observed in his newly invented cloud chamber.[26] These photographs made visible the idea that particles not only collided with gaseous molecules to serve as nucleation sites for droplet condensation, but that the particles could be "seen" to hit and ricochet from atoms and possibly other subatomic structures. In this same year, James Franck (1882–1964) and Gustav Hertz (1887–1975) published an experimental result that was designed to test the ionization of molecules in the vapor phase by accelerated electrons. Their results were surprising, even to them, and were the first new test of the Bohr model of the atom with bound electrons. The Franck-Hertz experiment also required a "projectile-target" interpretation.

The physicists studying atomic structure initially focused on the fate of the projectiles, that is, absorption and scattering, but by about 1914, Rutherford and others started to talk about transmutation experiments, indicating an interest in target physics. A next logical step in the bombardment experiments was to look at the fate of the target material. Atomic theory suggested that it might be possible to transmute one element into another by bombardment with the right particles with the right energy. As early as 1910 there was talk of needing high-voltage technology to accelerate particles for bombardment experiments to overcome the limitation on the energies of the natural sources of particles, that is, the alpha and beta from uranium. The physicists began to focus on the target. By 1919 Rutherford had clearly formulated these notions in terms of target-atom destruction, that is, "atom smashing."[27] In 1922 Rutherford's students James "Jimmy Neutron" Chadwick (1891–1974) and Etienne Bieler (1895–1929) described a model of the nucleus as an oblate spheroid calculated from collision theory and supported by particle bombardment and scattering experiments, the "target theory" of the physicists.[28]

Radiation and Biology

The earliest and most frequently cited paper that describes the "target theory" (at least in its current conception) is by James Arnold Crowther (1883–1950) and appeared in 1924 in the *Proceedings of the Royal Society of London*. While Crowther is credited as the originator of the target theory as it is known to biologists, the fundamental notions involved in the target theory had been employed by physicists for over a decade.

Thomson's first scattering paper appeared the same month that Crowther received his bachelor's degree at Cambridge and joined Thomson's department, where he took up the challenge of determining the number of electrons (n) in an atom of atomic weight A. Between 1908 and 1912 Crowther published the results of his scattering experiments, which significantly influenced Bragg and Rutherford in their thinking about the interior structure of the atom. In 1912 Crowther became demonstrator and lecturer in physics in Thomson's department at the Cavendish Laboratory, a position he held until 1924. Crowther's problem was to study the unseen, internal structure of the atom, and his approach initially focused on the observations of the fate of the projectile, that is, the incident and scattered particles and radiations, with apparently little attention to the effects on the absorbing or scattering material.

Crowther seemed to get on famously with Thomson and was apparently a skilled experimentalist. With the coming of the Great War, Crowther was detailed to serve in the Medical Radiography Unit at Addenbrooke's Hospital in Cambridge, where many war casualties were taken. As an adept experimental physicist, he was able to contribute to the war effort in a department that needed technological expertise with X-ray equipment. Such facilities for X-ray diagnosis were rare, and Crowther took this opportunity to collaborate with medical colleagues. These contacts had a lasting impact on Crowther, and after the war he became involved in the medical aspect of radiation studies. In 1921 he was appointed lecturer in medical radiology at Cambridge, concurrently with his position in the physics department. He wrote a text for the Cambridge diploma course in medical radiology, *Principles of Radiography*, and edited another book on applied topics, *The Handbook of Industrial Radiology*. Although this switch in interests may have been stimulated by financial and professional incentives, because, according to Heilbron, Crowther "had long wanted an independent, and needed a more remunerative position,"[29] Crowther maintained these interests even after he was appointed professor and chair of the physics department at the University of Reading in 1925.

Crowther's popular textbook, *Ions, Electrons and Ionizing Radiations*,[30] went through eight editions between 1919 and 1961. In this text he described the bombardment and transmutation work with increasing detail and enthusiasm as it went through rapid revisions from the first edition in 1919 to the fourth in 1924. For the fourth edition he added a new section entitled "Collision of α-Particles with Atoms" with a long discussion of different target and projectile models.

At the same time, very nearby in Cambridge, the biological effects of X-rays were being investigated. In the early 1920s the growth of cells and tissues outside the animal body was a major goal of many biologists. One of the pioneers in this effort was Thomas S. P. Strangeways (1856–1926), director of the Cambridge Research Hospital.[31] Since radiation was being used to treat a wide variety of difficult conditions, from cancer to infections to arthritis, Strangeways and others were interested in studying the effects of radiation on tissues and cells growing in isolation in culture flasks.

In 1923 Strangeways and H. E. H. Oakley published a paper entitled "The Immediate Changes Observed in Tissue Cells after Exposure to Soft X-rays,"[32] which caught the attention of Crowther. This paper gave qualitative observations on the frequency of tissue culture cells entering into mitosis as a function of exposure to X-rays. The highly qualitative nature of these data are such that it is very hard to understand why anyone would be tempted to base any sort of quantitative physical theory on them. However, two factors may be relevant: first, the experiments used radiation bombardment, one of Crowther's primary interests, and second, Strangeways and Crowther were both longtime fellows of St. John's College in Cambridge. As such, they undoubtedly had opportunities to discuss Strangeways's work beyond the vague and qualified descriptions in his publications.

Crowther wrote: "The authors [Strangeways and Oakley] find that after an exposure to the rays of 5 minutes the number of cells in mitosis was appreciably diminished, after an exposure of 10 minutes the number was still smaller, after 15 minutes only a few cells in mitosis were visible, and for still more prolonged exposures cells in mitosis were only seen occasionally in some of the cultures."[33] He goes on to state: "Although no numerical data are given these statements suggest *very strongly* that the number of cells capable of passing into mitosis is decreasing exponentially with the time of exposure; in other words that the action of the X-rays on the cells which produces the incapacity for entering into mitosis follows a probability law." Crowther went on: "It seemed interesting to consider whether this probability might not be due to the X-rays themselves, and represent the probability that a given structure in the cell would actually be affected by the incident radiation."[34]

From this point, Crowther went on to derive a simple expression relating the size of this "given structure" to its sensitivity to inactivation by the X-rays. For this derivation he assumed that the X-rays produce their effects by producing clusters of ions that must occur in the critical structure (in his words "hit") in order to produce an inactivating event.[35] From these calculations and assumptions, he estimated that the structure would be about the size of the centrosome, a cytologically observable structure in the cell related to chromosome partitioning, described by Theodor Boveri, its discoverer, as the central organ of cell division. For another observed effect reported in this paper, that is, cell disintegration, Crowther noted that a simple assumption of one hit per target was inadequate to account for the observations, but that a two-hit model sufficed. It seems clear that Crowther had accepted the quantum theory of X-rays and that he interpreted a hit as a local ionization in the target structure.

Over the next several years Crowther accumulated data in his own laboratory to test his approach to the target theory.[36] One objective was to test it on known intracellular targets to verify the calculations and assumptions of the model. Others soon picked up on this approach to investigate intracellular processes too complex to be studied biochemically or morphologically. It was particularly appealing to microbiologists because it gave a rational explanation for the exponential dose-response curves seen with heat, ultraviolet, and ionizing radiation.

Another approach to explanation of the biological effects of ionizing radiation was elaborated in Frankfurt am Main by Friedrich Dessauer (1881–1963) and his colleagues, Marietta Blau (1894–1970) and Kamillo Altenburger.[37] Dessauer was founder and director of the Institute for the Physical Basis of Medicine at the University of Frankfurt and an important figure in the development of the German school of radiation therapy. In his first paper in 1923 he reasoned from the quantum nature of light and X-rays and the recent physical studies on the ionization of gases, that in the cell, X-rays would be expected to interact with molecules, "electrolytes, colloids and membranes,"[38] and excite them and break them into fragments. The splitting and recombination of these molecules would, he reasoned, result in the very local dissipation of energy in the cell. This energy deposition in a small volume would cause "the small region (*Punkten*) to experience heating to a high temperature—I will call it 'point-heat' (*Punktwärmen*) in what follows."[39] Dessauer cited Wilson's cloud-chamber results as well as other physical studies in his attempts to calculate the magnitude of the heating he predicted. This calculation required not only information about the quantity of energy deposited per ionization event but also an estimate of the volume over which this energy was absorbed and dissipated. He arrived at the conclusion that the heating would be at least sufficient to coagulate enzymes inside the cell.

Dessauer constructed a model of the cell in which there were sensitive "particles" (*Teilchen*), and he realized that the fraction of particles that remained intact after irradiation with a certain dose was an exponential function of the dose. He then considered the case where a particle had to be hit more than once to be heated sufficiently to be inactivated. This assumption gave a single-target, multi-hit model. In search of biological data with which to test his theory, Dessauer cited experiments reported in September 1921 by F. C. Wood from Columbia University, in which Wood was observing the effects of radiation on transplanted mouse tumors. Wood's dose-response data, according to Dessauer, "did not go to zero, but on the contrary showed an exponential curve."[40]

Dessauer and his colleagues clearly applied the physical concepts of target and projectile, and in an addendum to his first paper he acknowledged a recent essay of Emil Warburg (1846–1931)[41] in which the photochemical effect of light was discussed in somewhat similar terms. He also acknowledged the contributions of J. J. Thomson's work to his thinking.[42]

While the model of radiation effects on biological material envisioned by Dessauer embodied the principles borrowed from atomic physics, the "point-heat theory" seemed aimed at explaining the nature of the physical event that killed cells, and then, from knowledge of that process, deducing what kinds of biological processes were responsible for the sensitivity of the cells to radiation effects. Blau and Altenburger, in the second paper on this subject from Dessauer's institute, elaborated further on the possible heating to be expected, and they gave detailed, step-by-step derivations of the exponential equation from Dessauer's first paper. They discussed the interpretation of each of the dose-response curves that were derived with different hypotheses: single-target, single-hit versus single-target, multi-hit. In all their arguments the number of particles in the cell was not treated as a variable that might be measured by this experimental approach, but instead this number was taken as more or less known from other work: "It is known that the cell has between 10^9 and 10^{10} such [protein] molecules, between 5,000 and 20,000 molecular weight."[43] In contrast to Crowther, Dessauer apparently did not realize that his concept of projectile and target might allow him to infer something more about the nature of the target itself (i.e., size and number), beyond the process of its physical destruction.

What Is a Hit?

The investigations of the biological effects of radiation were initially qualitative. While it is often taken for granted that dose-effect relationships are fundamental aspects of the understanding of causal effects, this was not a well-formulated concept, even in pharmacology, until the nineteenth century when quantitative methods because available. So too with radiation. "Doses" of radiation were initially measured by time of exposure, under specified conditions (distance and shielding), to a given amount of radioactive material, usually radium. With the development of X-ray generators, doses were described in terms of exposure times with applied voltages, tube design, and shielding being specified. In medical work, doses were sometimes stated in terms of multiples of the "minimal erythema dose," that is, the dose needed to produce an acute radiation "sunburn" on the skin of the experimenter (or his assistant).

One of the early known physical effects of X-rays was their ability to discharge a charged electroscope. This phenomenon was interpreted by the second decade of the twentieth century as the result of the ionization of the air by the X-rays. X-rays were recognized as "ionizing radiation," and a standard measure of X-rays in the physical laboratory was by an "iontometer"; the numerical values were reported in terms of time to discharge a standardized electroscope. Crowther's 1919 monograph was the first comprehensive textbook on this subject and gives a clear summary of the contemporary understanding of the way X-rays interact with matter.[44] The main evidence for this understanding came from Wilson cloud-chamber experiments where the X-ray tracks are invisible and show no water-droplet condensates; but only after the X-rays produce a primary ionization and eject secondary electrons are tracks visible. In 1919 Crowther concluded, "The X-rays produce no ions directly, but only through the medium of the secondary electrons ejected by them from the atoms of the gas."[45] Straddling the wave and particle concepts, he described X-rays as "ether pulses" of wavelength about 10^{-8} centimeters.[46]

By 1927 a more detailed description of the events in the region of the primary ionization was provided by Edward U. Condon (1902–74) and Harold M. Terrill (1890–?) from the Physics Department and the Institute of Cancer Research, respectively, at Columbia University: "The quantum of X-ray energy, when absorbed, is taken up by one atom of the absorbing substance and a high-speed photo-electron is liberated. This electron moves about in the neighborhood of the place of its liberation, losing energy by collisions with atoms and causing a good deal of local ionization. It is presumably the disturbing effect of this ionization on certain colloid equilibria which causes biological action, but that question is outside the realm of this paper."[47] Condon and Terrill noted that about 30–40 electron volts were expended to create an ion pair, so "it follows that such an absorption of one quantum of 160 KV X-ray is like a highly localized burst of ionic shrapnel in which about 4,000 ion pairs are liberated in less than a millionth of a cubic centimeter."[48] They were clearly aware that the effects of ionizing radiation depend on the size of the volume in which the ionizations occur, and they estimated the sensitive volumes for several biological endpoints as well as showing that some data were best described by the single-hit model and other data best fit the multi-hit model.

Ten years later, in the fifth edition of his textbook, Crowther described the progress in understanding X-rays by including a new section on Compton scattering and noted that "Compton made a somewhat daring, but very successful application of the quantum hypothesis. He assumed that the quanta of energy in the primary X-ray beam were so highly localized that they might be regarded as particles of negligible size, and further, that the collision between one of these quanta and an electron might be treated by the ordinary laws of dynamics, in exactly the same way as the collision between two perfectly elastic particles."[49] Between 1919 and 1929, X-rays had evolved from "ether pulses" of short wavelength to classical "particles of negligible size." Crowther resisted the new term, "photon," for such particles, preferring instead his own term: "quant."[50]

The interaction of radiation with matter was of crucial importance in the evolving understanding of its biological effects. While it was known that absorption of radiation involved deposition of energy in matter and that energy could lead to chemical changes, it was unclear just how such energy absorption should be measured. Three basic approaches were in use: a biological measurement used by early radiation therapists, that is, the minimal exposure needed to give a reddening of the skin, the "minimal erythema dose" or MED; a dose measurement based on the chemical changes caused by the absorbed radiation that employed standardized chemical reactions such as the oxidation of ferrous iron to ferric iron in solution, the so-called Fricke dosimeter; and the amount of ionization produced in air by exposure to X-rays as measured by discharge of an electroscope.

The Bohr model of the atom, the Franck-Hertz experiments, and the Compton scattering experiments provided some basis for thinking about these matters. Hans Bethe (1906–2005) developed theoretical approaches to calculate the energy required to produce one ion pair, from an atom of atomic number N, the so-called ionization potential. His theoretical calculations yielded a value of 32 electron volts as the amount of energy absorption needed to produce one ion pair in air. The experimental determination of this value, usually called W, was reviewed in 1930 by Rutherford, Chadwick, and Ellis, and they concluded that the experimental values for W were in close agreement with Bethe's theoretical value. However, because the ionization potential varies with the atomic number, it was unclear whether Bethe's theory held for all matter, for example biological tissues, and for all energies of X-rays. Working independently, both William H. Bragg and Louis Harold Gray (1905–65) provided the theoretical approach to this problem by 1928, and by 1936 Gray had investigated this matter experimentally.

The understanding of the ionization potentials provided the basis for standardization and quantization of X-ray exposure doses. In 1918 Bernhard Krönig and Walter Friedrich published *Die physikalischen und biologischen Grundlagen der Strahlentherapie* to lay out the underlying principles for radiation medicine.[51] In particular, they also emphasized the work on the nature of X-rays of Friedrich and Paul Knipping in Max von Laue's Munich laboratory (for which Laue won the Nobel Prize in 1914). Krönig and Friedrich described several biological systems for the study of radiation effects (the frog embryo, their own skin, human cancer nodules, and human ovaries) and related those effects to the dose of radiation as measured by an ionization meter. Much of this text is devoted to the study of ways to standardize X-ray doses. They investigated both the quantification of the "intensity" of the X-rays as well as their "hardness." They compared energy absorption in water, tissue, air, and metals. Probably their most important study was to characterize the properties of the "iontoquantimeter," their name for a device that measured the quantity of ions produced in air by the X-rays. All their biological experiments were reported in terms of doses related to air-ionization measurements.

Friedrich proposed that doses of X-ray exposure be standardized, based on the quantity of ions produced in air under standard conditions. In this approach he was following Bela Szilard, who had proposed that the dose be expressed in terms of the number of ions produced in one cubic centimeter of air, with a unit Szilard called the "megamegaion."[52] For economy of terminology, Friedrich preferred a unit based on the production of one electrostatic unit (esu) of charge in one cubic centimeter.

Even though it was understood that X-rays caused ionizations in air, which could be measured quantitatively, there was no systematic attempt to relate that fact to the quantitative responses to irradiation. In Muller's famous 1927 paper on X-ray mutagenesis in *Drosophila melanogaster*, for example, he indicated various dosages as "t2, t4," presumably referring to length of exposure without any additional details.[53] Pais noted that "how slow quantitative methods penetrated [into biological and medical uses of radiation] is seen, for example, from the fact that the roentgen [Friedrich's unit] was adopted internationally only in 1928, when an International Commission on X-ray units convened—for the first time."[54]

Because the roentgen as a unit of dose is a measure of the total number of ions produced per unit volume, and, importantly, the energy required to produce an ion pair (W) is constant (i.e., about 32 eV), by the early 1930s it became possible to think about the effects of radiation of different energies in more precise and mechanistic terms. Both experimental biologists and physicians using radiation began to be more concerned about precise dosimetry, both for theoretical reasons and for the practical aspects of reproducible treatment regimens.

Against this background of evolving understanding of X-rays and their measurement, it is interesting to examine the work of Karl Zimmer, the biophysicist in our trio of authors. Zimmer's attention, too, was focused on problems of X-ray dosimetry, and in 1936, a year after the 3MP, he published a small monograph on *Radiumdosimetrie*[55] and in 1937 a book *Strahlungen: Wesen, Erzeugung und Mechanismus der biologischen Wirkung* [Radiations: Nature, production, and mechanism of biological action].[56] In the latter work, he described in detail the way X-rays produce ions and how they can be collected and measured in a typical air-ionization chamber. After reviewing the production of photoelectrons and Compton electrons, he provided a chapter called "The General Theory of Biological Radiation Action: 'Exact Radiation Biology.'" This short chapter (seven pages) provides a contemporary view of Zimmer's concept of quantitative radiation biology. Zimmer's general treatment of dose-response curves ("The Damage Curve and Its Possible Interpretations") seems to approach the subject empirically rather than theoretically; that is, he noted that the general form of dose-response curves is sigmoid when the effect is plotted on linear coordinates versus dose. This sigmoid relationship can be transformed into a more-or-less symmetric curve with a maximum at some mean value of dose, if the damage is expressed as damage per dose increment, that is, the first derivative of the sigmoid curve. He also considered what might be the effect of varying the intensity (dose/time) on the biological effects and included data on mutations in *Drosophila melanogaster* irradiated to the same final dose (measured in roentgens) but with the irradiation carried out over several orders of magnitude of time, showing a negligible effect of dose rate on biological effects. The most interesting aspect of Zimmer's exposition is the section called "Statistical Theory of Damage Kinetics." He used the language of the target theory in his reference to target number, he provided hypothetical sigmoid dose-effect curves for cases of a single hit per target and 5.6 hits per target, and he noted that these are exponential relationships but did not adopt the point of view of Crowther's target theory and interpret the exponential curves in terms of Poisson statistics. Indeed, the relationship between the random distribution of hits and the shapes of the survival curves was left mystifyingly unanalyzed.

Zimmer compared two cases for effects plotted against dose (measured in roentgens) for radiation of two different wavelengths and argued that the expected differences result from the difference in the distribution of ions in the two cases, with the very low energy radiation being relatively ineffective in producing a biological effect. This approach, based on comparison of radiation of different wavelengths (energies), played a significant role in the biophysical analysis in the 3MP.

When we turn to the experimental results in the 3MP, we find emphasis on the two key biophysical points made by Zimmer in his exposition of radiation biology: measurement of dose in roentgens and the study of wavelength effects. Timoféeff-Ressovsky's experiments on production of mutations after exposure to X-rays were initially qualitative, but in one section of the 3MP (251 [II.3.b, 220]) he described the quantitative dose-response relationships between frequency of mutation and X-ray doses. He explicitly stated that the doses were measured in roentgens and italicizes this statement for emphasis (253 [II.3.c, 222]) The casual reader might wonder why this fact required emphasis. As noted above, X-ray dosimetry was in its early stages and standardization of doses was not yet routine. Further, and most important, *the use of the roentgen as a measure of dose allowed a direct interpretation of the dose in terms of the number of physical events (ionizations) and the biological events (mutations)*. Without such a quantization, Delbrück's calculations at the end of the paper would have been impossible.

The interpretation of these experiments was made more secure by Timoféeff-Ressovsky's experiments with radiation of differing wavelengths (253 [II.3.c, 222]). As Zimmer noted, because the dose was measured in roentgens, which measures the number of ions produced in air, it was a way to normalize for the fact that the quanta of short-wavelength radiations carried more energy per quantum than radiation of longer wavelength. The fact that they observed no significant wavelength effect when the doses were measured in terms of number of ions produced suggested that the biological effect (i.e., mutation) was dependant on the number of ions produced, rather than on some direct effect of the photons on the gene. If one photon caused one mutation, for example, a high-energy and a low-energy photon might both have equal effects in terms of mutation, but in terms of the dose measured in roentgens, the high-energy photon would produce more ions and hence give a higher dose per photon.

Zimmer's analysis of "The Problem of the Primary Processes," section 4 in his chapter called "Exact Radiation Biology," provides us with his conception of a "hit" only a year later than the 3MP. He summarized the two prevailing notions: first, secondary and tertiary ionizations and, second, localized thermal heating (point heat). He then offered a third, more original, possible mechanism, somewhat reminiscent of the disruption of "certain colloid equilibria" of Condon and Terrill: "Finally, there is yet a third possible reaction mechanism; namely, through the ionizing action of radiation, the degree of dispersion of colloids can change because of their excited state." But he concludes: "A final decision for one or the other of these three mechanisms of the primary event [direct ionization, point heating, and colloidal dispersions] cannot yet be ascertained."[57]

The experiments and analysis in the 3MP seem to have been a confluence of viewpoints already held by the three authors prior to their collaboration. Timoféeff-Ressovsky had been a champion of the point-mutation hypothesis as contrasted with L. J. Stadler's conception of mutation as a chromosome rearrangement.[58] Zimmer's interest in dosimetry with the air-ionization chamber would seem to complement a view of radiation mutation experiments in terms of interactions of ions with genes as individual physicochemical events rather than as cytologically observable rearrangements.

The central importance of dosimetry based on ionization methods is corroborated by Delbrück's memory many years later: "As I recall, and I have not reread the paper, the experimental conclusions were that the number of recessive mutations that you find in the X chromosome was proportional to that dose, *if one measures the dose in terms of ion pairs produced, or small clusters of ion pairs*."[59] Delbrück acknowledged Zimmer's contribution to this program in his recollection: "In order to do this [measure radiation mutagenesis] quantitatively, we had to have quantitative dosimetry of the ionizing radiation, and the person responsible for that was [K. G.] Zimmer."[60] Without appreciation for Zimmer's role in quantitative dosimetry, his participation in this three-way collaboration would seem a bit superfluous.

The Rise of Biophysics

After Crowther's seminal paper on the target theory, the next most frequently cited work on this subject again came from the Cavendish Laboratory and was published in 1936 by Douglas E. Lea (1910–47), Raymond B. Haines (1905–43), and Charles A. Coulson (1910–74). Entitled "The Mechanism of Bactericidal Action of Radioactive Radiations,"[61] it reviewed Crowther's derivation, but added extensive laboratory data demonstrating the exponential "probability law" for "disinfection" of different kinds of bacteria under different conditions. Interestingly, this paper was submitted to the Royal Society by Frederick Gowland Hopkins (1861–1947), professor of biochemistry at Cambridge, perhaps an indication of the increasing influence of this target-theory approach.

The use of the target-theory concept seemed to mark the user as something special, not quite a biologist, not quite a biochemist. It may even be the case that the use of the target-theory approach was one of the defining characteristics of the new discipline of biophysics. The words of the novelist may capture the essence of the era: Sinclair Lewis provided a contemporary perspective on young Martin Arrowsmith's X-ray inactivation data: Rippleton Holabird, director of the McGurk Institute, "was as much bewildered as Tubbs would have been by the ramifications of Martin's work. What did he think he was anyway—a bacteriologist or a biophysicist?"[62]

This target-theory approach had great appeal to physicists who were looking for interesting problems in biology during the 1930s and 1940s. Erwin
Schrödinger, Max Delbrück, Ernest Pollard, and Fernand Holweck were
among the well-known physicists to whom the conceptual and experimental
simplicity of the target theory held great attraction. The application of
these ideas from the atomic physics of Thomson and Rutherford to study
the biology of the gene seems to have depended in the first instance on
a contingent conjunction of events and individuals with specific interests
and knowledge. The wider application of the target theory would then
serve the interests and accord with the scientific style of a generation of
physicists who were developing the new field of biophysics. Initially, the
target theory became a central tool of biophysics to study the size and
number of functional molecules and structures in complex biological systems. Later, the target theory became accepted by many biologists as a
valid approach to biological structure and function, perhaps because biology as a field became more and more dependent on chemistry and physics. Indeed, almost as a follow-up to the studies of the fictional Martin
Arrowsmith, the PhD thesis of one of our current icons of biology and an
acknowledged disciple of Max Delbrück, James D. Watson (1928–), was a
study of the detailed application of the target theory to the X-ray inactivation of bacteriophage.[63]

Conclusions

While it seems logical today that radiation was "the right tool for the
job"[64] of studying gene mutation and gene structure in the 1930s, this was
not the only choice available at the time. The particular collaboration of a
physicist and a biophysicist both interested and well-versed in the recent
physics of X-rays and particle physics with a geneticist whose particular
bias was toward mutation as a singular "point" event led to the specific
outcome in this landmark paper. They were not the first to appreciate the
power of the target-theory approach, but had a particularly clear vision of
its application and the logical consequences that followed from the recent
physical understanding of X-ray physics. The conception of X-rays as particles (photons) that produced ionizations localized in space and involving
molecular changes was key to this work. The quantitative relations between the dose-response relationships and the ionizations produced made
possible the detailed calculations that give the work much of its authority.

With a richer context for the X-ray physics of the time, the 3MP can be more fully appreciated.

Notes

1. G. S. Stent, *Molecular Biology of Bacterial Viruses* (San Francisco: W. H. Freeman and Co., 1963), 18.

2. E. P. Fischer and C. Lipson, *Thinking about Science: Max Delbrück and the Origins of Molecular Biology* (New York: W. W. Norton, 1988).

3. N. V. Timoféeff-Ressovsky, K. G. Zimmer, and M. Delbrück, "Über die Natur der Genmutation und der Genstruktur," *Nachrichten von der Gesellschaft der Wissenschaften zu Göttingen, mathematisch-physikalische Klasse, Fachgruppe VI: Biologie,* 1 (1935): 189–245.

4. Erwin Schrödinger, *What Is Life? The Physical Aspect of the Living Cell, with Mind and Matter and Autobiographical Sketches,* with forward by R. Penrose (Cambridge: Cambridge University Press, 2000), 57.

5. K. G. Zimmer, "The Target Theory," in *Phage and the Origins of Molecular Biology,* ed. J. Cairns, G. S. Stent, and J. D. Watson (Cold Spring Harbor, NY: Cold Spring Harbor Press, 1966), 37.

6. Stent, *Molecular Biology of Bacterial Viruses,* 18.

7. H. J. Muller, "The Production of Mutations by X-Rays," *Proceedings of the National Academy of Sciences of the U.S.A.* 14 (1928): 714–26.

8. Examples include Yale's Department of Biophysics, the University of California's Donner Laboratory of Biophysics and Medical Physics, and the University of Rochester's Department of Radiation Biology.

9. H. J. Muller, "Artificial Transmutation of the Gene," *Science* 66 (1927): 84–87.

10. L. J. Stadler, "Mutations in Barley Induced by X-rays and Radium," *Science* 68 (1928): 186–87.

11. The concepts and terminology subsumed under the rubric of the "target theory" derive from two main pathways of investigation. Both approaches assumed that there is a critical structure or region in the cell, the "target," in which the deposition of energy, the "hit," is able to cause some change in function, such as inhibition of mitosis, cell death, or, later, mutation. The exact nature of the "hit" was debated. Crowther, in the English-language literature, interpreted a hit as the ionization of some component of the target that in itself caused the biological effect; Dessauer and his colleagues, in the German literature, thought of the hit as local heating by the ionization and that the biological effects resulted from thermal effects, "Punktwärme" [point heat]. That the primary event was ionization was not in question.

12. Richard P. Setlow and Ernest C. Pollard, *Molecular Biophysics* (Reading, MA: Addison-Wesley, 1962).

13. Douglas E. Lea, *Actions of Radiation on Living Cells* (Cambridge: Cambridge University Press, 1946); second edition, 1955.

14. John Heilbron, "The Scattering of α and β Particles and Rutherford's Atom," *Archive for History of Exact Sciences* 4 (1967): 247–307.

15. M. Curie, "Radium and Radioactivity," *Century Magazine* (January 1904): 461–66.

16. Heilbron, "Scattering," 253.

17. Heilbron, "Scattering."

18. Heilbron, "Scattering," 269; The paper in question is J. J. Thomson, "On the Number of Corpuscles in an Atom," *Philosophical Magazine* 11 (1906): 769–81.

19. Abraham Pais, *Inward Bound: Of Matter and Forces in the Physical World* (Oxford: Oxford University Press, 1986), 94.

20. A. Einstein, "Über einen die Erzeugung und Verwandlung des Lichtes betreffenden heuristischen Gesichtspunkt," *Annalen der Physik* 322 (1905): 132–44.

21. Einstein, "Heuristischen Gesichtspunkt," 133.

22. Quoted in Abraham Pais, *Subtle Is the Lord: The Science and Life of Albert Einstein* (Oxford: Oxford University Press, 1982), 382.

23. Gilbert N. Lewis, "The Conservation of Photons," *Nature* 118 (1926): 874–75.

24. Roger H. Steuwer, *The Compton Effect: Turning Point in Physics* (New York: Science History Publications, 1975).

25. Quoted in Pais, *Subtle Is the Lord*, 414.

26. C. T. R. Wilson, "On an Expansion Apparatus for Making Visible the Tracks of Ionizing Particles in Gases and Results Obtained from Its Use," *Proceedings of the Royal Society, Section A* 87 (1912): 277–92.

27. Ernest Rutherford, "Collision of α-Particles with Light Atoms," *Philosophical Magazine* 37 (1919): 537–87.

28. James Chadwick and Etienne Beiler, "The Collision of α-Particles with Hydrogen Nuclei," *Philosophical Magazine* 42 (1922): 923–40.

29. Heilbron, "Scattering," 297.

30. James Arnold Crowther, *Ions, Electrons, and Ionizing Radiations* (first edition, London: Longmans, Green and Co., 1919; eighth edition, London: Edward Arnold, 1961).

31. The Cambridge Research Hospital was a "scheme" devised by Strangeways that consisted of a free hospital devoted to the "systematic investigation of some of the important diseases the pathology and treatment of which are as yet undetermined." http://www.srl.cam.ac.uk/history/thomas_strangeways.html. The focus of the research was on rheumatoid arthritis and allied diseases. When the hospital first opened in 1905, it had two wards of three beds each and a laboratory in a converted coal shed. It subsequently became well known for its pioneering work in cell and tissue culture.

32. T. S. P. Strangeways and H. E. H. Oakley, "The Immediate Changes Observed in Tissue Cells after Exposure to Soft X-rays," *Proceedings of the Royal Society, Section B* 95 (1923): 373–81.

33. J. A. Crowther, "Some Considerations Relative to the Action of X-Rays on Tissue Cells," *Proceedings of the Royal Society, Section B* 96 (1924): 207.

34. Crowther, "Considerations," 207–8, emphasis mine.

35. See note 11.

36. J. A. Crowther, "The Action of X-Rays on *Colpidium Colpoda*," *Proceedings of the Royal Society, Section B* 100 (1926): 390–404.

37. Friedrich Dessauer, "Über einige Wirkungen von Strahlen, I," *Zeitschrift für Physik* 12 (1922): 38–47; Marietta Blau and Kamillo Altenburger, "Über einige Wirkungen von Strahlen, II," *Zeitschrift für Physik* 12 (1922): 315–29; Dessauer, "Über einige Wirkungen von Strahlen, IV," *Zeitschrift für Physik* 20 (1923): 288–98; Dessauer, "Zur Erklärung de biologische Strahlenwirkungen," *Strahlentherapie* 16 (1924): 208–21.

38. Dessauer, "Über einige Wirkungen, I," 40.

39. Dessauer, "Über einige Wirkungen, I," 41.

40. Dessauer, "Über einige Wirkungen, I," 45.

41. E. Warburg, "Über die Anwendung der Quantenhypothese auf die Photochemie," *Die Naturwissenschaften* 30 (1917): 67.

42. Dessauer, "Über einige Wirkungen, I," 46–47.

43. Dessauer, "Über einige Wirkungen, IV."

44. Crowther, *Ions* (1919).

45. Crowther, *Ions* (1919), 179.

46. Crowther, *Ions* (1919), 5–6.

47. E. U. Condon and H. M. Terrill, "Quantum Phenomena in the Biological Action of X-Rays," *Journal of Cancer Research* 11 (1927): 324–33.

48. Condon and Terrill, "Quantum Phenomena," 325.

49. Crowther, *Ions* (1929), 177.

50. Crowther, *Ions* (1929), 177.

51. Bernhard Krönig and Walter Friedrich, *Die physikalischen und biologischen Grundlagen der Strahlentherapie* (Berlin: Urban and Schwarzenberg, 1918). English translation: *The Principles of Physics and Biology of Radiation Therapy* (New York: Rebman, 1922).

52. Krönig and Friedrich, *Principles*, 60.

53. Muller, "Artificial Transmutation of the Gene."

54. Pais, *Inward Bound*, 94.

55. Karl G. Zimmer, *Radiumdosimetrie: Verfahren und bisherige Ergebnisse* (Leipzig: Georg Thieme Verlag, 1936), 40.

56. Karl G. Zimmer, *Strahlungen: Wesen, Erzeugung und Mechanismus der biologischen Wirkung* (Leipzig: Georg Thieme Verlag, 1937), 72.

57. Zimmer, *Strahlungen*, 62.

58. James F. Crow, "Sixty Years Ago: The 1932 International Congress of Genetics," *Genetics* 131 (1992): 764–66.

59. Max Delbrück interview by Carolyn Harding, Pasadena, CA, July 14–September 11, 1978, Oral History Project, California Institute of Technology Archives, retrieved February 18, 2008, from http://oralhistories.library.caltech.edu/16/01/OH_Delbruck_M.pdf, 49, emphasis mine.

60. Ibid.

61. D. E. Lea, R. B. Haines, and C. A. Coulson, "The Mechanism of Bactericidal Action of Radioactive Radiations: I, Theoretical," *Proceedings of the Royal Society, Section B* 120 (1936): 47–76.

62. Sinclair Lewis, *Arrowsmith* (New York: Harcourt Brace, 1945), 420.

63. James Dewey Watson, "The Properties of X-Ray-Inactivated Bacteriophage: I, Inactivation by Direct Effect," *Journal of Bacteriology* 60 (1950): 697–718); and "II, Inactivation by Indirect Effects," *Journal of Bacteriology* 63 (1952): 473–85.

64. A. E. Clarke and J. H. Fujimura, eds., *The Right Tools for the Job: At Work in Twentieth-Century Life Sciences* (Princeton, NJ: Princeton University Press, 1992).

CHAPTER THREE

Biophysics in Berlin:
The Delbrück Club

Phillip R. Sloan

It brings back to memory the idyllic and enthusiastic sessions at your house and at our house where we delighted in our first adventures at bringing genetics and physics together. Like the young Ladies and Gentlemen of Boccaccio's Decamerone [*sic*], we were brought together by withdrawing from the terrors of a great plague to jointly consider some of the riddles of life.
—Max Delbrück to Nikolai Timoféeff-Ressovsky, October 1, 1970[1]

It is the Three-Man Paper (3MP) itself that tells us briefly about its genesis in its opening paragraphs; in the words of the authors, it originated from "lectures [*Vorträgen*] and discussions in a small, private circle of representatives from genetics, biochemistry, physical chemistry and physics."[2] But this tells us very little about how these discussions originated, the nature of the personnel involved, or the reasons why genetics was the focus of these discussions. This chapter illuminates these circumstances and details the close interaction of inquiries into two areas of research, genetics and photosynthesis, that provided the scientific context for the genesis of the 3MP.

The specific social context from which the 3MP emerged must also be clarified. In Max Delbrück's retrospective oral account of 1978, he speaks of having organized a group of five or six "exiled, internal [*sic*] exiled, theoretical physicists" in 1934 to participate in some informal discussions at his family home in Grünewald near the main institutes of the Kaiser Wilhelm Society in the southwest Berlin suburb of Dahlem. Delbrück described this group as follows:

This little club which started out as theoretical physics, and then brought in genetics, also brought in biochemists and photosynthesis physiologists. The photosynthesis man was Hans Gaffron, and he and Kurt Wohl lived together [with their families] in the same house in Dahlem. As a result of the talks that we had in our club on photosynthesis, they published a series of papers on the kinetics of photosynthesis . . . . Wohl and Gaffron discussed these experiments, and essentially already described what is now accepted; namely, that photosynthesis is done in photosynthetic units, which consist of about 1000 molecules of chlorophyll all funneling their energy into one photosynthetic reaction center.[3]

In this retrospective oral interview, Delbrück enumerates the other members of this group: the physicists Gert Molière, Werner Bloch, Ernst Lamla, and Karl Walter Kofink;[4] the radiation physicist Karl G. Zimmer; and the *Drosophila* geneticist Nikolai Timoféeff-Ressovsky.[5] Detail on these known participants is provided in the introduction to this volume or later in this chapter.

As we look more closely at this list of participants and the possible reasons for organizing such a discussion, several puzzles emerge. At least two questions can be posed. First, in view of Delbrück's intellectual biography prior to 1935, why was he the one physicist in the group to participate in a collaborative paper on genetics? As will be argued below, his more plausible entry into biological discussions would seem to have been through the discussions of photosynthesis in this group. Second, what is the basis for the linkage of quantum, and more specifically atomic, physics to genetics that emerges in the 3MP, and what made the linkage novel?

There is some uncertainty about the sequence of events that generated this unusual paper. From Karl Zimmer's retrospective account of its origins, appearing in a volume dedicated to Max Delbrück, the 3MP did not originate from the Grünewald discussions, but from intensive meetings between Timoféeff-Ressovsky, Zimmer, and Delbrück that sometimes lasted up to ten hours at Timoféeff-Ressovsky's apartments on the campus of the Kaiser Wilhelm Institute for Brain Research in the northeast suburb of Berlin-Buch, a ninety-minute train journey from Berlin-Dahlem.[6] To quote Zimmer,

At about the time these studies reached completion, Delbrück became interested in our line of work: however hard I try, I cannot remember exactly how the contact was established, but I do remember vividly the discussions that followed. Two or three times a week we met, mostly in Timoféeff-Rossovsky's [*sic*] home in Berlin, where we talked, usually for ten hours or more without any break, taking some food during the session. There is no way of judging who learned most by this exchange of ideas, knowledge and experience, but it is a fact that after some months Delbrück was so deeply interested in quantitative biology, and particularly in genetics, that he stayed in this field permanently. As an outcome of these discussions, a joint paper had been completed.[7]

To clarify these circumstances, this chapter first summarizes some relevant background discussions generated by Niels Bohr, dealing with the relations of physics and biology in the 1930s, and pursues in detail some issues discussed more generally in the Nils Roll-Hansen and Daniel McKaughan chapters in this volume. This is followed by brief biographical sketches of the participants in the discussion group. An analysis of the context provided by the photosynthesis discussions in the Delbrück meetings for interpreting the genetics work then follows. Finally, the novel dimensions of Delbrück's contribution to the 3MP as a theorist interested in drawing physics and biology together constitutes the closing discussion. The general goal of this chapter is to detail the importance of the Delbrück discussion group for its role in the development of a new style of doing "biophysics" that Delbrück was later to make famous at his Cold Spring Harbor summer workshops and in his California Institute of Technology research groups. As Delbrück reports, there was "no secretary of the society, no record keeping or anything. We were just a handful of people."[8] The reconstruction here must for this reason contain some speculation, as none of the participants is still alive and archival documentation remains elusive and fragmentary.

Delbrück's "club" lasted from approximately the fall of 1934 until Delbrück's departure for the United States in August 1937. It was situated against a broader backdrop of intellectual, scientific, and sociopolitical transformations of the late Weimar Republic. Generally in the background were the detailed discussions taking place more broadly within biophysics in Germany at the time, with a prehistory of exploration of issues involved in the relations of radiation and medicine in German experimental biology.[9] It was also situated within the umbrella of the Kaiser Wilhelm Society institutes at their primary location in Dahlem. As it was to bear on the specific content of the 3MP, the landmark work of Hermann Muller is always in the background. On the philosophical side, there were the active discussions related to issues of vitalism, mechanism, and the intermediate position of "holism" that formed a prominent feature of the interwar discussions in theoretical biology, especially in Weimar Germany, and particularly the new turn in these discussions initiated by Niels Bohr.[10] The genesis of the 3MP was also situated against the rapid nazification of Germany taking place in 1933–37 that affected several participants individually, and the Kaiser Wilhelm institutes generally.[11] Exploration of this full context is outside the scope of this chapter. The focus is instead restricted to an examination of the way in which physics impacted on biological discussions in these unusual discussions that marked Delbrück's first venture into biology.

The Quantum Enters Biology

The impact of new developments in physics after 1925 on theoretical debates within German biology forms a point of entry into the specific circumstances that created the 3MP. Although there is little reason to expect a priori that special and general relativity and the development of quantum mechanics after Planck were logically connected with issues in embryology and cell biology that were being discussed more broadly in biological circles in Germany at this time, the role of specific personalities served to create such a connection. Central to creating these connections was Niels Bohr (1885–1962), the nodal figure in the research into quantum and nuclear physics in the late 1920s and early 1930s. Bohr's philosophy of biology and his discussion of biological "complementarity," described in detail in the chapters by Nils Roll-Hansen and Daniel McKaughan in this volume, affected two individuals associated with him who drew different conclusions from some of his speculations, Ernst Pascual Jordan (1902–80) and Max Delbrück. Following brief remarks on Jordan, whose role in the biophysical discussions of the 1930s has been developed in depth by Richard Beyler, Norton Wise, and Finn Aaserud in several studies,[12] the relevance of some of Jordan's speculations for creating the context of the Delbrück discussions will be developed in more detail.

Pascual Jordan was born and first educated in Hannover, initially studying mathematics and physics at the Technische Hochschule in Hannover. Moving to the University of Göttingen in 1922, he completed his doctoral dissertation in 1924. As a result of his collaboration with his *Doktorvater* Max Born and with his contemporary Werner Heisenberg that created one of the foundational papers in matrix mechanics,[13] he was awarded a postdoctoral fellowship to study at Bohr's Institute for Theoretical Physics in Copenhagen during the summer and fall of 1927. During his attendance at the annual Easter meeting of theoretical physicists at Bohr's institute in 1929, he began discussions with Bohr about biological topics. More explicit conversations on these topics took place at the March 1931 theoretical physics gathering at the same time that Max Delbrück was also in Copenhagen on a Rockefeller fellowship. These discussions initiated a sustained correspondence between Jordan and Bohr on the interrelations of biology and physics in May and June 1931.[14] This correspondence resulted in Jordan's submission to Bohr of a manuscript in May 1931 that was eventually to appear as Jordan's article "Quantum Mechanics and the Foundational Problems of Biology and Psychology" in *Die Naturwissenschaften* for November 4, 1932.[15] In this paper, Jordan put forth his "amplifier" (*Verstärker*) theory. As he explained the main point of this theory,

> according to this hypothesis the structure and mode of function of an organism would be wholly the same as that of an *amplified arrangement*, as is used by physicists, in order to amplify the a-causal fluctuations of a stationary process, relative to the *atomic individual processes*, into *macroscopic* effects.
>
> According to this conception—which we call for short the *amplifier theory* of the organism—there is a conspicuous difference in the behavior of organisms compared to the entities of inorganic nature, without need for further explanation.[16]

Jordan's amplifier theory also made a direct extension of physics to psychology. The subjective experience of inner freedom is genuine, and it is a manifestation of the intrinsic causal indeterminism that extended from the atomic level upward. In a brief closing section to the paper, Jordan then extended this indeterminacy to claim a general indeterminism in all biological action.

The publication of Jordan's paper in the main journal of general science in the German language set off a debate over the relations of physics and biology that reverberated through a series of discussions in the 1930s. In his analysis of the events surrounding the 1932 paper, Richard Beyler has detailed the arguments of both the supporters and critics of Jordan's theory.[17] Of particular relevance is the critique of Jordan's paper in a public address to the Naturforschende Gesellschaft in Zurich in June 1933 by the elder statesman of Swiss psychiatry, Paul Eugen Bleuler (1857–1939), best known for his work on schizophrenia. In this talk, published subsequently as a forty-five-page article, Bleuler attacked Jordan and others who were trying to extend physical concepts, such as acausality, outside physics.

Much of Bleuler's essay dealt with the concept of causality, particularly engaging Erwin Schrödinger on this question. In the third section of the paper, he turned his attention to issues of psychology and biology in relation to microphysics, and here he directly addressed Jordan's 1932 essay. In developing his arguments against the claim that microphysics somehow undermines causality in psychology, he made brief mention of an issue that had not been raised by Jordan—genetics—as a strong counterexample to Jordan's arguments.

In support, he appealed to the experimental work on the cause of mutation through irradiation being carried out, in the wake of Hermann Muller's pathbreaking research, by workers such as the Americans F. B. Hanson and F. Heys.[18] As Bleuler interpreted their work, "mutation" can no longer be considered "spontaneous": "most recently Hanson and Hey [sic] actually have confirmed that the genetic mutation rate generated in Drosophila through radium rays is in direct proportion to the applied dose." He concludes with the question: "Is this not causal connection?"[19]

The introduction of genetics into the debate by Bleuler was, however, a side issue used to shore up his defense of traditional deterministic causation in science in general. But in Jordan's extended follow-up, published in the August 30, 1934, issue of *Erkenntnis*, the main journal of the Berlin and Vienna branches of the Unity of Science movement, Jordan gave genetics a more prominent place in his application of his amplifier theory to biology.[20] The 1932 and 1934 papers by Jordan, along with Bohr's earlier papers on the relations of physics and biology, republished in Berlin in 1931,[21] generated among a group of physicists, philosophers of science, and biologists a concern to develop a common response to Bohr and Jordan. The interpretation of Bohr's views on these matters was further complicated by the publication of his "Light and Life" lecture in *Die Naturwissenschaften* in March 1933, which put forth Bohr's epistemological, rather than ontological, interpretation of the relations of biology and physics through "complementarity."[22]

In developing his argument in his 1934 essay, Jordan opened with a long introductory section summarizing the developments in atomic physics since Planck's 1900 paper. He also expounded briefly on such topics as the statistical nature of scientific law, a point Bleuler had questioned. Jordan also highlighted the importance of the Heisenberg uncertainty principle for microphysics. Bohr's concept of complementarity was also summoned in support. Only after an extended discussion of this expansive theoretical framework did Jordan turn to the main topics suggested by the title—a quantum-physical interpretation of biology and psychology.

As a direct response to Bleuler's appeal to deterministic causality in genetics, Jordan for the first time extended his amplifier theory, which had up to this point concerned itself with the general theory of the organism and psychological issues, to the explanation of inheritance. Responding to the brief paragraph in Bleuler's paper, Jordan developed in some detail the claim that quantum physics can give an explanation of the stability and discontinuity of genetic inheritance as revealed by Mendelian laws without abandoning the underlying statistical nature of the microphysical world. This stability is interpreted as consistent with his amplifier theory and his conception of the statistical nature of scientific causality. Quantum discontinuity and genetic discontinuity are not simply *analogous* phenomena. They are based on the same physical foundations. Furthermore, quantum mechanics explains how the underlying causation can be fundamentally statistical rather than classically deterministic, yet result in the discontinuity observable in genetics:

> The absence of continuous transitions in the discontinuous increments of hereditary factors doubtlessly must signify that the elementary, nondecomposable [*zerlegenbaren*] gene is demonstrated to be an *individual molecule* (if also a very large one). *Hence we see in the theory of heredity the most conclusive and broadest empirical basis for the thesis that the organism is not a macrophysical, but a microphysical, system.* The instability [*Unstetigkeiten*] of the hereditary factors (wholly analogous to the atomic-quantum instability in physics) requires the adoption of *statistical concepts*. Inherent in the Mendelian laws of inheritance are statistical laws.[23]

Responding to Bleuler's claim that radiation genetics represented an example of strict causal relationships in biology, Jordan replied that the situation in genetics "corresponds apparently exactly to the above mentioned physical example of light absorption through atoms."[24] It implies that in radiation genetics, there is no predetermined (*Vorherbestimmung*) causality holding between radiation and mutation, but only a statistical relationship between dosage and effect.[25]

In expanding on these points, Jordan claimed that a statistical interpretation of quantum physics provided the basis for the statistical causality of genetics:

> *Without* our modern knowledge of the unstable [*unstetigen*] variability of hereditary factors, one could then assume that a wholly slight and insignificant modification could be inherited by a lineage. Since, however, in fact, from the

foundational instability [*Unstetigkeitsgründen*], a change of the genotype cannot happen that is arbitrarily "small," actually only two possibilities exist: either
that of *no* alteration, or that of an inherently *significant* one. Therefore the
weak effect in the form must show that even the *probability* of such a mutation
is very small; a more positive result of the experiments *must be statistically very
infrequent.*[26]

In the elaboration of his arguments on the applicability to biology of
quantum indeterminacy and the underlying statistical nature of natural
laws, Jordan sought to undermine the "mechanistic" conception of biology that was locked in struggle with vitalism and holism in the Weimar
debates. The discussion in the final portions of the paper also invokes the
authority of Bohr in numerous places, as Jordan discussed issues of organic life, freedom of will, and even Freud's psychoanalytic theory.[27] The
concluding claim of the paper summarizes his ambitious program:

> To be sure, it is scarcely necessary to extend our beginning inquiry into the
> domain of atomic physics to these most distant questions, in order to justify the
> belief that the new physics has created *new forms of natural scientific thought*
> whose clarifying power extends beyond the boundaries of physics, for the first
> time allowing certain fundamental problems and states of affairs in further di-
> vergent domains of natural science to be understood as *related* and *linked.*[28]

Jordan's extension of quantum physics to biology in his 1932 and 1934
papers, and the claimed support of his positions by Bohr, formed the subject of critical discussion at a preconference gathering of members of the
nascent Unity of Science movement on September 1 and 2 that preceded
the Eighth International Congress of Philosophy, held in Prague, September 2–7, 1934. Attended by Rudolf Carnap, Hans Reichenbach, C. I.
Morris, Otto Neurath, Philip Frank, Alfred Tarski, Ernest Nagel, Moritz
Schlick, Edgar Zilsel, and Carl Hempel, the September preconference
gathering represented almost a *Who's Who* of the early logical-empiricist
movement.[29] Jordan's claims were particularly the focus of critical presentations at this meeting by Moritz Schlick and by the Viennese philosopher
of social science Edgar Zilsel.[30] Schlick's paper, entitled "Concerning the
Concept of Wholeness," took particular aim, without naming individuals,
at the claims advanced by Hans Driesch, Ludwig von Bertalanffy, Adolf
Meyer-Abich, Jakob von Uexküll, and Jordan, who were arguing in some
way for the autonomy of biology from physics on the basis of notions of

holism and organicism.[31] Schlick's point in his short essay is that a careful analysis of the language of holism indicates that the metaphysical claims that seemed to be behind such questions were without warrant and the problem was only a fictitious issue. Following the restriction of philosophy to the analysis of language and "questions of sense" (*Sinnfragen*) and fact (*Tatsachenfragen*), the "holism" question could simply be dissolved.

In his contribution to the discussion, Zilsel took specific aim at Jordan's extension of quantum mechanics to biology. Interpreting him as endorsing vitalism, Zilsel attacked Jordan's "amplifier" theory. In the latter portion of the paper, he took up Jordan's recent extension of the theory to genetics and Jordan's claim that mutation studies supported his interpretation of quantum indeterminacy in an amplified stable organic structure—the chromosome—at the boundary of the microscopic and macroscopic. In response, Zilsel appealed to the recent work of Hermann Muller on the causal connections of mutation and radiation. The demonstration in Muller's studies of the predictable relation between both X-radiation and temperature increase with mutation rates was, for Zilsel, a clear counterexample, a point that Bleuler had also made. Furthermore, the process of development was sufficiently predictable and determinate, implying that "in most and even in the most characteristic phenomena of life, the amplification analogy remains completely superfluous."[32]

Zilsel admitted that, unlike Driesch, Jordan was not grounding the "peculiar *stability*" of the organism on vital causes, but on microphysics.[33] But the effect was seen to be the same. Both options—Drieschian vitalism and Jordanian "amplification" theory—had the same goal, that being "to rescue in some way for science, the unpredictable striving, free will, the incalculable 'ensoulment' of the organism, which are naturally considered in prescientific observations."[34] Zilsel suggested in summary that it is more important to understand the psychological and sociological reasons that would lead one to embrace such views after a half century of scientific inquiry into the physical conception of life, rather than see these as valid arguments.

Zilsel's critique of Jordan generated a set of short written responses by Hans Reichenbach, Otto Neurath, Moritz Schlick and Philip Frank, along with a longer response by Jordan to his critics, published in a special issue of *Erkenntnis*. In his reply, Jordan argued emphatically that he was not suggesting an "autonomy" of biology from physics, but only questioning the ability to "attribute biological lawfulness to *presently known* physical laws."[35] Instead, appealing directly to Bohr's arguments concerning the impossibility of explaining biology through classical—that is, deterministic, strong-causal—physics, Jordan argued that the new quantum physics and Bohr's interpretation of it implied for biology and psychology "a dissolution and overhaul of all previous conceptions."[36]

As a consequence of these public discussions, the arguments of Jordan and Bohr on the relationships of physics and biology were very much in the air in the fall of 1934 when Delbrück decided to bring together a group of physicists and biologists to discuss the theoretical relationships of these domains.

Convening the Club

The relevance of the issues discussed in the previous section becomes apparent from the report of events surrounding a lecture given in November 1934 by Pascual Jordan at a meeting of the Berlin Society for Empirical Philosophy (Gesellschaft für empirische Philosophie)—the "Berlin" circle—held at Berlin-Dahlem.[37] As described in the introduction to this volume, Delbrück had been interested in relating biology to physics at least since the spring of 1932 and became deeply engaged with the theoretical relations of biology and physics following Bohr's lecture in August 1932, predating the publication of any of Jordan's papers on this topic. By November 1934 the claims made by Jordan in the November 1932 and August 1934 papers, and the responses by the logical empiricist community in September 1934, were surely well known to those associated with the sponsors of the lecture.[38] In attendance at Jordan's lecture were Max Delbrück and several biologists, including the codirector of the Kaiser Wilhelm Institute for Biology, Max Hartmann (1876–1962), who seems to have been the biologist at the KWI most interested in the philosophy of science.[39] Jordan's talk evidently ranged over the same topics treated in his long *Erkenntnis* paper. Delbrück's reaction to Jordan's lecture, recorded in a letter to Bohr, illuminates his differences from Jordan.[40]

Berlin-Dahlem, November 30, 1934

Dear Professor Bohr,

A few days ago, Jordan gave a lecture here before the Society of Empirical Philosophy and spoke about quantum mechanics and biology, to which the biologists showed up. In the part dealing with physics, the lecture was very poor

[*dürftig*], and it described the state of the discussion in the year 1927. In the biological part, which was even poorer [*durftiger*], he completely distorted each of
your arguments, to the extent that he mentioned them. In the discussion which
followed, the biologist Hartmann complained bitterly about the confusion
[*Verwirrung*] in the biological literature caused by your and Jordan's papers.
Thereby he distorted also those of your arguments [*Sätze*] which Jordan had
not even mentioned. The result was that subsequently, all the biologists scolded
all the physicists. I have now taken the liberty of giving Mr. Hartmann the enclosed survey [*Zusammenstellung*] of what we assert* ([footnote] i.e. what I
believe you assert), and what we do not assert. I should like to know how far
you endorse the formulation. It is important to express oneself very briefly because biologists are wont to read long papers very superficially. Therefore they
can never comprehend a new subtle thought [*Gedanken*] correctly, [and] they
must always force it into the ready-made scheme of their concepts. From my
survey of what we do *not* assert, you can gather the kind of misunderstandings
that arose. I believe that it would be very useful if you were to publish a short
explanation of this kind.

This semester I have arranged a private seminar [*Privatseminar*] with biologists, biochemists and physicists, in which rather able people participate, so that
all of us learn a lot [*recht tüchtige Leute hinkommen*].

With best wishes,
yours
Max Delbrück

Delbrück's appended explanatory document, drawn up for Hartmann's
clarification, emphasized the following points: First, contrary to Jordan,
Delbrück denies that "the laws of atomic theory can clarify *specific* life
phenomena." Second, he claims that it is impossible to give simultaneously
a causal description of the relationship of the domains of the living and
nonliving, a point he interprets to be Bohr's meaning of "complementarity": "Precisely *because* in a living organism physical and chemical phenomena are interwoven far into the atomic realm, the common root of
biology *and* physics *and* chemistry *must* be found in the *atomic* [level]. Just
for this reason, however, a *causal* description of the relationship *cannot* be
based on physical *and chemical concepts alone.* For in the atomic domain,
physics and *chemistry* allow no common causal description."[41]

Similar again to Bohr's "Light and Life" arguments, Delbrück asserts
that because biologists study *living* organisms, their inquiries into genetics,

developmental mechanics, and biophysics "cannot penetrate to the investigation of *individual atomic* elementary processes, from which they are
indeed far removed; there is complete agreement on this point . . . [F]or
this reason these domains are not causally reducible to one another, as
physics and chemistry are not causally reducible to one another."[42]

We see that in the late fall of 1934, Delbrück had revealed himself, like
Jordan, to be an antireductionist, and like Jordan, he was appealing to
Bohr for legitimation of these views.[43] But he was also rejecting Jordan's
reading of Bohr, and was instead relying on his own understanding of
Bohr's complementarity arguments, particularly as Bohr had developed
these in the "Light and Life" lecture.[44] There is no simple derivation of
biological properties from physical principles, and no extension of quantum acausality on up the chain. Consequently, one can give no explanation
of properties at the macrolevel through a simple "amplification" of those
at the microlevel.

Delbrück's letter to Bohr also confirms that by November 1934 his informal discussion colloquium was under way. In his interview with Carolyn
Harding in 1978, Delbrück supplied some scanty detail on these meetings,
commenting that the group met "sometimes every week, sometimes once
a month, and so on."[45] It is evident that these meetings were more than
informal discussions, and they included presentations by some of the at-
tendees of important papers that were later published.

The sequence of inquiry and the recruitment of members into the discussion group is of some importance for situating the genesis of the 3MP.
The core group of the club was, according to Delbrück's retrospective
report, initially dominated by physicists, but then "at my suggestion we
soon brought in also some other people, some biologists and biochemists."[46] But from other reports, the photosynthesis expert Hans Gaffron
(1902–79) from the KWI for Biochemistry was involved from the first,
with Timoféeff-Ressovsky and Zimmer added further downstream.[47]

The physicists included several interesting individuals. One was Paul
Friedrich Gaspard Gert Molière (1909–64), trained at the University of
Berlin.[48] Affiliated in this period with the Kaiser Wilhelm Institute (KWI)
for Physical Chemistry, he was located in quarters adjacent to the KWI
for Chemistry with which Delbrück was associated as a research assistant
to Lise Meitner. In these years Molière was particularly engaged with a
quantum-mechanical interpretation of the X-radiation of crystals. This
resulted in a multipart paper in 1939 and further explorations of X-ray
theory in 1940.[49] One of Delbrück's early scientific papers, published with

Molière in 1936 in the *Proceedings* of the Berlin Academy of Sciences, was a detailed mathematical discussion of statistical quantum mechanics and thermodynamics, and it is possibly an outcome of the Delbrück club discussions.[50]

Two other attending physicists present challenges in detailing their biographies. Werner Bloch (1890–1973) was a headmaster of a gymnasium and had been an attendee at Einstein's 1916/17 relativity lectures and his 1917/18 lectures in Berlin on statistical mechanics.[51] His short popular presentation of relativity, *Introduction to Relativity* (1918), was an early exposition of Einstein's theory that Einstein praised.[52] In 1930 he translated Paul Dirac's text on the principles of quantum mechanics into German.[53]

Another individual presents difficulties because of his misleading designation by Delbrück as "Werner."[54] This is Karl Walter Kofink (1907–75), who typically went by the name of Walter. He was a theoretical physicist who completed his PhD under Erwin Schrödinger and Max von Laue at the University of Berlin in 1934. He was appointed that same year as an assistant at the Institute for Theoretical Physics at Berlin-Charlottenburg Technische Hochschule before moving to the University of Frankfort am Main in 1935, where he became professor *ordinarius* in 1940. During the period of the Delbrück club, he was publishing on statistical mechanics and on Paul Dirac's theory of electron spin.[55]

The physicist Ernst Lamla (1888–1986) received his doctorate from the University of Berlin in 1912 under Max Planck, with a dissertation on hydrodynamics and relativity. He taught physics at the gymnasium level from 1924 to 1933, until he was dismissed for political reasons when the National Socialists came to power.[56] He was also the author of a textbook, *Foundations of Physics for Natural Scientists, Physicians and Pharmacists*.[57] Lamla was one of the "internally exiled" physicists who did not leave Germany during the Nazi era, and he remained active in scientific circles during the war years.[58] In 1947, along with his mentor Max Planck, he was given an honorary professorship at the University of Göttingen. Three years later he was made editor of Germany's main journal of general science, *Die Naturwissenschaften*, which he edited until 1966.

Berlin-born physical chemist Kurt Wohl began his academic career in 1914 as a chemical engineering student at the Technical University of Danzig, with his education interrupted by three years of service on the front lines in World War I, and he completed his engineering and chemistry diploma in 1920. He then studied physics for a semester at the University of Heidelberg before transferring to the University of Berlin, receiving his

PhD at the University of Berlin under Walter Nernst in 1923, with a dissertation on the specific heat of diatomic gases. Beginning in the same year,
he was appointed assistant to the KWI for Physical Chemistry, a position
he held until 1935. From 1929 to 1933 he was also a lecturer in chemistry
at the University of Berlin. Forced into partial retirement in 1933 from his
position at the university because of his partial Jewish ancestry, he was formally terminated at the KWI in 1935, and he formed one of the "internally
exiled" members of the group spoken of by Delbrück. To support himself,
Wohl gave private lectures and was employed by I. G. Farben Industry
in Berlin from 1936 to 1939 as a physical and electrochemist before his
departure for England in 1939.[59]

Hans Gaffron, the biochemist in the group, was born in Lima, Peru, and
lived there until 1912, when his family returned to their native Germany.[60]
He studied chemistry at the Universities of Heidelberg and Berlin from
1921 to 1924, and received his doctorate in chemistry from the University
of Berlin in 1925. He accepted a position as assistant to the biochemist
Otto Warburg (1883–1970), then at the KWI for Biology. In 1931 Gaffron
accepted an invitation for a year-long research position at the Biological
Laboratory of the California Institute of Technology in Pasadena to work
with photosynthesis expert Robert Emerson (1903–59), the professor of
biophysics at Caltech from 1929 to 1946. Returning to Germany in 1932
on the death of his father, he then spent six months in research on marine
photosynthesis issues at the famous Stazione Zoologica at Naples before
accepting a research position at the KWI for Biochemistry, working under
Carl Neuberg (1877–1956) until Neuberg's dismissal by the Nazis in 1936.
He continued his research as a guest researcher under Fritz von Wettstein
(1895–1945), the director of the KWI of Biology, until he left Germany
with his wife in 1937, both for conscience reasons and to accept a fellowship from the Rockefeller Foundation to work with Cornelius van Niel
at the Hopkins Marine Station in Pacific Grove in California. He partici-
pated in van Niel's research group until the fall of 1939.[61]

This multidisciplinary group of researchers, connected through their
acquaintance with Delbrück and proximity to the Dahlem institutes, and
all sharing some interest in the interaction of physics and biology, generated from their discussions at least two fundamental theoretical developments in biophysics, one in photobiology and the other in genetics. Since
evidence suggests that it was photosynthesis, rather than genetics, that
formed the original focus of the biophysical discussions,[62] a brief examination of the photosynthesis inquiries allows us to see the 3MP in a context
larger than genetic research. Delbrück's enduring interests after this time

in photosynthesis and photobiology can be followed in some detail through his correspondence with Hans Gaffron between 1944 and 1977, into his final research project on phototropism in the fungus *Phycomyces*.[63]

From Photosynthesis to Genetics

The importance of photobiology for the arguments of the 3MP emerges in several places in a close reading of the text, particularly in sections clearly authored by Delbrück. One reason for an initial focus on these questions in the discussion group was geographical location. Delbrück, Wohl, and Gaffron were all located close to the Dahlem institutes, with Delbrück in the KWI for Chemistry, Wohl in the adjoining KWI for Physical Chemistry, and Gaffron in the KWI for Biochemistry.[64] Timoféeff-Ressovsky's genetic unit, by contrast, was located a considerable distance away in east Berlin. Photosynthesis was a hot research topic at the KWI in this period, and it provided the most obvious linkage between quantum physics and biology, as explained below.[65] Otto Warburg had been engaged with quantum theory and photosynthesis since the 1910s and had already published seminal papers in the area when Gaffron joined him as his assistant in 1925. Warburg's revolutionary work was made possible by his development of novel microtechniques that allowed intensive study of photosynthesis through his invention of a special gas manometer able to measure gas exchange in the microliters. He also developed these techniques around a model organism, the microscopic green alga *Chlorella*.[66] With this he was able to develop the means to study the so-called quantum yield of the photosynthesis reaction, discussed in more detail below.

Delbrück's engagement with photosynthesis grew logically from his previous interest in Bohr's interpretation of quantum mechanics and the linkage of physics and biology in a research program inspired by Bohr's "Light and Life" lecture. Such interest suggests why he would have quickly sought out the acquaintance of Hans Gaffron, who was actively working on this topic, when he returned to Berlin in the autumn of 1932.

Delbrück's publications before 1935 dealt with issues in group theory, chemical bonding, and nuclear physics, with his most substantial early work a short joint monograph on the structure of the atomic nucleus, co-authored with his research director, Lise Meitner, that appeared in 1935.[67] Some insight into Delbrück's transition into biology in this crucial period can be detected from this joint publication. In this monograph, the two authors collaborated in an analysis of nuclear transmutations. Delbrück's section, entitled "Application of Quantum Mechanics to the Atomic Nucleus," gives us an insight into the way Delbrück was attempting to draw some connections between the issues of nuclear physics and biological problems in this period.[68]

The closing theoretical section of the text argues that the knowledge of the structure of the nucleus not only has meaning for atomic processes, but also has implications "in the domain of biology and chemistry."[69] The new quantum understanding of nuclear structure supplies a way of interpreting chemical reactions, "and also opens up new ways for the biological investigation of the conversion and transformations of different materials in the organism."[70] In other words, the connections of physics with biology are through metabolism and biochemistry, not genetics.

In the discussion in the second half of the monograph, the concern is specifically with radiation and radioactivity, with discussion of such issues as alpha and beta decay, the neutrino hypothesis, and the nature of X-rays at different wavelengths. The intent of this monograph, as the authors summarized in closing, was to show "the remarkable implication, that the conceptual schema of quantum theory is completely sufficient for the qualitative ordering of nuclear phenomena, even if we can give quantitative proof only for a few exceptional cases."[71]

A link between Delbrück's preexistent interests in nuclear physics with the inquiries in photosynthesis at the KWI is made through the concept of the "photosynthetic" unit that was developed by Gaffron and Wohl in an important theoretical paper that emerged from the Delbrück discussion club. In many respects this was to be a more lasting theoretical contribution to biology than the work in genetics itself.

The Concept of the Photosynthetic Unit

The 3MP analyzes the structure of the gene and the way in which radiation research can help determine this structure as well as explaining the causes of gene mutation. But the specific analysis of this question in the 1935 paper seems to owe considerable insights to the work on photosynthesis that was taking place in parallel in the Delbrück discussions. These researches led to the concept of a discrete structural unit that interacted with the input of specific quanta of energy. The explanation of the relation between quantum theory and photosynthesis has several similarities to the "target theory" developed in the 3MP. In the hands of the Kaiser Wilhelm physical chemists and biochemists with an interest in photosynthesis, quantum mechanics formed the theoretical foundation for analyzing the photon theory of light, a theory that had been hotly debated since Einstein first put it forward in 1905.[72] By 1930 the photon theory had achieved a general consensus endorsement within physics, at least within the discussions of interest here. The photon theory conceived light to be transmitted in discrete and measurable packets of energy with a discrete value, symbolized typically in chemistry and experimental physics by $h\nu$, where h is Planck's constant and ν (the Greek nu) the photon's specific frequency.[73] The theoretical question resolved in part by the work of Gaffron and Wohl in the context of the Delbrück club concerned the way that light functioned in the photosynthesis reaction.

The international research community pursuing photosynthesis generally in the 1930s was largely developing the methods instituted by Warburg.[74] Following the publication of his foundational papers in the early 1920s,[75] presenting his new analytic techniques, and his proposal of a standard model organism, Warburg had created the theory and tools with which to open up the complex black box of plant physiology to gain sophisticated understanding of how plants accomplished photosynthesis. Hans Gaffron began his work on photosynthesis with Warburg at the KWI, and then added to this training work with Caltech biophysicist Robert Emerson and interchanges with Stanford microbiologist Cornelius van Niel (1897–1985). This series of interactions supplied a background for Gaffron's theory, which he then worked out in company with physicist Kurt Wohl in the context of the Delbrück seminars.

Gaffron drew both on the work of Warburg and the revolutionary work of Cornelius van Niel. The new layer of inquiry that van Niel added to the Warburg work concerned the new understanding of the photosynthetic reaction itself. In important papers published in 1931 and 1932, van Niel presented the results of his extensive studies on photosynthesis in purple and green anaerobic sulfur bacteria, rather than on photosynthesizing algae. His work revised the traditional understanding of the photosynthetic reaction, which up to then was characterized by the general formula

$$6\,CO_2 + 12\,H_2O \rightarrow C_6H_{12}O_6 + 6\,O_2 + 6\,H_2O.$$

Van Niel instead reinterpreted this as a reducing reaction involving a hydrogen donor in which the oxygen is derived from water rather than from the splitting of CO_2, in an energy-driven reaction summarized by the general equation

$$CO_2 + 2H_2A \rightarrow CH_2O + 2A + H_2O,$$

where A can be oxygen, sulfur, or another electron receiver.[76] The carbohydrate product is then formed by polymerization of CH_2O.

To explain the energetics of this reaction was the point where quantum
physics entered the picture. Warburg and his collaborator Erwin Negelein
(1897–1979) determined that the reaction required four quanta of energy
from red light to reduce one molecule of carbon dioxide to carbohydrate
in the photosynthesis reaction. But the mechanism of this reaction was
unknown.[77]

Gaffron adapted van Niel's insights to the investigation of photosynthetic phytoplankton under low light intensities. This allowed him to give
an alternative account of photosynthesis that also explained in greater
detail how the input of quantum radiation was captured and used. This
was worked out in the Gaffron-Wohl collaboration in the period of the
Delbrück discussions.[78]

Quantized Units

The Gaffron-Wohl theory is still today acknowledged as a foundational
development in the history of photosynthesis.[79] It helped unravel the puzzling way in which chlorophyll, light, and carbohydrate production were
related. Furthermore, this work directly incorporated principles from
quantum physics in its analysis.

On the Gaffron-Wohl theory, energy supplied by light was actually captured by the chlorophyll molecule without appreciable loss, and the CO_2
molecule simply attached to the chlorophyll without chemical bonding. The
captured energy in the chlorophyll molecule was then held in a "reducing
center," X, presumably the chlorophyll molecule, in which light quanta
of a measurable specific energy are stored without loss of energy, and to
which the CO_2 molecule becomes attached. For the reaction to proceed, the
Gaffron-Wohl theory assumed, following Otto Warburg, that four quanta
of light are necessary, which are then expended in a cascade of reactions
schematized by Wohl in 1940 as a feedback loop taking the form[80]

$hv = hv$

$$X_0 \xrightarrow{hv} X_1 \xrightarrow{hv} X_2 \xrightarrow{hv} X_3 \xrightarrow{hv} X_4 \xrightarrow[+ H_2O + CO_2]{-O_2 - CH_2O} X_0,$$

with X_4 symbolizing the final photoreaction products, oxygen and formaldehyde ($-CH_2O$), from which the carbohydrate sugar, for example glucose

$(C_6H_{12}O_6)$, is synthesized by polymerization.[81] Important in this quantum analysis of the process was the evidence that the reaction involved no "induction" period as required by the competing model developed in 1935 by the physicist James Franck.[82] The relation of light intensity to the rate of photosynthesis was linear, once sufficient quanta had been absorbed, rather than sigmoid in shape. The photosynthetic reaction only occurred when the necessary four quanta had been stored up by the reducing unit, after which the energy was released in a linear relation to further irradiation. This unit was assumed to be located in the chloroplast, and involved approximately 2,500 chlorophyll molecules. It also formed a "solid structural unit, having at one place a reducing center for carbon dioxide."[83] On the Gaffron-Wohl theory, the captured photons resonate "without loss through the body of the photosynthetic unit." Furthermore, this unit, conceived as a "crystalloid structure," would be "activated as a whole as soon as a light quantum is absorbed at any point."[84]

There are several levels of synergy between these reflections on the photosynthetic unit and speculations on the nature of the gene that emerged from the same Delbrück discussions. First, radiant energy is conceptualized as transmitted in discrete packets that are captured by concrete structural units that can retain this energy essentially undiminished for a measurable period of time. This is similar to assumptions of the target theory of the 3MP. Second, there is no "induction" period. The photosynthetic unit is activated immediately in a discontinuous way, even though it can function only when the necessary four quanta have been absorbed. The reaction, once initiated, is in a linear relation to the radiant energy received rather than one gradually induced over a period of time. Finally, the photosynthetic unit is conceptualized as a finite structural unit of molecules, possibly of crystalline form, that is stable in structure and functions as a physiological unit.

The similarities between this concept of the quantum-activated photosynthetic unit and the "gene" of the 3MP are significant, and it is unclear in which way the interaction was moving during these conversations from which both the Gaffron-Wohl theory and the 3MP emerged. It is evident that Delbrück was contributing directly to these discussions of photosynthesis as well as to those on genetics, and offering suggestions about the underlying physics involved.[85] At the same time, there were important differences between the concept of the photosynthetic unit and the gene as conceptualized in the 3MP. The incoming quanta remain external to the photosynthetic unit, whereas they are conceptualized as penetrating the gene and even causing rearrangements of its molecular structure. The gene of the 3MP is also conceptualized as a much simpler structure than the photosynthetic unit, and Zimmer even estimates the target-strike area to be a unit with a radius of approximately one micron, leading Schrödinger later to attribute to the gene the probable size of one thousand atoms.[86] But the notion of the photosynthetic unit as a complex of molecules affected by radiant energy also gives us some insight into the otherwise puzzling final claim of the 3MP that "genes are physical-chemical units; perhaps the whole chromosome . . . consists of such a unit, a large assemblage of atoms, with many individual, largely autonomous sub-groups."[87] This interplay between photosynthesis and genetics illuminates some of the subtle nonreductive claims of the 3MP.

Genes and Quanta: The Biophysics of Mutation

The connections of quantum theory and biology were plausibly first developed in the Delbrück discussions through a route that led Delbrück from work with Meitner into discussions of photosynthesis, encouraged by his preexisting contacts with Wohl and Gaffron. It was then that genetics entered into the discussions, with the invitation of Timoféeff-Ressovsky and Zimmer to join the small club.[88]

Assuming Jordan did claim in his Dahlem lecture of November 1934 that quantum theory had considerable implications for the understanding of genetics and, in particular, that it had bearing on the stability of the gene and on the relation of mutation to X-radiation, the latter would have been specifically relevant to the research program of Timoféeff-Ressovsky then in progress at the KWI genetics unit within the Institute for Brain Research in Berlin-Buch. This is a plausible point that may date the moment when Delbrück then "brought in," as he later reports, Timoféeff-Ressovsky and his young assistant Zimmer, still officially attached to the Cecilienhaus Hospital in nearby Charlottenburg, to participate in the conversations.[89] By professional background, Timoféeff-Ressovsky, trained in population biology and experimental genetics, would have been the least prepared to interact with the interests of the existing group. Zimmer, on the other hand, was the individual with the specific expertise that could link Timoféeff-Ressovsky's fruit-fly genetics with current experimental radiation research, and he was by training most able to deal with the more theoretical aspects of photosynthesis theory of interest to Gaffron, Wohl, and Delbrück.[90] However the collaboration occurred, the involvement of Timoféeff-Ressovsky and Zimmer in these discussions brought the group into immediate contact with practical experimental genetics coupled with the ongoing research on radiation and mutation taking place at Timoféeff-Ressovsky's laboratory in Berlin-Buch.

The novel features of the 3MP that emerged by April 1935 can be seen by comparing it to Timoféeff-Ressovsky's work immediately predating the initiation of the Delbrück club. A lengthy review paper by Timoféeff-Ressovsky, published in English 1934 in the Cambridge-centered *Biological Reviews*,[91] provides a window into the nature of his research before the input of Zimmer and Delbrück is in evidence.[92] Much of the content of this paper, including its diagrams, would reappear in altered form in Timoféeff-Ressovsky's section of the 3MP with the new insights provided by the collaboration with Zimmer and Delbrück. The focus of the 1934 paper is a survey of the research since the 1920s on radiation and mutation.

After supplying an overview of the research on the experimental production of mutations by radiation, Timoféeff-Ressovsky argued for the following claims:[93]

1. The relationship between quantity of radiation and mutation rate is linear, and demonstrates no "induction" effect that might indicate, as one might find with some poisons, a lack of effect at low dosage. Instead there was a simple proportionality of dose and effect.[94]
2. Within broad limits, the effect remained linear for all wavelengths of radiation. Particularly within the range of X-rays, there is little difference between the effects of "hard" and "soft" rays.
3. Different genes show different degrees of stability under radiation. For example, within the white-eyed series in *Drosophila melanogaster*, "the darker allelomorphs mutate more frequently under the influence of the same X-ray dosage than do the lighter allelomorphs."
4. The effect of X-radiation is direct, rather than working through its effect on intermediate biochemical systems. This is demonstrated by the existence of "reverse" mutability.
5. The effect of chemical or temperature variation does not show the same relation to mutation production as that produced by irradiation.

Viewed against the later 3MP, noticeably absent from this paper is any substantial attempt to develop a causal account of mutation, and there is no effort to relate this to quantum physics. We also find no mention of a prominent feature of the 3MP, the "target theory," developed particularly by Zimmer. The closest the paper comes to dealing with the issue of causation is in a section entitled "The Nature of the Effect of Rays on the Process of Mutation." Drawing on some of Hermann Muller's suggestions, whose "unpublished research" is mentioned in the paper, Timoféeff-Ressovsky comments that "the most plausible assumption would be that gene mutations are reconstructions of the gene, *i.e.* some physico-chemical changes of its structure."[95] This suggests to Timoféeff-Ressovsky that some conclusions on the nature of the gene itself might be drawn from such research. The material theory of the gene, worked out particularly by Muller, lies in the background of much of this discussion.

Timoféeff-Ressovsky outlines in the 1934 paper two options for understanding the gene: first, genes can be conceived as "fixed quantities of specialised matter (consisting of several or even many equal physico-chemical units)"; or second, they can be conceptualized as "physico-chemical units (molecules, micellae, or colloid particles of specific structure)." The former would support Richard Goldschmidt's quantitative theory of inheritance in which the chromosome is conceived as a biochemical bag or "sausage" that expresses its contents quantitatively, rather than involving atomic genes located on chromosomes as assumed in the Morgan theory.[96] The second option, which Timoféeff-Ressovsky supports, was better able to explain reverse mutations—that is, the ability to reverse artificially produced mutations by applications of additional radiant energy. The issue of reverse mutation was later seen by Delbrück as Timoféeff-Ressovsky's most important theoretical contribution, and it plays an important role in the development of the explanatory theory of the 3MP. But all of this is tentative at best for Timoféeff-Ressovsky in this 1934 review: "our present empirical knowledge is far too insufficient to build up more detailed theories of the structure of the gene."[97]

The new dimensions incorporated into the 3MP in the ensuing six months were the contributions of Zimmer and the target theory, and Delbrück's own theoretical contributions. Since the target theory is discussed in detail in the Summers and Beyler chapters, I focus here on the nature of Delbrück's contribution.

Delbrück's negative reaction to Jordan's talk seems to have restrained any penchant he may have had for engaging in philosophical speculations on implications of quantum physics for biology similar to Jordan's. Neither the name of Bohr nor the term "complementarity" is ever mentioned in the 3MP, nor is Bohr cited in the bibliography to the paper. Indeed, in sending Bohr the manuscript of the 3MP in April 1935, Delbrück explicitly wrote that the paper contains "no sort of complementarity argument. To the contrary, it shows that one can lay out a unified atomic-physical theory of mutation and the stability of the gene."[98] Although, as argued below, the paper implicitly appeals to Bohr's complementarity thesis, Delbrück's analysis proceeds on another theoretical plane. Illustrative of this philosophically modest theoretical stance, Delbrück begins his section of the paper by presenting the arguments for and against any attempt to deal with the issues of genetics and mutation from the standpoint of theoretical physics, and the main conclusion of this discussion even makes an argument for asserting the autonomy of genetics from physics:

> On the basis of such considerations, one may assert that genetics is autonomous and that it should not be muddled with physical-chemical notions. In particular, one could think this because, in the places in biology where the use of physical and chemical concepts has so far been met with real success, no convergence with the phenomena of genetics can be seen. Rather, while genetics always deals provisionally with the isolation of processes that do indeed have, physically-chemically, an unambiguous character, they appear as only parts of processes when viewed biologically, and their relation to the whole life process remains problematic, unless their coordination is viewed as arising on the basis of some heuristic scheme in which the life process is postulated in principle as physical-chemical machinery.[99]

With such a cautious statement opening Delbrück's section, allowing a "mechanistic" conception of life only as something akin to a Kantian regulative hypothesis,[100] what is most evident is his rejection of Jordan's claim that one can *derive* crucial biological phenomena or claims about living states *directly from* microphysics, reiterating a point he made to Bohr in his letter of November 1934. The novelty of his approach to these questions as a physicist lies on two levels, one related to his expertise in nuclear physics, and the other to a special methodological stance related to his background as a theoretical physicist.

To the first point, we observe in the opening section of the paper how Delbrück parallels photochemical changes with those likely producing mutations, but in this case they take place by entry of quanta of energy directly into the gene, resulting in atomic rearrangement of its molecular structure. This allows him to develop a theory of the material structure of the gene, and a physical explanation of mutation, building on the target-theory account. This leads to the definition of the gene near the end of the second part of the multi-authored conclusion of the paper: "we view the gene as an assemblage of atoms within which a mutation can proceed as a rearrangement of atoms or dissociation of bonds."[101] This implies some kind of concrete structure to the gene that can be rearranged by the input of sufficient radiant energy. It is in some ways similar to the photosynthetic unit, but in other ways quite different.

On the methodological level, Delbrück's contribution proceeds by construction of a kind of *Gedankenexperiment*,[102] an unusual methodological stance for biologists to adopt at that time. His account envisions a hypothetical scenario in which genes are conceptualized as material molecules that are able to "retain their identity [*Gleichartigkeit*] unchanged in the face of all surrounding influences."[103] His explanation of this molecular stability through physical principles forms the framework for developing a theory of gene structure that is then employed in an explanation of mutation. This explanation is put forth in terms of an explicitly hypothetical model that discusses the gene structure by means of simplifying assumptions. The gene is conceptualized as an arrangement of atoms. This conception of the gene he then relates to the observation of both Muller and Timoféeff-Ressovsky, summarized in the first part of the 3MP, that mutation rates increase with increases of temperature in accord with the van't Hoff rule governing the rate of chemical reactions.[104] This increase implies that the reaction rates at every ten-degree increase in temperature will be altered stepwise by the fluctuations in energy needed to overcome the stability of molecular structures. These fluctuations are created either by thermal changes in the system or by the input of external ionizing radiation in the form of specific packets of energy that are either dissipated internally as heat, or captured to create specific rearrangements of atoms at a specific point.[105] Since the organism exists within very narrow thermal limits, the main cause of mutation is presumed, therefore, to be that effected by external ionization.

On this basis, Delbrück reasons that the changes in the material gene "can occur only in jumps [*nur springweise*]."[106] The "gene" is a unit that, like the photosynthetic unit, is acted upon by a specific quantity of radiant energy delivered in specific packets. Unlike photosynthesis, however, "the amount of matter primarily affected is determined not by the number of absorbed light quanta, but rather by the absorbed energy per unit volume."[107] The fact that reversible forward-and-back mutations— Timoféeff-Ressovsky's great contribution to this issue—could be caused by X-radiation striking a specific volume—Zimmer's contribution from the standpoint of the target theory—allowed Delbrück to sidestep the possibility that the gene might be involved in a complex biochemical process subsequent to the strike by radiant energy.[108]

Between its rearrangements, the gene was stable because of the nature

of the chemical structures and the energy holding these together. Mutation is thus conceptualized as a "rearrangement of a stable assemblage
of atoms" due to external energy input.[109] Delbrück's approach through
fundamental physics pushes the explanation of these events to the level of
atomic physics, offering a plausible explanation for how the ionizing radiation and the target theory could actually function in relation to the gene.

Some of the claims that result from the 3MP could suggest that Delbrück was advocating a strong reductionism of life to the action of material genes, the Mullerian interpretation of Delbrück's theory advocated
later by Schrödinger in his *What Is Life?* lectures. This may have been
Timoféeff-Ressovsky's own view.[110] But as developed in the introduction to
this volume and in the Roll-Hansen and McKaughan chapters that follow,
this is to misread the argument of the paper. In a subtle way, Delbrück's
analysis was still grounded on his underlying defense of the "autonomy"
of biology and his understanding of Bohr's notion of "complementarity,"
which remains subliminal in the paper. By framing these reflections in the
language of a hypothetical model to be tested for fit against experience,
Delbrück avoids strongly realist claims about the definitive nature of his
model of the gene and his account of mutation.

But if the 3MP avoids a "bag of genes" reductionism, there is still the
sense that the "genome" is the controlling factor, and we are left in the
end with some ambiguity in the reading of Delbrück on the issue of reductionism that seems difficult to resolve decisively. In the background
is always his complex meaning of "complementarity," which he absorbed
from Bohr but in important respects misunderstood.

Entering the domain of genetics as a theorist migrating from physics,
Delbrück was able to make some new kinds of connections between several strands of discussion taking place in the Grünewald conversations.
Comparisons and differences between photosynthesis and genetics are
drawn out; a style of theorizing is employed that is not typical of the biology of his day; he has linked together the concept of genes and mutations
with issues in atomic physics. Delbrück's contribution is not at the level
of empirical research, but on the plane of theoretical unification through
simplifying assumptions.

Conclusion

The Delbrück gatherings continued to meet until Delbrück left for the
United States on a Rockefeller fellowship in August 1937. The participants

then scattered in various directions under the pressure of the growing political turmoil in Europe. Wohl emigrated to England in 1939, and to the United States in 1942. Gaffron left about the same time as Delbrück in 1937 to work with van Niel and remained in the United States. Timoféeff-Ressovsky and Zimmer would spend the war years working together at the Subunit for Genetics in the KWI in Buch and, as detailed in the Beyler chapter, continued to explore radiation and genetics to some extent unhindered by external events.[111] Together they published in 1947 a monograph on target theory as the first of an intended three-volume work on biophysics that appeared after they had both been transported to Russia.[112] The impact of the war years on Gert Molière, Ernst Lamla, Walter Kofink, and Werner Bloch requires additional study. The primary theoretical products of the Delbrück club—the Gaffron-Wohl theory of the photosynthetic unit and the target theory of gene mutation—only in part represent the result of these interactions. By bringing into direct conversation this interdisciplinary association of scientists, unified by an interest in exploring the nature of the relationships between biology and physics, this ephemeral group ranged broadly in its interdisciplinary discussions into areas of life science and physics. In orchestrating these gatherings, Delbrück's first effort at transporting what he had absorbed as the "Copenhagen spirit" of collaborative and open inquiry into theoretical questions under the guidance of a charismatic leader,[113] Delbrück displayed the style of organizational management that was to make him famous at Cold Spring Harbor and Caltech. The Delbrück gatherings were informal and intentionally interdisciplinary, combining intensive intellectual discussions with the formation of an interacting social group. In its later California setting, this tacit social bond of an interdisciplinary group was created by Delbrück through his famous desert camping trips. The Grünewald meetings were obviously more restrained, and seem to have confined their social dimensions to group ventures to Timoféeff-Ressovsky's laboratory in Berlin-Buch to look at fruit flies.[114] But through the character of the Delbrück discussions, we can see in its early phases one aspect of shift from a form of "biophysics" as practiced in many contexts by the mid-1930s, to what became the "style" of a self-fashioned identity of an interconnecting body of interdisciplinary researchers who by the 1960s designated themselves "molecular" biologists. To be sure, there are multiple dimensions related to the creation of "molecular biology" that were not present in the Delbrück club; the participants certainly did not see themselves as forming some new disciplinary structure, and their association was, as we have seen, the product of many contingent factors that lay outside science. Only Timoféeff-Ressovsky and Zimmer attempted to develop institutional structures for biophysical research that extended immediately beyond these discussions,[115] and other research groups at the KWI carried on cooperative researches among biologists and biochemists that were also biophysical in character.[116] What Delbrück created through his discussions was a social structure that brought together both theoretical and experimental physicists and biologists and biochemists to explore jointly the relations of biology and physics, orchestrated by Delbrück in accord with his own particular and even idiosyncratic vision of the relations of biology to the physics in which he was originally trained.

Notes

1. Max Delbrück to Nikolai Timoféeff-Ressovsky, 1 October, 1970, Delbrück Papers, California Institute of Technology, box 22, folder 3. Hereafter cited in the form DP 22.3. All citations and quotations from the Delbrück papers appear by permission of the Archives of the California Institute of Technology and the Delbrück family.

2. N. V. Timoféeff-Ressovsky, K. G. Zimmer, and M. Delbrück, "Über die Natur der Genmutation und der Genstruktur," *Nachrichten von der Gesellschaft der Wissenschaften zu Göttingen, mathematisch-physikalische Klasse, Fachgruppe VI: Biologie,* 1(1935): 189–245; 3MP 222 (19). Hereafter cited as 3MP. References to specific pages in this paper are given by page number as found in the translation in this volume, followed by the pagination in the original paper in parentheses.

3. Max Delbrück interview by Carolyn Harding, July 14–September 11, 1978, California Institute of Technology Archives, 55, available online at http://oralhistories.library.caltech.edu/16/01/OH_Delbruck_M.pdf. Hereafter Harding interview. All quotations from this document used with permission of the Archives of the California Institute of Technology. Delbrück's various use of terms like "club" or "seminar" suggests something more formal than a simple informal gathering at his home, but it was nothing as formal as a traditional German research seminar. The existence of informal discussion groups surrounding the Kaiser Wilhelm Dahlem institutes was not itself unusual, and many groups had been meeting for informal colloquia at the Harnack House in Dahlem after its opening in 1929. The changes wrought by the rise of National Socialism, and its impact on the Kaiser Wilhelm Society, altered this world and made Delbrück's "off-premises" discussion a new kind of association. See M. J. Nye, "Historical Sources of Science-as-Social Practice: Michael Polanyi's Berlin," *Historical Studies in the Physical and Biological*

Sciences 37 (2007): 411–36. I will use the designator "club" or "discussion group" with these complexities in mind.

4. The Harding interview (56) identifies him as *Werner* Kofink. See note 54 below.

5. Ibid., 50, 56. He also mentions one other participant, whose name he could later not recall, who was interested in alcoholic fermentation.

6. Ibid., 49.

7. K. G. Zimmer, "The Target Theory," in *Phage and the Origins of Molecular Biology,* ed. J. Cairns, G. Stent, and J. D. Watson (Cold Spring Harbor, NY: Cold Spring Harbor Press, 1966), 36–37.

8. Harding interview, 55.

9. See Summers and Beyler chapters, this volume. Extensive information on this context is given in Alexander von Schwerin, *Experimentalisierung des Menschen: Der Genetiker Hans Nachtsheim und die vergleichende Erbpathologie, 1920–1945* (Göttingen: Wallstein, 2004), especially chap. 3.

10. On this broader context see Anne Harrington, *Reenchanted Science: Holism in German Culture from Wilhelm II to Hitler* (Princeton, NJ: Princeton University Press, 1996). See also Richard Beyler, "From Positivism to Organicism: Pascual Jordan's Interpretations of Modern Physics in Cultural Context" (PhD dissertation, Harvard University, 1994); and C. Lawrence and G. Weisz, eds., *Greater than the Parts: Holism in Biomedicine, 1920–1950* (New York: Oxford University Press, 1998).

11. The multiyear Max Planck Society study of the Kaiser Wilhelm Society under nazification, *Geschichte der Kaiser-Wilhelm-Gesellschaft im Nationalsozialismus*, edited by Reinhard Rürup and Wolfgang Schieder, is now in its fourteenth volume. For an overview see C. Sachse and M. Walker, eds., *Politics and Science in Wartime: Comparative International Perspectives on the Kaiser Wilhelm Institute, Osiris,* 2nd ser., 20 (2005). See also S. Heim, C. Sachse, and M. Walker, *The Kaiser Wilhelm Society under National Socialism* (Cambridge: Cambridge University Press, 2009); and K. Macrakis, *Surviving the Swastika: Scientific Research in Nazi Germany* (New York: Oxford University Press, 1993), chap. 6. On the impact on the biological sciences with specific reference to the Kaiser Wilhelm institutes, see Bernd Gausemeier, *Natürliche Ordnungen und politische Allianzen: Biologische und biochemische Forschung an Kaiser-Wilhelm-Institute, 1933–1945* (Göttingen: Wallstein, 2005) in the Rürup and Schieder series. See also K. Macrakis, "The Survival of Basic Biological Research in National Socialist Germany," *Journal of the History of Biology* 26 (1993): 519–43; and U. Deichmann, *Biologists under Hitler,* trans. T. Dunlap (Cambridge, MA: Harvard University Press, 1996).

12. On Jordan's larger project and his construction of "quantum" biology, see especially the Beyler chapter, this volume; and Beyler, "Positivism to Organicism"; Beyler, "Targeting the Organism: The Scientific and Cultural Context of Pascual Jordan's Quantum Biology, 1932–1947," *Isis* 87 (1996): 248–73; Beyler, "Exporting the Quantum Revolution: Pascual Jordan's Biophysical Initiatives,"

in *Pascual Jordan 1902–1980* (Max-Planck-Institut für Wissenschaftsgeschichte, Preprint 329, 2007, http://www.mpiwg-berlin.mpg.de/en/research/preprints.html); N. Wise, "Pascual Jordan, Quantum Mechanics, Psychology, National Socialism," in *Science, Technology, and National Socialism*, ed. M. Renneberg and M. Walker (Cambridge: Cambridge University Press, 1994), 224–54; F. Aaserud, *Redirecting Science: Niels Bohr, Philanthropy, and the Rise of Nuclear Physics* (Cambridge: Cambridge University Press, 1990), chap. 2.

13. M. Born, W. Heisenberg, and P. Jordan, "Zur Quantenmechanik," *Zeitschrift für Physik* 34 (1925): 858–88; 35 (1926): 557–615.

14. Aaserud, *Redirecting Science*, 82–92. See also Bohr to Jordan, January 25, 1930, in "Selected Correspondence," trans. D. Favrholdt, in *Niels Bohr: Collected Works*, vol. 10, ed. D. Favrholdt (Amsterdam: Elsevier, 1999), 515. All subsequent citations to the Bohr correspondence will be made to this edition as Bohr, "Selected Correspondence," wherever possible.

15. The original title is "Statistik, Kausalität und Willensfreiheit." This was sent by Jordan with a letter to Bohr on May 20, 1931. In Bohr, "Selected Correspondence," 317.

16. P. Jordan, "Die Quantenmechanik und die Grundprobleme der Biologie und Psychologie," *Die Naturwissenschaften* 20 (1932): 820. My translation here and elsewhere. For extensive discussion of this concept, see Beyler, "Positivism to Organicism," chap. 2.

17. Jordan, "Die Quantenmechanik," 820.

18. F. B. Hanson and F. Heys, "An Analysis of the Effects of the Different Rays of Radium in Producing Lethal Mutations in *Drosophila*," *American Naturalist* 63 (1929): 201–13; Hanson and Heys, "Radium and Lethal Mutations in *Drosophila*: Further Evidence of the Proportionality Rule from a Study of the Effects of Equivalent Doses Differently Applied," *American Naturalist* 66 (1932): 335–45. The work of Hanson and Heys built on the radiation studies by Muller discussed in the introduction to this volume.

19. E. Bleuler, "Die Beziehungen der neueren physikalischen Vorstellungen zur Psychologie und Biologie," *Viertel Jahrschrift der naturforschenden Gesellschaft in Zürich* 78 (1933): 176–77.

20. P. Jordan, "Quantenphysikalische Bemerkungen zur Biologie und Psychologie," *Erkenntnis* 4 (1934): 215–52.

21. N. Bohr, "Wirkungsquantum und Naturbeschreibung," *Die Naturwissenschaften* 17 (1929): 483–86; Bohr, "Die Atomtheorie und die Prinzipien der Naturbeschreibung," *Die Naturwissenschaften* 18 (1930): 73–78. These were republished in the volume *Atomtheorie und Naturbeschreibung: Vier Aufsätze mit einer einleitenden Übersicht* (Berlin: Springer, 1931), translated as *Atomic Theory and the Description of Nature: The Philosophical Writings of Niels Bohr*, vol. 1 (Cambridge: Cambridge University Press, 1934; reprint, Woodbridge, CT: Ox Bow Press, 1987).

22. N. Bohr, "Licht und Leben," *Die Naturwissenschaften* 21 (1933): 245–50. The English version appeared in two parts (*Nature*, March 25, 1933, 421–23; April 1, 1933, 457–59). By "epistemological," I mean that as applied to biology, Bohr was developing aspects of Kant's solution to the antinomy of teleological judgment, particularly as this had been interpreted by Harald Høffding. I am developing this point in more detail in my manuscript "How Was Teleology Eliminated in Early Molecular Biology?" (in press, *Studies in History and Philosophy of the Biological and Biomedical Sciences*). For additional details, see Roll-Hansen and McKaughan chapters, this volume.

23. Jordan, "Quantenphysikalische Bemerkungen," 238, emphasis in original.

24. Ibid., 238–39.

25. Ibid., 239.

26. Ibid., 240.

27. Ibid., 248. "Freude hat die hier auftretende '*psychologische Komplementarität*' in einer Formulierung geschildert, welche in geradezu frappanter Weise an Bohrsche Sätze errinnern."

28. Ibid., 251–52.

29. See Ernest Nagel's report on this meeting in *Journal of Philosophy* 31 (1934): 589–601. Jordan's manuscript had been sent to the *Erkenntnis* editor Hans Reichenbach in June 1934, and his views on biology and causality had been under discussion with Reichenbach since January 1934 and were obviously well known to the organizers and participants in this symposium prior to its August 30 publication date. Jordan-Reichenbach correspondence, Special Collections, University of Pittsburgh Libraries.

30. Both papers are published in 1935 as M. Schlick, "Über den Begriff der Ganzheit," 52–53; and E. Zilsel, "P. Jordans Versuch, den Vitalismus Quantenmechanisch zu Retten," *Erkenntnis* 5 (1935): 56–63.

31. On this context see Harrington, *Reenchanted Science*, and M. G. Ash, *Gestalt Psychology in German Culture, 1890–1967* (Cambridge: Cambridge University Press, 1995). On the complex relation of early logical empiricism to biology, see Gereon Wolters, "Wrongful Life: Logico-Empiricist Philosophy of Biology," in *Experience, Reality, and Scientific Explanation*, ed. M. C. Galavotti and A. Pagnini (Dordrecht: Kluwer, 1999), 187–208.

32. Zilsel, "P. Jordans Versuch," 62.

33. The views of Driesch were immediately relevant. He was scheduled to give one of the opening plenary addresses at the impending international conference.

34. Zilsel, "P. Jordans Versuch," 64.

35. P. Jordan, "Ergänzende Bemerkungen über Biologie und Quantenmechanik," *Erkenntnis* 5 (1935): 348–52, emphasis in original.

36. Ibid., 352.

37. The Berlin circle was jointly responsible with the Wiener Kreis for the publication of *Erkenntnis*. For a short overview of the Berlin circle, see N. Rescher,

"The Berlin School of Logical Empiricism and Its Legacy," *Erkenntnis* 64 (2006): 281–304.

38. The relevant issue of *Erkenntnis* had been published on August 30.

39. See Max Hartmann, *Die Kausalität in Physik und Biologie* (Berlin: De Gruyter, 1937) and *Philosophie der Naturwissenschaften* (Berlin: Springer, 1937). Hartmann wrote several treatises on philosophical issues between 1937 and 1956 (*Gesammelte Vorträge und Aufsätze*, 2 vols. [Stuttgart: Fischer, 1956]). The other unspecified biologists in the audience could logically have included Otto Warburg, Alfred Kühn, and Fritz von Wettstein, all with important leadership positions in the Kaiser Wilhelm biological institutes at the time. See Macrakis, "Survival of Basic Biological Research," 520–24.

40. Max Delbrück to Niels Bohr, November 30, 1934, slightly modified from Bohr, "Selected Correspondence," 466–69.

41. Ibid., emphasis in original.

42. Ibid.

43. Bohr agreed with Delbrück's summary in his response of December 8, 1934, while commenting that this is not a "comprehensive account of the viewpoints but only a correction of the misunderstandings that are unfortunately widespread among the biologists" (ibid., 470). Bohr continues that he had prepared a paper in the summer of 1934 on these issues that he was intending to publish and that he would send this to Delbrück. The closest thing to such a published paper on these issues was his lecture to the Second International Congress for the Unity of Science held in Copenhagen, June 21–26, 1936, and published as "Causality and Complementarity," *Philosophy of Science* 4 (1937): 289–98 (reprinted in Bohr, *Causality and Complementarity: Philosophical Writings*, vol. 4, ed. J. Faye and H. J. Folse [Woodbridge, CT: Ox Bow Press, 1998], 83–91). In this he comments that his position "stands far removed from every attempt to exploit in a spiritual sense the failure of causal description in atomic physics" (90), denying Jordan's interpretation.

44. See note 21.

45. Harding interview, 55.

46. Ibid., 49.

47. See note 62.

48. Biographical details on Molière have been difficult to locate. This sketch has been derived from the short account in the university archives page of the University of Tübingen, available at http://www.uni-tuebingen.de/UAT/j2009g1.htm, and from descriptions of his later career. Molière retained his position at the KWI during the war and then was appointed to the Max Planck Institute for Physics at Göttingen. From 1952 to 1957 he served as the second director of the Institute for Theoretical Physics at São Paolo, Brazil, and from 1957 to 1959 as a scientific investigator at CERN in Geneva. He then became professor of theoretical physics at the University of Tübingen until his death.

49. G. Molière, "Quantenmechanische Theorie der Röntgenstrahlinterferenzen

in Kristallen, I–II," *Annalen der Physik*, ser. 5, 35 (1939): 272–313; Molière, "Zur Strahlungstheorie I," *Annalen der Physik*, ser. 5, 37 (1940): 415–20.

50. G. Molière and M. Delbrück, "Statistische Quanten-mechanik und Thermodynamik," *Akademie der Wissenschaften, Berlin: Physisches-Mathematisches Klasse Abhandlungen* 1 (1936): 1–42.

51. Bloch's notebooks for these courses survive. See comments on Bloch in M. Klein et al., eds., *The Collected Papers of Albert Einstein*, vol. 3: *The Swiss Years* (Princeton, NJ: Princeton University Press, 1993), 7, 600. Useful information on Bloch was also supplied by Ernst Peter Fischer of the University of Konstanz (personal communication).

52. W. Bloch, *Einführung in die Relativitätstheorie* (Leipzig: Teubner, 1918). See letter of Albert Einstein to Werner Bloch, June 27, 1917, in *Collected Papers of Albert Einstein*, 10:94.

53. P. Dirac, *Die Prinzipien der Quantenmechanik*, trans. W. Bloch (Leipzig: Hirzel, 1930).

54. Although the Harding interview (56) and one other document in the Delbrück archives (DP 40.3) identify him as "Werner" Kofink, extensive research seems to prove decisively that the person in question was the theoretical physicist Karl Walter Kofink. A crucial piece of evidence in resolving this problem is from comments of the German physicist Bernhard Gross, of the Institute of Physics of the University of São Paolo, Brazil, and a former fellow student with Kofink at the University of Berlin. He speaks of "Werner" Kofink as having invited him for a visiting professorship at the University of Karlsruhe in 1968 (B. Gross, "Recollections," *IEEE Transactions on Electrical Insulation* E1-21 [June 1986]: 245–47). This invitation has been confirmed by subsequent research to have been issued by Karl Walter, then holder of the chair in the Institute for the Structure of Matter at Karlsruhe. Walter was in Charlottenburg from October 1934 to March 1935, and relocated to Frankfort am Main from June 1935 to December 1944. Following the war, he held positions at the University of Strasbourg and the University of Stuttgart until appointed to the Institute of Theoretical Physics at Karlsruhe in November 1950. He was named to the chair in the structure of matter in 1961. From 1955 to 1956 he was a visiting professor at the California Institute of Technology. Biographical details have been constructed from documents supplied by the Archives of the Karlsruhe Institute of Technology and by Humboldt University of Berlin, and assistance from Princeton University Library. We wish to express our appreciation to Wulf Wulfhekel, Klaus Nippert, Antje Kreienbring, Ben Primer, and Peter Fischer for assistance in unraveling this mistaken identity.

55. W. Kofink, "Statistische Thermodynamik unter Zugründelegung der Quantenmechanik in Hartreescher Näherung," *Annalen der Physik*, ser. 5, 28 (1937): 264–96. His *Habilitationsschrift* is published as Kofink, "Zur Diracschen Theorie des Elektrons," *Annalen der Physik*, ser. 5, 38 (1940): 428–35, 436–55, 562–82, 583–600.

56. Details on Lamla have been drawn primarily from H. Autrum and F. L. Boschke, "Obituary for Ernst Lamla," *Die Naturwissenschaften* 73 (1986): 689.

57. E. Lamla, *Grundriss der Physik für Naturwissenschaftler, Mediziner und Pharmazeuten* (Berlin: Springer, 1925).

58. For more details on the scientific work of scientists who stayed in Germany in this period, see Macrakis, *Surviving the Swastika*.

59. After the period of the seminars, and with the support of the British Society for the Protection of Science and Learning, organized to assist dismissed faculty from German universities, Wohl moved in February 1939 to the Department of Botany at Oxford, and there continued his fundamental research on photosynthesis. In November 1942 he emigrated to the United States, taking temporary positions at the New School for Social Research in New York and at Princeton University's Palmer Physical Laboratory before joining the chemical engineering faculty of the University of Delaware in 1943, where he remained until his death. He was well known in his later years for his pioneering work on the theory of combustion. Biographical details are based on Jaime Wisniak, "Kurt Wohl: His Life and Work," *Educación Química* 14 (2003): 36 – 46; a personal narrative and faculty publicity document in the Wohl archives of the University of Delaware; a narrative curriculum vitae submitted by Wohl for a faculty position at the University of Chicago (James Franck Papers, Regenstein Library, University of Chicago, box 10, folder 6); and a file of obituary materials supplied by the archivist of the University of Delaware.

60. Biographical details have been drawn from a curriculum vitae drafted by Gaffron around 1970, and an autobiography written for naturalization purposes in 1944. Both were supplied by Hans's daughter-in-law, Barbara Gaffron, and are cited by permission (personal communication). See also biographical remarks in R. Rürup and M. Schüring, *Schicksale und Karrieren: Gedenkbuch für die von den Nationalsozialisten aus der Kaiser-Wilhelm-Gesellschaft vertriebenen Forscherinnen und Forscher* (Berlin: Wallstein, 2008), 199–201. Information on his doctoral research was supplied by Antje Kreienbring of Humboldt University.

61. After the period of the seminars, Gaffron accepted an invitation from fellow German émigré James Franck to join the Department of Biochemistry at the University of Chicago as a member of Franck's research group working on photosynthesis at the university's Fels Research Laboratory. He remained at the University of Chicago from 1938 until 1960, when after mandatory retirement he moved to the new Institute of Molecular Biophysics at Florida State University, from which he retired in 1972.

62. The original emphasis on photosynthesis is confirmed by the report of Bentley Glass, who knew Delbrück and several of the other participants in these discussions. He was a student in Berlin in this period and claims that Hans Gaffron's interests in photosynthesis brought the group together. B. Glass, "Two Prize Nobelists," *Quarterly Review of Biology* 64 (1989): 325. The strong photosynthesis emphasis is

also confirmed by the report of C. Stacy French, who worked in Berlin in 1935–36 with Otto Warburg on photosynthesis. She reports having attended "a few seminars on photosynthesis at Max Delbrück's house with Hans Gaffron and Eugene Rabinowitch" ("Fifty Years of Photosynthesis," *Annual Review of Plant Physiology* 30 [1979]: 6). This is the only evidence I have found that Eugene Rabinowitch may have attended these seminars after his dismissal from Göttingen in 1933.

63. DP 8.17. The correspondence covers February 20, 1944, to October 19, 1977. Delbrück introduced photosynthesis theory and biophysics into his lecture courses in physics at Vanderbilt in 1944, and was requesting advice and references to literature from Gaffron. Gaffron was also reporting to Delbrück on his photosynthesis research with James Franck at Chicago in this period. On Delbrück at Vanderbilt, see R. T. Lagemann and W. G. Holladay, "Max Delbrück at Vanderbilt," in *To Quarks and Quasars: A History of Physics and Astronomy at Vanderbilt University* (Nashville, TN: Vanderbilt University Department of Physics and Astronomy, 2000), chap. 6, accessed at http://www.vanderbilt.edu/delbruck/documents/Lagemann_Delbruck_Chapter.pdf, April 23, 2009. See also remarks in D. T. Zallen, "Redrawing the Boundaries of Molecular Biology: The Case of Photosynthesis," *Journal of the History of Biology* 26 (1993): 77n26.

64. Personal contacts between Delbrück and the Wohl and Gaffron families began in the early 1930s, and these warm personal friendships were maintained when all three emigrated to the United States in the late 1930s by different routes. See E. P. Fischer and C. Lipson, *Thinking about Science: Max Delbrück and the Origins of Molecular Biology* (New York: Norton, 1988), 70.

65. For a discussion of the wider background of botanical and photosynthesis research at the KWI, see Claus Schnarrenberger, "Botany at the Kaiser Wilhelm Institutes," *Englera* 7 (1987): 105–46; and Macrakis, "Survival of Basic Biological Research."

66. D. T. Zallen, "The 'Light' Organism for the Job," *Journal of the History of Biology* 26 (1993): 269–79.

67. L. Meitner and M. Delbrück, *Die Aufbau der Atomkerne: Natürliche und künstliche Kernumwandlungen* (Berlin: Springer, 1935).

68. On Delbrück's authorship of this section, see "Autobiography Dictated by Delbrück, Winter 1981," DP 45.7, p. 25.

69. Meitner and Delbrück, *Die Aufbau*, 37.

70. Ibid., 38.

71. Ibid., 61–62.

72. See Summers chapter, this volume.

73. This is defined by the equation E = $h\nu$, with E the quantum of radiant energy, ν the frequency of its source, and h the Planck constant with the approximate numerical value of 6.626×10^{-34} joules/second.

74. For an overview see K. Nickelsen, "The Construction of a Scientific Model: Otto Warburg and the Building Block Strategy," *Studies in History and Philosophy of Biological and Biomedical Sciences* 40 (2009): 73–86; J. Myers, "Conceptual

Developments in Photosynthesis, 1924–1974," *Plant Physiology* 54 (1974): 420–26; Zallen, "Redrawing the Boundaries."

75. O. Warburg and E. Negelein, "Über den Energieumsatz bei der Kohlensäureassimilation," *Zeitschrift für physikalischen Chemie* 102 (1922): 235–66; Warburg and Negelein, "Über den Einfluß der Wellenlänge auf den Energieumsatze bei der Kohlensäureassimilation," *Zeitschrift für physikalischen Chemie* 102 (1922): 191–218.

76. C. B. van Niel, "On the Morphology and Physiology of the Purple and Green Sulphur Bacteria," *Archiv für Mikrobiologie* 3 (1931): 108. The complexities of this relationship, and particularly the way in which the carbon in the CO_2 ends up as sugar, was clarified only by intensive work in the next two decades, culminating in Melvin Calvin's and Andrew Benson's Nobel Prize work in 1961 on the path of carbon. Calvin's conclusion was that the intermediate product was phosphoglyceric acid, rather than formaldehyde, from which carbohydrates were formed. The "free formaldehyde" theory was still assumed by Wohl in 1940 ("The Mechanism of Photosynthesis in Green Plants," *New Phytologist* 39 [1940]: 33–64), but was already questioned by Gaffron by 1939. See H. Gaffron, "Chemical Aspects of Photosynthesis," *Annual Review of Biochemistry* 8 (1939): 497; and his article "Photosynthesis," *Encyclopaedia Brittanica*, 1949 edition (unchanged from 1943), 17: 848a–848b.

77. On the issues surrounding this debate, see Nickelsen, "Construction," and Govindjee, "On the Requirement of Minimum Number of Four versus Eight Quanta of Light for the Evolution of One Molecule of Oxygen in Photosynthesis: A Historical Note," *Photosynthesis Research* 59 (1999): 249–54.

78. H. Gaffron and K. Wohl, "Zur Theorie der Assimilation," *Die Naturwissenschaften* 24 (1936): 81–90, 103–7. I am drawing on Gaffron's own summary of this theory in his "Chemical Aspects."

79. On its importance see J. M. Anderson, "Changing Concepts about the Distribution of Photosystems I and II between Grana-Appressed and Stroma-Exposed Thylakoid Membranes," in *Discoveries in Photosynthesis*, ed. J. T. Beatty, H. Gest, and J. F. Allen (New York: Springer, 2005), 729–36; P. Joliet, "Period-Four Oscillations of the Flash-Induced Oxygen Formation in Photosynthesis," in ibid., 371–78; P. H. Homann, "Hydrogen Metabolism of Green Algae: Discovery and Early Research—A Tribute to Hans Gaffron and His Coworkers," in ibid., 119–29; A. Melis and T. Happe, "Trails of Green Alga Hydrogen Research: From Hans Gaffron to New Frontiers," in ibid., 681–89.

80. Wohl, "Mechanism of Photosynthesis," 37. Experimental attack on the four-quantum theory was carried out by the research group at the University of Wisconsin in the late 1930s (W. M. Manning, J. F. Stauffer, B. M. Duggar, and F. Daniels, "Quantum Efficiency of Photosynthesis in *Chlorella*," *Journal of the American Chemical Society* 60 [1938]: 266–74). See also Myers, "Conceptual Developments," 421. By the time he wrote an article on photosynthesis for the *Encyclopaedia Brittanica* in 1943 (unchanged in 1949), Gaffron assumed that the correct value was more like ten quanta. Current values confirm this conclusion.

81. On the formaldehyde theory, see E. Rabinowitch, *Photosynthesis*, vol. 1 (New York: Interscience, 1945), 255ff.

82. J. Franck, "Beitrag zum Problem der Kohlensäure-Assimilation," *Die Naturwissenschaften* 14 (1935): 226–29; J. Franck and H. Levi, "Zum Mechanismus der Sauerstoff-Aktivierung durch fluoreszenzfähig Farbstoffe," ibid., 229–30. Franck's model required a significant induction delay.

83. Wohl, "Mechanism of Photosynthesis," 37.

84. Ibid.

85. As Gaffron and Wohl report in the 1936 paper: "Aber erst aus Diskussionen in einem von Herr M. Delbrück veranlassten Kolloquium ergab sich, dass man genötigt ist, diese Vorstellung einer Theorie der Assimilation zugrunde zu legen. Die folgenden Erörterungen haben ihren Ausgangspunkt in jenen Diskussionen" (Gaffron and Wohl, "Zur Theorie der Assimilation," 81).

86. 3MP, 252 (235). See E. Schrödinger, *What Is Life? The Physical Aspect of the Living Cell, with Mind and Matter and Autobiographical Sketches* (Cambridge: Cambridge University Press, 2000), 44. This compares with the photosynthetic unit, which is estimated to be around 2,500 chlorophyll *molecules*. As the Beyler chapter in this volume develops in detail, the assumption that the target-strike volume corresponded with the actual gene is never actually asserted in the 3MP and was the most readily revised feature of the 3MP. Schrödinger's reading of this as the conclusion of the paper was only one example of his misleading interpretation of its results.

87. 3MP, 270 (240).

88. On the uncertainty about the form of Zimmer's participation, see his recollection quoted at the beginning of this chapter.

89. Harding interview, 49.

90. His doctoral research was on photochemistry. See the introduction to this volume.

91. N. V. Timoféeff-Ressovsky, "The Experimental Production of Mutations," *Biological Reviews* 9 (1934): 411–57. The paper was submitted for publication on April 7.

92. Timoféeff-Ressovsky describes his hiring a new young assistant (Zimmer was twenty-two) to help with this work on radiation sometime after Muller departed for the Soviet Union in September 1933 (Timoféeff-Ressovsky to Muller, undated, H. J. Muller Papers, Lilly Library, Indiana University, LMC 1899, box 30).

93. Timoféeff-Ressovsky, "Experimental Production," 425.

94. As Zimmer pointed out subsequently, this meant that there is no allowable "tolerance" for human radiation exposure. See H. D. Griffith and K. G. Zimmer, "The Time-Intensity Factor in Relation to the Genetic Effects of Radiation," *British Journal of Radiology*, n.s., 8 (1935): 40–47.

95. Timoféeff-Ressovsky, "Experimental Production," 439. On Muller's contribution, see the introduction to this volume.

96. See E. A. Carlson, *The Gene: A Critical History*, reprint edition (Ames: Iowa State University Press, 1989), chap. 15; J. Harwood, *Styles of Scientific Thought: The German Genetics Community, 1900–1933* (Chicago: University of Chicago Press, 1993), 41–45.

97. Timoféeff-Ressovsky, "Experimental Production," 440.

98. Delbrück to Bohr, April 5, 1935, in "Selected Correspondence," 472.

99. 3MP, 255–56 (224–25).

100. See Roll-Hansen and McKaughan chapters, this volume.

101. 3MP, 268 (238).

102. On the concept of *Gedankenexperimenten* in physics, see John Norton, "Thought Experiments in Einstein's Work," in T. Horowitz and G. Massey, eds., *Thought Experiments in Science and Philosophy* (Savage, MD: Rowman and Little-field, 1991), available at http://philsci-archive.pitt.edu/archive/00003190/ (accessed May 5, 2008). Delbrück explicitly uses the term in his section (3MP 256 [225]).

103. Ibid.

104. The van't Hoff rule is given by the equation

$$\frac{W_{T+10}}{W_T} = e^{+\frac{10U}{kT^2}},$$

where T is the absolute temperature in degrees kelvin, and U is the energy needed to overcome a state of stability of an atomic configuration.

105. On the relationship of point heat and the target theory, see Summers chapter, this volume.

106. 3MP, 257 (226).

107. 3MP, 263 (232).

108. On the subsequent criticism of the Target theory by Demerec and Fricke on these grounds, see Beyler chapter, this volume.

109. 3MP, 262 (231).

110. See further discussion of this point in the introduction to this volume.

111. See also discussion in the introduction, note 17.

112. N. V. Timoféeff-Ressovsky and K. G. Zimmer, *Biophysik*, vol. 1, *Das Tref-ferprinzip in der Biologie* (Leipzig: S. Hirzel, 1947). Only one volume of this intended series was published. See details on their biophysics program in the Beyler chapter, this volume.

113. The "Copenhagen spirit" has been used to designate several things. As John Heilbron characterizes it, the designation means the extension of quantum interpretations of causality and complementarity as a universal epistemological position that Heisenberg had termed the *Kopenhagener Geist der Quantenphysik* (John Heilbron, "The Earliest Missionaries of the Copenhagen Spirit," *Revue d'histoire des sciences* 38 [1985]: 201). Finn Aaserud's interpretation is more multifaceted and includes institutional, theoretical, and attitudinal aspects. He uses it to characterize

Bohr's combination of theoretical research, experimentalism, and institutional support for science that he developed at the Institute for Theoretical Physics in Copenhagen (Aaserud, *Redirecting Science*, 6–15). As Delbrück might show some heritage from Copenhagen in his own work, it characterizes his personal leadership style and the encouragement of orchestrated, collaborative interdisciplinary research carried out in intensive seminar discussions.

114. Delbrück later recalled that "we also went to [Timoféeff-Ressovsky's] place just to see some flies, and talked about fly genetics and mutation research" (Harding interview, 49).

115. See Beyler chapter, this volume.

116. Macrakis, "Survival of Basic Biological Research," 531–37; Gausemeier, *Natürliche Ordnungen und politische Allianzen*.

CHAPTER FOUR

Exhuming the Three-Man Paper: Target-Theoretical Research in the 1930s and 1940s

Richard H. Beyler

In an interview many years after the event, Max Delbrück asserted that his 1935 paper cowritten with Nikolai V. Timoféeff-Ressovsky and Karl G. Zimmer, "Über die Natur der Genmutation und der Genstruktur," had received "a funeral first class" due to its publication in a relatively obscure forum, the proceedings of the Göttingen Academy of Sciences. Timoféeff-Ressovsky, the lead author of this "Three-Man Paper" (hereafter 3MP) sought to overcome the effects of this burial, Delbrück explained, by sending out a multitude of reprints. In Delbrück's view, even this action did not result in much understanding of the 3MP in the scientific community, since scientists at Caltech needed him to explain it to them upon his arrival there in 1937.[1] Delbrück's statement did not reveal why, exactly, the 3MP was buried in this fashion. Perhaps the paper was too long for publication in a conventional journal but too short to appear as a conventional monograph.

According to an orthodox narrative that seeks to identify a series of right turnings in the development of the discipline now known as molecular biology—a narrative promulgated not least by Delbrück and his protégés—his contemporaries' purported ignorance of the paper was (fortunately) not of much real consequence. The real consequence of the paper was its connection to the entry of its coauthor Delbrück and various associated phage researchers into the story.[2] The 3MP marked Delbrück's move from physics to biology. Likewise, the recruitment into biology of

Delbrück's Nobel Prize corecipient, Luria, arose from his reading of the 3MP.[3] Moreover, and perhaps most significantly, the 3MP formed a large part of the inspiration for Erwin Schrödinger's lectures *What Is Life?* The publication of these lectures in 1944, in turn, induced a cohort of energetic and enthusiastic workers into the new field of molecular biology just as it was working out the structure of DNA.[4] Thus, in this historical narrative, the significance of the 3MP lies in its shining example of an attempt to apply physicists' methods to biological problems and in the motivation it gave for physicists to move into biology. More precisely, this significance is attached primarily not to the paper itself but rather to its reinterpretation by Schrödinger almost a decade after its original publication.

Within this retrospective search for the "origins of molecular biology,"[5] what appeared exciting about the 3MP was not so much its content as simply the illustration it offered that a physicist might do interesting work in biology. Indeed, a retrospective evaluation of the 3MP's argument was tricky. Timoféeff-Ressovsky, Zimmer, and Delbrück were judged correct in identifying genes as molecules—or, more precisely, assemblages of atoms (*Atomverbände*)—and in this sense the 3MP could serve as a landmark on the triumphal road of molecular biology. Embarrassingly for a linear understanding of scientific progress, however, by the late 1940s evidence had accumulated that confirmed some initial skepticism voiced already in the 1930s by Hermann Muller, Milislav Demerec, Douglas E. Lea, and others about the 3MP's methodology. Complex chemical processes rendered inadequate the simplifying assumptions underlying the 3MP's analysis of the effect of ionizations in organic tissues. (I will return to some of these objections below.) For the orthodox perspective, ultimately, the 3MP emerged as a right turning on the "path to the double helix" but one taken for the wrong reasons, or (in another formulation) as a prime example of a "successful failure."[6]

We might then unpack Delbrück's statement regarding the "funeral" of the 3MP as acknowledging its significance for his personal development as a scientist—an experience that pushed him in directions that later led to triumph—but derogating its importance for the actual work done in and with the paper at the time. Indeed, the paper's obscurity might even seem deserved, since its argumentation proved to be flawed. In Delbrück's retrospective assessment the 3MP constituted a meteoric (i.e., ultimately failed) contribution, the appreciation of which was (for the time being) limited to a few cognoscenti, particularly in his native country.[7]

The orthodox interpretation looks at the 3MP from a post-double-helix viewpoint. But viewed within its own historical horizon—that is, situated within the landscape of biophysics research in Germany in 1935, a horizon that necessarily excludes prescience of later events and subsequent consensus about right approaches in genetics—the situation of the 3MP is somewhat different from that which Delbrück wanted to explain retrospectively. As this chapter will show, far from having been buried, the 3MP was cited—often with approval, occasionally with disapproval—in the work of biophysicists in at least a dozen or so locales of research in radiation biology. As might be expected, these users of the 3MP were found primarily in Germany, but also occasionally in various other countries including Great Britain, the United States, and France. In other words, in its own day the 3MP was a prominent contribution to a vigorous discourse among an international and interdisciplinary network of scientists, a discourse with its own problemata that do not necessarily correspond to those of later molecular biology. Moreover, though the 1935 paper was a landmark, taken either as a signal success or target of critique within this discourse, it was by no means a *founding* document. Rather, it was clearly understood to be part of an ongoing research trend integrating the experimental and mathematical techniques of modern physics into radiation biology. This discourse had been under way for at least a decade before its publication—that is, well before Delbrück's involvement—and it continued after his departure for the United States in 1937 in the burgeoning field of radiation biology, though in the end it was not central to the unraveling of genetic structure.

This body of research was the investigation of the structure and properties of organisms at the submicroscopic level through the statistical analysis of the results—namely, death, mutation, or other damage—of bombarding the organisms with X-rays, ultraviolet light, alpha particles, neutrons, or other forms of radiation. In German-language publications from the 1930s, the term for this approach—or rather, the collective term for a variety of related approaches—was *Treffertheorie*. In English, the most common term was "target theory," though "hit theory"—a closer translation of the German term—was sometimes also used. The French physicist Fernand Holweck sometimes used the term "statistical ultramicrometry."[8] These names suggest the basic premises and goals of the approach. The governing idea was that organisms at the subcellular level possessed sensitive zones or "targets" whose being "hit" by the impinging radiation resulted in significant damage. A statistical analysis of the results—death, mutation, and so forth—compared with models drawn from more or less

plausible assumptions about what biophysical, biochemical processes constituted a hit, resulted in conclusions about the structure of the supposed
sensitive zones—at least their size, and perhaps something about their
composition. For the target theorists, radiation-induced death and mutation were signs pointing toward the basic physiological structures of life,
provided they could be decoded using the experimental tools of radiation
physics and the mathematical tools of statistics. At least, that was their
hope. The 3MP was part of the development of target theory. As discussed
elsewhere in this volume, it combined a target-theoretical analysis of X-
ray mutagenesis in fruit flies (*Drosophila melanogaster*) with thermodynamic arguments about the rate of mutation under various conditions to
conclude that a gene is an assemblage of atoms (*Atomverband*).[9]

Many target theorists were indeed motivated by the question of the
nature and action of the gene—certainly this was the case for Timoféeff-Ressovsky, Zimmer, and Delbrück in 1935. As Lily Kay puts it, as
of the early 1930s "the gene problem was considered . . . to be the cutting
edge of science."[10] Not just the approximate size of the gene, but even its
very existence as a physical entity was a subject of interest and disputa-
tion. Milislav Demerec, at the Carnegie Institution's genetics laboratory
in Cold Spring Harbor, stated definitively in a 1935 *tour d'horizon* that
the "gene is a minute organic particle, probably a single large molecule."
Hitherto, the argument for this was based on dividing the microscopically
observed volume of a chromosome (of, say, *Drosophila*) by the estimated
total number of genes; however, Demerec suggested that more direct experimental evidence was on the way, based on examination of the chemical changes induced by "electron hit[s]" in artificially induced mutations.[11]
For the American geneticist Hermann Muller, who had discovered X-ray
mutagenesis in 1927, the nature of the gene had been a favorite topic since
the 1920s. From his new position at the Soviet Academy of Sciences, after
an extended visit to Timoféeff-Ressovsky's laboratory in 1932–33, Muller
and his colleague A. A. Prokofyeva published a paper trying to address
"to what extent the chromosome is subdivisible into its constituent genes
(or perhaps still further?), [and] the cytological basis of the constituents
thus separated out."[12] But as Evelyn Fox Keller notes, Muller did not
reach a satisfactory answer to such questions. Above all in Germany, the
particulate theory of the gene was far from unanimously held in the early
1930s. Richard Goldschmidt, codirector of the Kaiser Wilhelm Institute
(KWI) for Biology in Berlin-Dahlem, strenuously argued against it, holding what we call the "gene" instead to be the manifestation of a hierarchy
of structure within the chromosome as a whole.[13]

Clearly, tracking down the nature of the gene, using methods hitherto unconventional in the life sciences, was a large part of the agenda of the 3MP. Does this mean that the paper was part of the history of molecular biology? That depends, in part, on how one defines the term. Strictly speaking, the 3MP could not have been part of this field, since the term was evidently first used—such is the historical consensus—by Rockefeller Foundation officer Warren Weaver in 1938. And it was above all after the end of World War II that institutions, institutes, journals, and other disciplinary structures really began to transform themselves under this label. Apart from this literal usage, historians differ on the extent to which various forms of research prior to that period share essential characteristics with subsequent molecular biology.[14] In any event, neither subsequent developments of molecular biology, such as the discovery of the structure of DNA, nor disputations over the division of credit for that discovery can serve as a guide to target theorists' interest in genes in the early to mid-1930s. To make only one obvious point: at that time the chemical identity or internal structure of the genetic substance was by no means clear. Linking genes to any specific biochemical structure, still less to DNA, was simply not on the target-theoretical agenda in 1935. In a sense, the target theorists stopped at the outer boundary of the putative gene particles, not attempting to get inside that structure. Insofar as this is commonly asserted to be a key characteristic of molecular biology, the 3MP clearly stands prior to or outside of the history of that scientific field.

In other respects, if one posits a gradual development that extends beyond the formal labels and postwar institutional structures, the 3MP and the target-theory project do share some characteristics that some historians have ascribed to molecular biology. As Demerec's 1935 survey made clear, garnering direct experimental evidence of the particulate nature of genes required a novel move into submicroscopic territory. This focus on the submicroscopic region has been identified by Lily Kay, particularly, as a salient feature of the molecular biological enterprise.[15]

Similarly, target-theoretical discourse as it developed during the 1930s and into the 1940s entailed several bold assumptions—which could be and were strongly contested—about the nature of the organic, which were somewhat at odds with the traditional disciplinary culture of biology. To put it into a rather stark dichotomy: Can living organisms be usefully, albeit incompletely, understood in terms of an abstract model with homogeneous components that respond at the submicroscopic level in a uniform yet probabilistic fashion to physical events? Or is the life scientist, by definition, essentially concerned with a world of diverse, heterogeneous, unique forms and individuals, each of whom responds in an individually characteristic yet causally deterministic way to the stimuli of its environment? To have taken target theory as even plausible reflected a willingness to accept a radically new, abstract, even simplified conception of life itself, a theme that historians have sometimes identified with molecular biology as a whole.[16]

Target-theoretical radiation biology was, consequently, seldom something that established biologists or medical scientists undertook by themselves. As we will see in examining the various institutes where the work was done, target theory was usually done by physicists who had taken up research on organic objects, by scientists who made a disciplinary shift from physics to biology, or by collaborative teams of one or more biologists or medical scientists together with one or more physicists. There was apparently something about the experimental apparatus—the X-ray tubes and so on—in combination with the conceptual apparatus—for example, the mathematical subtleties of Poisson distributions and other statistical models—that needed to be run by someone familiar with physics. As noted above, Delbrück's move from physics to biology has become something of a founding myth for (at least some) molecular biologists, and to be sure, numerous commentators have testified to a certain style of doing science that seemed distinctively physicist-like, or even more specifically in the "Copenhagen spirit" inspired by theorist Niels Bohr.[17] But certainly much physics—and much good, successful physics—was not necessarily done under the Bohrian Copenhagen aegis. As the examples we will consider below indicate—Friedrich Dessauer and his collaborators, James A. Crowther, Edward U. Condon, Douglas E. Lea, Fernand Holweck, Richard Glocker, Kurt Sommermeyer, Karl G. Zimmer—there were plenty of precedents before 1935 for migrations of physicists into biology, or collaborations between physicists and biologists, over the conduit of target theory.

What motivated this kind of transdisciplinary collaboration or interdisciplinary movement, which was by no means obvious or simple for the parties involved? In many cases—particularly at its beginning—target theory emerged from what might be called an instrument-driven agenda. Here is an exciting new tool: what can it do? The primary tool in question here was, of course, the X-ray tube; later, other sources of radiation such as neutron generators also came into play. But target theory was also driven by its use of conceptual tools: statistical analysis, the notion of the quantized energy transfer, the tendency toward abstraction in the approach to the organic. For some biophysicists, there was a fascination in applying a given statistical model to a wide range of phenomena. For others, there was an excitement generated by the sense that a new fundamental principle of the cosmos had been discovered and a quest to discover the widest ramifications of this principle. Of course the X-ray machines of the 1920s were simple compared to the sophisticated and complicated apparatus that pervaded, and arguably dominated, later molecular biology; nevertheless, instrumentality is a common thread across these episodes.[18]

The target-theory or statistical-ultramicrometry approach emerged more or less independently in several places in the 1920s. Though the research from these several groups eventually coalesced into a common international literature, the context, origins, and orientation of each center were unique and illustrative of different aspects of the themes suggested above. I will consider several of these, but concentrate on developments in Germany that had the closest connections to the 3MP team.

Let us first consider a qualitative, somewhat schematized overview of the theory, disregarding for the time the diachronic dimension and glossing over several variants, modifications, and amendments of it that emerged.[19] If we expose cultures of bacteria or collections of some other organism to radiation—say, X-rays of a given wavelength and intensity—and then observe the proportion of the organisms that are damaged in some way—die, fail to reproduce, and so forth—we can then plot the survival or damage rates versus the dosages of radiation at the various wavelengths. We assume that each organism has a sensitive "target" volume at the subcellular level. Likewise, we assume—consonant with the new findings of quantum physics—that the energy from the radiation is not absorbed in a continuous fashion, but rather in a series of discrete deposits. The minute size of these deposits makes them insignificant at the macroscopic level, but in the subcellular realm they appear as a series of distinct and discrete "hits" distributed in a (presumably) random fashion. Increasing dosage results in an increasing density of "hits." Hits occurring outside the sensitive target area remain without significant effect. There is also a chance that a sensitive target in any given cell will receive more than one hit, the probability of which increases with increasing dosage. In the terminology of statistics, we are dealing with a Poisson distribution. If one hit on the sensitive zone suffices to cause the effect of interest in the organism—say, the death of a bacterium—then the relationship between increasing dosage and proportion of survivors is given by an exponentially decreasing curve. But if more than one hit to the target is required to kill the organism, the dose-response curve takes on a sigmoid (S-shaped) form, with the mathematically determinable contours of the curve corresponding to the number of hits required.

We are therefore interested in the "dose-response" curve: the proportions of death, mutation, or the like, that occur under designated conditions—type of radiation and its dosage, temporal and spatial distribution, and so forth. Moreover, we are interested in comparing the various dose-response curves obtained under various experimental conditions with an appropriate family of theoretical curves drawn under a set of more or less plausible assumptions about the nature of the process. If our assumptions are correct, then the form of the curve tells us how many hits are required for the effect—death, mutation, and so forth—to occur. Specifically, the hit number can be determined from the inflection point of the curve. Furthermore, once we know the number of hits on the target required to produce the effect in question, from the average overall density of hits and the absolute value of the dose-response ration we can calculate the effective volume of the sensitive zones or target areas.

Target theory therefore rests on a number of assumptions about the physical structure of the organism and its interaction with the impinging radiation. Above all, the theory rests on a quantized understanding of radiation: the effective hit represents in some fashion or other a localized deposition of a quantity of energy, and indeed the theory makes sense only if the transfer of energy from radiation to tissue takes place in discrete, localized packets. Researchers are then confronted with a series of questions. What physical process constitutes a hit: the complete absorption of the energy of the radiation (say, in the form of heat), the passage of a scattered electron, an ionization, or a cluster of ionizations? Can we actually determine the average density of hits? Are they actually randomly distributed, or are there underlying patterns to the distribution? Does the target zone indicate the dimensions of a physical structure, or is it an abstract coefficient expressing the likelihood of a given physicochemical process irrespective of where it occurs, or some combination of the two? Either way, what, exactly, is the nature of the interaction between the effective "hit" and the relevant portion of the living cell?

The original home of target theory in Germany was Friedrich Dessauer's Institute for the Physical Foundations of Medicine in Frankfurt. As noted elsewhere in this volume, Dessauer moved from a highly successful career as an entrepreneur of medical X-ray equipment to research in the physical aspects of medicine and biology. He also became politically active as an outspoken member of the Catholic-based Center Party, eventually winning election to the German parliament in 1924. In 1920, he was appointed an honorary professor at the new University of Frankfurt, which he had been instrumental in founding. The following year the Institute for the Physical Foundations of Medicine was created around Dessauer's professorship.[20]

It was at this institute, in 1922, that Dessauer began his forays into the statistical analysis of radiation biology. Dessauer modeled the absorption in organic tissues of energy from radiation in organic tissues—or rather from the secondary electrons produced—as proceeding in discrete, localized points of heat that in turn resulted in the localized denaturing of proteins. Assuming that these deposits occur randomly in time and space, he introduced an "effectiveness coefficient" to relate the dosage to the observed effects. Statistically, the number of biological units (e.g., cells) that remained unaffected after bombardment would be related to the dosage by an exponentially decreasing function.[21] Dessauer's junior collaborators, nuclear physicist Marietta Blau and mathematical physicist Kamillo Altenburger, extended the statistical analysis to cases in which there must be more than one energy deposit or "hit" per object to produce the effect, producing a sigmoid dose-response curve.[22] Comparison of the experimentally observed dosage-response curve with one of these theoretically predicted curves would give an indication of how many energy hits were entailed for the biological process in question to occur.

Dessauer's "point-heat" hypothesis about the physical nature of a hit was not necessary for the target theory to work. The theory rested on an abstract statistical analysis; as long as one accepted the assumptions that the energy of the radiation was transferred in a series of discrete deposits, one could remain (as it were) agnostic about the actual physiological mechanism while generating a formalism that produced a curve matching the experimental data. Such agnosticism did not satisfy all biologists or medical researchers, however. Soon a search was on for a way of associating Dessauer's theoretically generated "effectiveness coefficient" with a real region or structure within the organism or cell. In other words, the abstraction of the effectiveness coefficient became conceptualized as a physiologically identifiable spot or structure; an event of the required kind, such as an ionization, taking place within the target region then constituted a hit.

While Dessauer seems to have been the first to develop this theoretical structure, similar developments took place quite soon thereafter, and apparently independently, in several other institutes. In Cambridge, England, beginning in 1924, experimental physicist and medical radiologist James A. Crowther drew on his experience with scattering experiments in the laboratory of J. J. Thomson to undertake a statistical analysis of the effect of radiological "hits" in the protozoan *Colpidium*; his experimental statistical analyses proved to correspond to the theoretical curves produced by Dessauer's group.[23] Crowther's approach was soon applied by Edward U. Condon and Harold M. Terrill at Columbia University to experiments by Francis Wood on cancer cells and by Charles Packard on *Drosophila* mutations.[24] Again in Britain, at the Strangeways Research Laboratory in Cambridge, Douglas E. Lea and a cluster of collaborators, including Raymond B. Haines and Charles A. Coulson, took up the lead provided by Crowther, applying various kinds of radiation to bacteria and later to viruses.[25]

Another parallel and apparently independent appearance of the target theory occurred in Paris, where Fernand Holweck—an experimental physicist who was *chef de travaux* and *maître de recherches* at the Radium Institute[26]—and oncologist Antoine Lacassagne—director of physiology at the Radium Institute and simultaneously *chef de service* at the Pasteur Institute—began research in 1928 into the topic in frequent interaction with each other. Their original articles simply pointed to the notion of a quantized, localized transfer of radiation whose random distribution accounted for the differential response to irradiation of bacteria within the culture.[27] Holweck concluded that the photoelectrons produced by the impinging X-rays were localized in an area no more than one-tenth of the diameter of *Bacillus pyocyaneus*. The next year, they elaborated their statistical analysis, and Marie Curie herself provided the mathematical functions that defined the underlying statistical models.[28]

For the target theorists, the new physical instruments of X-rays and other forms of radiation, plus the statistical method of analysis, thus promised access to a new level of biology. But there were also new assumptions—new to the life sciences, that is—about the nature of biological entities. Above all, the target theory presupposed a certain uniformity among its objects of study. This was standard procedure in the physical sciences. In a 1931 Festschrift commemorating the tenth anniversary of his institute, Dessauer described the conceptual basis of his work in radiation biology. He employed, he said, "a conception which works with necessary physical simplifications. In the place of the quite complicated field of effect of, say, a human or animal body, I imagine, as a physicist, a cube filled with homogeneous, single-phase biological material."[29] Or in the words of a later textbook by the Freiburg biophysicist Kurt Sommermeyer, with the contrast in disciplinary assumptions made explicit: "We . . . proceed from the *presupposition*, that *all the objects* which are subjected to the statistics *behave completely similarly with respect to the radiation*, as we are accustomed to with physical objects . . . This assumption is quite familiar to the physicist. Conversely, the biologist takes it as self-evident that biological objects always exhibit certain differences, a *variability*."[30] The targets, in other words, all had to act alike with respect to the impinging randomly distributed radiation if the theory was to work.

In the eyes of many life scientists, Dessauer's assumption of uniformity ignored what was for them the most significant object of biological inquiry: the evident uniqueness of individual organisms. This complaint motivated, for example, a caustic critique of Dessauer's theory leveled by Hermann Holthusen—a lifelong skeptic about the uncritical use of radiation in biomedical contexts—in the mid-1920s. Rather than assuming that organisms would act the same under similar circumstances, Holthusen argued, it was the task of the life scientist to explain mutually differing reactions and behaviors. He charged the target theorists with avoiding what was for him the really crucial phenomenon: the variations among individual organisms in their responses to radiation. Holthusen carried out his own series of experiments on seedlings, the results of which, he maintained, could only be explained in terms of a variable susceptibility.[31] Subsequent mutually critical exchanges between Holthusen and Dessauer and one of the latter's collaborators, Yoshisada Nakashima, revolved around the former's contention that Dessauer and Nakashima had neglected possible photochemical phenomena in the absorption of radiation, as well as possible cumulative effects over time.[32]

As Arne Hessenbruch has pointed out, objections such as Holthusen's were closely linked to his sensitivity to the potential *human* implications of the theory in the medical context: conceptualizing organic responses to radiation in a way that neglected the variability among the organisms might be the first step on the slippery slope to denying individuality in human beings more generally. Another prominent physician, Lothar Heidenhain, attacked Dessauer on exactly this question, which, in turn, had particular urgency due to the concerns raised by the ongoing industrialization and rationalization of medicine, and of life in general, in the Weimar era.[33] But while in this scientific initiative Dessauer appears in a distinctly modernizing and (to his critics) mechanizing guise, in many other works he was intensely concerned with the broader philosophical implications of technology and science, seeking, within the framework of Catholic social political thought, a reconciliation between technical rationality and human values.[34] The controversies between Dessauer, on the one hand, and Holthusen or Heidenhain on the other, ultimately were more about differing cultural, political, and philosophical values, but they had their point of origin in profoundly differing conceptions of the relationship between physical and biological inquiry.[35] Thus the incursion of physics into biology was provoking both excitement and concern well before the 1930s.

Despite criticisms, the target-theoretical approach developed apace in a variety of contexts, though the various results did not always agree. In the United States, researchers deploying the method included Ralph Wyckoff at the Rockefeller Institute for Medical Research, who examined the effects of ultraviolet light as well as X-rays upon *Bacterium coli* [*sic*] and *Bacterium aertryke*. He concluded that the "sensitive area" of a bacterium was no more than 0.05 of its volume.[36] John Gowen and E. H. Gay, using Wyckoff's facilities at the Rockefeller Institute, published in early 1933 an essentially target-theoretical analysis, inspired by Crowther, of X-ray-induced lethal mutations in *Drosophila* sperm. Considering the gene to be the sensitive volume, they arrived at a maximum size of 1×10^{-18} cubic centimeter; based on a current hypothesis that the chromosome was a folded protein thread, this would mean that "the gene could be composed of as many as 15 protein molecules of average atomic size."[37] Their dosimetry, however, was challenged by Demerec soon thereafter.[38]

In Germany, several research centers also quickly elaborated the Dessauer-Blau-Altenburger theory. A network of target-theoretical research grew up around Richard Glocker, director of the X-ray laboratory at the Stuttgart Institute of Technology who began work in the field partly in collaboration with researchers at the Katharinenhospital in Stuttgart, including chief surgeon Otto Jüngling and the couple Hanns and Margarethe Langendorff of the hospital's Department for Radiation Biology.[39] Glocker had long been concerned with the dosimetry of X-rays in their metallurgical applications, and in this context became interested in quantifying biological effects.[40] Glocker, Jüngling, the Langendorffs, and several assistants studied X-rays' effects on bean, mustard, and sunflower seedlings, axolotl eggs, yeast, and algae.[41] For theoretical explanation of the results, Glocker drew on the target theory as developed by Dessauer and colleagues, by Crowther, and by the Holweck-Lacassagne-Curie collaboration. Confronting also Holthusen's critique of Dessauer, Glocker

designed more complex models of interaction between radiation and target in order to account mathematically for what appeared to be individual
bean seedlings' variable susceptibility to radiation. He introduced statistical formulas based on the hypothesis that each of several targets in each
biological object must receive n hits for the effect—such as the stunted
growth of seedlings—to result. Crucial in deciding the question, Glocker
argued, was to ascertain if the shape of the dose-response curve was
dependent on wavelength with overall dosage of radiation held constant:
if there was a dependence, then this was probably due to the quantized
character of the radiation, and various experimenters had indeed found
this to be the case.[42]

As the work of Glocker, the Langendorffs, and their collaborators
suggested, the austere simplifications at the basis of target theory were the
theory's source of power, but also its most vulnerable point. The approach
could be elaborated, and defended, by taking a more complex set of physical factors into consideration. Starting with Dessauer's one-hit, one-target
model, Blau and Altenburger added the possibility of multiple hits per
event, and Glocker added the possibility of multiple targets per organism.
More generally, the notion of variation among individuals could be subsumed under the concept that a hit (e.g., ionization) in a target region produced only a certain *probability* that the observed biological effect would
take place. From the target-theoretical point of view, all the biological
objects still behaved exactly the same with respect to a particular law, but
this law was deemed to be probabilistic. Holthusen's critique still applied,
however: was this not an unnecessarily abstruse way of conceptualizing
the familiar notion (familiar to biologists, that is) of individual variability?
As we will see below, proponents responded by creating more and more
complex forms of the theory. By taking more factors affecting the physical
nature and response of the supposed targets into account, more refined
statistical dose-response curves could be created, which could be matched
with more precision to experimental results. Moreover, despite disagreements about specific details, researchers such as Holweck and Lacassagne,
Wyckoff, and Gowen and Gay in the United States had used the techniques of target theory to venture educated guesses about the physical
size of "sensitive zones"—sometimes now identified with genes—within
living cells.

Thus by 1935 there were several precedents for the application of target-
theoretical methods to the problem of the physical nature of the gene.
In its constellation of resources, interests, and talents, N. V. Timoféeff-
Ressovsky's laboratory in Berlin-Buch was an ideal location for their even

Leaders of the Kaiser Wilhelm Institute for Brain Research, mid-1930s. The front row from left to right: Oskar Vogt, L. Minor, Cecile Vogt, Elena Timoféeff-Ressovsky, and Nikolai Timoféeff-Ressovsky. Others not identified. Archives of the Max Planck Society, Berlin-Dahlem (Kaiser Wilhelm Institute for Brain Research, no. III/4).

more assiduous application.[43] The Russian geneticist, who had been a protégé of Nikolai K. Kol'tsov, came to Berlin in 1926, along with his wife Elena—also a talented geneticist—at the invitation of Oskar Vogt, director of the KWI for Brain Research. Timoféeff-Ressovsky's Department of Genetics operated as a quasi-independent unit within the KWI.

In the ensuing years, the Timoféeff-Ressovskys were instrumental in integrating the perspectives of population genetics with exhaustive experimental work on *Drosophila*. After Muller's discovery of X-ray mutagenesis in 1927, Timoféeff-Ressovsky became one of the leading exponents of that technique, which provided one of the main motivations for Muller's visit in 1932–33.

As described in more detail in the chapter by Sloan in this volume, about the same time an informal circle of biophysical researchers formed in Berlin and made connections with Timoféeff-Ressovsky. This group included radiation physicist Karl G. Zimmer—then working at the Cecilienhaus Hospital in Berlin-Charlottenburg, later with Timoféeff-Ressovsky—and the young theorist Max Delbrück, who was at that time an assistant to Lise Meitner at the KWI for Chemistry in Berlin. Zimmer's growing interest in target-theoretical research seems to have been a smooth, pragmatic transition from his previous work in biomedical applications of physics.[44] Delbrück's interest seems to have been motivated, in large part, by questions of a methodological or philosophical character, including his response to theoretical physicist Pascual Jordan's move to export the quantum revolution into biology in the name of Bohr's correspondence principle and through what Jordan described as a "radical empiricist" epistemology.

Jordan, starting with his habilitation lecture in 1927, developed in a multitude of publications and lectures the theme that the new physics was a signal example of the triumph of empiricism. Strict attention to observables had enabled quantum theorists to break through the preconception of causal determinism. Among other consequences, according to Jordan, quantum theory had thereby revealed that a materialist understanding of the basic phenomena of life was a metaphysical dogma. When Jordan learned that Niels Bohr was also exploring the relations between biology and quantum physics, the result was an exchange of letters and conversations at conferences in 1930 and 1931. The result was a 1932 essay in *Die Naturwissenschaften*, in which Jordan took Bohr's "complementarity principle" to mean that we must be on the lookout for processes at the microscopic or submicroscopic level that exhibit indeterministic properties, inexplicable by classical deterministic physics, which have observable consequences for the living organism as a whole. This notion that certain indeterministic phenomena at the atomic or near-atomic level are somehow "amplified" into effects at the macroscopic, organismic level was the subject of protracted debate among Jordan and several critics. In a lengthy article in the logical-empiricist journal *Erkenntnis* in 1934, Jordan turned an objection from psychologist Eugen Bleuler into an argument in favor of the amplifier theory: the mutability of genes under irradiation suggested not a causal-deterministic relationship but, because of its statistical nature, just the opposite.[45]

Among the critics of Jordan was Delbrück, and (as described in Sloan's chapter) generating a response to Jordan evidently became part of the agenda of Delbrück's biophysical seminar. Yet though Delbrück perceived himself as defending the integrity of Bohr's complementarity principle against Jordan's too reductionist approach, there is good reason to doubt that he and Bohr were actually in philosophical agreement, either.[46] In any event, out of quite different motives, both Jordan and Delbrück brought to the preexisting network of target theorists a new philosophically charged *Problematik*.

While the use of radiation had been integral to Timoféeff-Ressovsky's work on *Drosophila* mutation genetics for several years, the target-theory approach promised to offer a new power and precision to the research. Zimmer and Delbrück, in particular, became collaborators with Timoféeff-Ressovsky, resulting in the 3MP in 1935.[47] As we have seen, target theorists already held that their method revealed the size and—perhaps—certain physical characteristics of the gene. But the 3MP provided the clearest evidence to date, combining the respective fields of expertise of the three authors, that the gene was an entity whose size was of a molecular order of magnitude, if not indeed a molecule per se. Part 1 of the paper summarized over a decade's worth of observations on spontaneous mutations in *Drosophila*, as well as X-ray mutagenesis experiments on this model organism, both over twenty separate studies by Timoféeff-Ressovsky himself as well as (by my count) work by at least eighteen other individual scientists or teams. He paid particular attention to the appearance of so-called back mutations—indicating the reversibility of whatever physico-chemical process was underlying the phenomenon—and the effects of radiation in varying temporal distributions, under different temperatures, and so forth. Zimmer's second section was a physical and statistical analysis of the model of a "hit" as comprising a single ionization in the effective "target" zone. Delbrück's third section was primarily an analysis of the thermodynamic aspects of the single ionization model, in which the experimental results on back mutations and his observations on spontaneous mutation (providing, so to speak, a theoretical baseline against which the effects of radiation could be analyzed) were crucial. The final section of the paper drew on three separate lines of experimental and theoretical evidence for a "discussion of the mutation process" and "theory of gene structure"—namely, that the former was a rearrangement (*Umlagerung*) of an arrangement of atoms (*Atomverband*) in the latter.[48]

The apparent success of the 3MP in defining the gene as a molecule-like structure was a signal for radiation biologists to redouble their efforts in this direction. In the target-theoretical literature of the late 1930s and early 1940s, the 3MP became a point of reference for ongoing research in Timoféeff-Ressovsky's department. The 3MP was thus not a static artifact, but rather subject to both elaboration and amendment. Of course it was members of the Timoféeff-Ressovsky circle who made most use of the 3MP, but as we will note below, other target theorists both in Germany and elsewhere took note of the paper or the extension of its arguments in other publications.

The 3MP confirmed ideas about the molecular nature of the gene that Muller—whose discovery of X-ray mutagenesis had brought Timoféeff-Ressovsky into radiation biology in the first place—had been advancing since the 1920s, in the face of some controversy. Moreover, Muller had pronounced himself enthusiastic about the application of physical methods and theories to biology.[49] Nevertheless, he had doubts about Timoféeff-Ressovsky, Zimmer, and Delbrück's target-theoretical argument. At a conference in Copenhagen in 1936, as well as in several subsequent papers, Muller cautioned Timoféeff-Ressovsky that the analysis was based on simplifying premises, all of which were possible but far from proven. Above all, Muller raised concerns about the notion that there was a one-to-one correlation between ionizations in the sensitive region and mutation and the notion that there was no significant spatial transfer of energy in the mutation process.[50] A 1936 article by Timoféeff-Ressovsky and Delbrück, following up on the 3MP, responded to these cautions. Here they clarified that their result of the previous year only established a minimum bound for the size of the gene, and they conceded that the target region was not necessarily identical with the gene itself.[51]

Demerec and Hugo Fricke voiced concerns similar to Muller's. They interpreted their experiments on lethal X-ray-induced mutations in *Drosophila* as showing a process that involved the transfer of energy from activated molecules. Such energy transfers, they asserted, could occur across distances of up to 100 microns, rendering highly problematic any identification of a mathematically abstract target with a given physiological area or entity. This was now a line of argument at odds with the views Demerec had expressed (admittedly rather speculatively) two years earlier.[52]

In 1937 Timoféeff-Ressovsky published a comprehensive survey of experimental genetics, *Experimentelle Mutationsforschung in der Vererbungslehre*, which came to be one of the leading German-language texts in the field. Here he gave a brief exposition of the general principles of the target theory in which the primary references were the 3MP as well as the work of Glocker, the British geneticist W. V. Mayneord, and Boris Rajewsky—Friedrich Dessauer's erstwhile student and his successor after Dessauer's forced emigration.[53] Timoféeff-Ressovsky then summarized the model developed in the 3MP, namely, that a mutation was a "rearrangement" (*Umlagerung*) of atoms in an otherwise stable, well-defined group or structure of atoms. Timoféeff-Ressovsky admitted that uncertainties remained: whether this assemblage of atoms should be understood "as a very large or smaller molecule, as a micelle, or as a part of a molecule or micelle can so far not be decided." Nevertheless, the 3MP model had much to offer by way of plausible and useful consequences: it accounted for the general stability and permanence of genes; it was much more consonant with the phenomenon of replication than alterative models that ascribed a more amorphous structure to the gene; it readily accounted for back-mutation.[54] Evidently Timoféeff-Ressovsky's bid to create a canonical survey of the state of the art of radiation genetics, this book thus includes the essential content of the 3MP, plus a couple of years' worth of refinements, as its capstone. If our concern is with the dissemination of ideas advanced in the 3MP, and not merely the document itself, this book, which incorporated the gist of the 3MP, should also be taken into account.

Experimentelle Mutationsforschung emphasized the way the 3MP model, and above all the thermodynamic arguments contributed by Delbrück, integrated with the known facts about spontaneous natural mutation.[55] For Timoféeff-Ressovsky, who in addition to being a leading exponent of experimental genetics was also a knowledgeable field biologist and evolutionary theorist, this illumination of the natural process of mutation by the experimentally derived model was evidently of considerable value for understanding problems of speciation and intraspecific variation. He cited the 3MP in this context in several publications from this period.[56]

Timoféeff-Ressovsky and Zimmer focused special attention in the late 1930s on one refinement of the target theory around the problem known as the "saturation effect." The effect arose when sufficiently energetic radiation produced more hits per target than were necessary to produce a lethal or mutagenic result. Jordan analyzed the phenomenon theoretically in detail, and argued from his mathematical models that the saturation effect could serve as an important test case for validity of the target theory.[57] To investigate this, Zimmer and Timoféeff-Ressovsky (with financial support from the German Research Association) procured in 1937 a neutron generator from the Philips electronics firm, which they used (among other things) to irradiate *Drosophila*.[58] An explicit goal of these neutron experiments was to elaborate the concept of the relationship between the target zone and the gene developed in the 3MP in the face of criticisms such as those of Fricke and Demerec.[59]

Their 1936 paper following up the 3MP, Zimmer and Timoféeff-Ressovsky now reminded their readers, had differentiated among three concepts that might or might not be physically the same entity: the sensitive target zone in which ionization occurred, the structure that undergoes the chemical or physical change composing the mutation process (possibly after a spatial transfer of energy from the spot of the ionization), and the gene abstractly defined as a locus of traits not susceptible to breaks or crossing-over in inheritance. In other words, the theoretical target zone should *not* automatically be assumed to be the gene itself. The target zone was, rather, the sensitive area within which ionizations effectively and necessarily led to mutations in the gene. The physical locales of mutation were, presumably, considerably smaller, and perhaps located elsewhere. Even using these more cautious definitions, however, the neutron radiation appeared to have a saturation effect. Neutrons produced ionizations at a density high enough to cause superfluous hits per target, and thereby seemed to confirm Timoféeff-Ressovsky, Zimmer, and Delbrück's original hypothesis, as well as Jordan's suggestions, rather than Fricke and Demerec's interpretation.[60] As Delbrück explained in an invited paper for the Pacific Division meeting of the American Association for the Advancement of Science in 1940, the neutron experiments "support the view that these sensitive volumes represent well-defined areas and that the effect is really a direct one on the gene."[61] Thus with the neutron experiments starting in 1938, the core argument of the 3MP seemed to be vindicated, and indeed made more precise.

Nevertheless, Zimmer's caution about possible limits of mathematical modeling in biology also increased. In contrast to the days of the 3MP itself, he was increasingly alert to the relevance of Holthusen's criticisms. A 1941 article in *Biologisches Zentralblatt*, the leading German-language biological review journal, surveyed the problem of biological variability from the perspective of the target theory. Zimmer cited as exemplars the 3MP, the collaboration between Glocker and Langendorff, and the work of Lea and his colleagues in Great Britain. Zimmer conceded that the assumption of homogeneity was indeed highly problematic, but argued that this problem did not invalidate the whole target-theoretical technique. Rather, coefficients for physical probability could be introduced into the mathematical models so as to formally account for biological variability.[62] Zimmer conceded that, in some cases, observed dose-response curves could best be explained by such an assumption of biological variability among individual organisms. This was the case in an exchange with the Frankfurt immunologist Richard Prigge over the validity of a target-theoretical interpretation of the process of immunization.[63] After Zimmer's own statistical analysis of Prigge's experiments on tetanus immunization showed that a Gaussian distribution of susceptibility provided the closest match—hence invalidating the prior assumed model—Prigge wrote to Zimmer that he had cooled on the prospect of target-theoretical analysis of this experimental results.[64]

Moreover, the difficulty of energy transfer within the cell or subcellular structures would not go away, despite the renewed confidence created by the saturation-effect experiments. Several new regular collaborators—namely the industrial physicists Friedrich Möglich (who was a scientific advisor at Osram, one of Germany's leading electrotechnical and lighting concerns), Robert Rompe (also at Osram), and Nikolaus Riehl (who worked for the Auergesellschaft, a company renowned for its expertise in nonferrous metals)—contributed their expertise on fluorescence, luminescence, crystallography, dipole quantum resonance, gaseous diffusion, and similar topics to a series of articles with Zimmer and Timoféeff-Ressovsky. The core issue was the relative dimensions of the "formal" and "true" target zones in view of the diverse mechanisms by which energy could be transported within the cell. These articles presented a more complex picture than the simple abstractions of the original target theory, and called for more investigation of the specific material structures—for example, crystals—possibly implicated in key biological processes, as well as the influence of environmental factors. Closer cooperation among the experts in various theoretical and experimental subjects would be required, concluded Möglich, Rompe, and Timoféeff-Ressovsky in 1942: "[W]ith the advance of our knowledge of the structure and chemical composition of the biologically reactive units, as well as of the target region under the most diverse conditions in radiobiological experiments, the analysis of immediate structures and energy propagation mechanisms and [the analysis] of the target zones can mutually complement and support each other, on the path to the clarification of the question of the nature of elementary biological units and their changes."[65] The Berlin-Buch group thus continued to elaborate a target-theoretical approach to radiation genetics, in which the 3MP served as a point of departure, but to which other subdisciplines of physics and biology contributed, well after Delbrück's departure for the United States in 1937.

To be sure, under the influence of criticisms from scientists such as Holthusen, Muller, Fricke, and Demerec, and following their own concerns regarding the specifics of energy transfer and organismic variability, Timoféeff-Ressovsky and Zimmer were increasingly nuanced in how they interpreted their results. This hardly represented a dismissal of the 3MP, however. Moreover, they were hardly peripheral or obscure figures within

German biology. As Ute Deichmann has shown, Timoféeff-Ressovsky and Zimmer (along with Rajewsky) were within the top seven in support for research in zoology provided by the German Research Council between 1934 and 1945.[66]

The Berlin-Buch institute was by no means the only place in which the methodology deployed in the 3MP was modified, elaborated, and extended. In 1936 in Frankfurt, the projected conversion of the university institute there into a new KWI for Biophysics constituted a kind of declaration of methodological independence for the simplification and regularity at the root of Dessauer's and Rajewsky's target-theory approach: "The powerful forward development of physics and physical technology [and] . . . the blossoming of atomic physics [open] fundamentally new paths in natural scientific knowledge . . . Transporting the new knowledge [of physics] over to living substance and researching the physical regularities which occur therein constitute urgent tasks for the immediate future."[67] The bid to create a KWI for Biophysics proved successful in 1937. Rajewsky's own interests moved away from target theory per se toward other subjects—for example, cosmic rays and their biological effects.[68] However, a collaboration with Timoféeff-Ressovsky in 1939 on the influence of cosmic rays on the rate of mutation reunited the two lines of investigation.[69]

In Stuttgart, Richard Glocker's attention likewise had turned increasingly to other subjects, especially after he became codirector of the KWI for Metals Research upon its reestablishment in Stuttgart in 1934.[70] Hanns Langendorff, however, became director of the Radiological Institute at the University of Freiburg and in the late 1930s began there a new collaboration with the experimental physicist Sommermeyer. Langendorff and Sommermeyer took Timoféeff-Ressovsky's work, including the 3MP, as inspiration for a lengthy series of experiments on *Drosophila* eggs.[71]

Meanwhile back in Stuttgart, in 1937 the theoretical physicist Ulrich Dehlinger of the KWI for Metals Research published a note in *Die Naturwissenschaften* suggesting that the gene had a crystalline structure. His starting point was the 3MP argument that mutations were alterations in a single atom group in a molecule. The process of cellular fission, however, suggested a "crystal-like repetition of identical atom groupings." Dehlinger consequently suggested that a mutation was, more precisely, an *Umklapp* (chain-reaction rearrangement) process within this crystal.[72] Sommermeyer juxtaposed this suggestion with Timoféeff-Ressovsky and Zimmer's neutron experiments, and found that it provided an alternative and perhaps more plausible interpretation of the latter.[73]

In the United States, influence of the 3MP was perhaps not as extensive as in Germany, but it did not escape notice, particularly if one takes into consideration works such as *Experimentelle Mutationsforschung* and the articles on neutron irradiation that elaborated the 3MP argument—an example of this being Fricke and Demerec's critique discussed above. Besides the 1940 lecture by Delbrück also discussed above, direct citations of the 3MP can be found (for example) in Charles Packard's paper at the 1935 meeting of the Radiological Society of America; in a survey of X-ray genetics by Milislav Demerec before the same forum two years later; in an article on autosomal mutations by the Amherst College Drosophilists Harold Plough and George Child; in a review of the state of gene theory by the Missouri biochemist Addison Gulick; in a survey of radiation as applied to botanical cytology by the botanist T. H. Goodspeed and the physicist Fred Uber; in an article on photosynthesis by the émigré Kurt Wohl (a former participant in Delbrück's seminar); and a theoretical article on the physiological action of genes by Sewall Wright.[74] The neutron studies were taken up in articles by Norman Giles, E. R. Dempster, Ugo Fano, and Alfred Marshak.[75] *Experimentelle Mutationsforschung*, which largely recapitulated the 3MP argument, appeared in the bibliography of several other articles.[76] A book review in *American Midland Naturalist* called the book an "admirable and up-to-date summary" of the field of radiation biology since Muller's 1927 discovery, and though not citing the 3MP by name did explicitly refer to the collaboration with Zimmer and Delbrück.[77] Conversely, a 1943 review of another biophysics text chastised it for ignoring Timoféeff-Ressovsky's work.[78]

In Britain, Douglas E. Lea and his collaborators at the Strangeways Research Laboratory were the main practitioners of target-theory research in the decade or so after 1935. The work of the Timoféeff-Ressovsky collaboration and of associated figures such as Jordan was evidently instrumental for Lea's work, though his view of the relevance of the 3MP changed somewhat over time. In 1937, Lea used the 3MP's approach as a starting point to analyze experiments by various scientists on the recovery of organisms from nonlethal radiation.[79] A 1940 article on determining the number of genes in a chromosome by the target-theory method takes the 3MP as its starting point, and adds the nuance of the saturation effect as found in the neutron experiments.[80] By 1945, Lea (together with D. G. Catcheside) wrote that "a re-examination of the position seems called for," noting that even Timoféeff-Ressovsky and Delbrück had after the 3MP itself cautioned that the "target" of target theory might or might not have anything to do with the physical size of the genes. While conceding that the theory rested on simplifying assumptions—that ionization in the gene and only within the gene causes mutation—Lea and Catcheside nevertheless contended that these assumptions were correct to within a small factor, at any rate for the maximum size of the gene.[81] In his classic 1947 overview of target theory, *Actions of Radiations on Living Cells*, Lea included the 3MP among nine items authored or coauthored by Timoféeff-Ressovsky in its bibliography.[82]

Following the lead of Holweck and Lacassagne, Paris also remained a center of target-theoretical biophysics throughout the 1930s. Timoféeff-Ressovsky's participation in an International Conference on Physical-Chemical Biology at the Palais de la découverte in October 1937 and subsequent publication of a monograph in the series *Génétique*, edited by Boris Ephrussi, brought German-language research more centrally to the attention of French biophysicists.[83] Subsequently, Eugène Wollman, who was *chef de service* at the Pasteur Institute, became another key member of the Paris constellation of target theorists.[84] Greater interest in the 3MP, specifically, arrived in the person of Salvatore Luria in 1938. Luria had initially studied medicine, but shifted to a career in natural science under the example and influence of his school friend Ugo Fano.[85] Luria became a student in Enrico Fermi's institute in Rome; it was there that Franco Rasetti—like Delbrück an erstwhile assistant to Lise Meitner—introduced him to the 3MP. Since Luria was of Jewish descent, he became eager to leave Italy as Mussolini increasingly emulated Hitler's anti-Semitic policies. His Italian fellowship for study in the United States was withdrawn, but Luria was able to obtain instead, through the agency of Holweck, a French National Research Fund fellowship for work at the Radium Institute.[86]

Luria published an article on bacteriological target theory in the *Comptes rendus* of the Academy of Sciences.[87] He then collaborated on a radiobiological study on phage with Holweck and Wollman, an English version of which appeared in *Nature*. These three authors concluded that their "results were closely similar to those obtained by Timoféeff-Ressovsky on gene radio-mutations in Drosophila" in the 3MP.[88] Luria also wrote a systematic overview of the topic, which appeared in the weekly *Paris médical* on June 29, 1940, and which cited the 3MP as one of twenty-three papers in its bibliography. Luria explained the target theory by making a metaphorical connection—taken "unhappily from current events" he noted—between mutagenesis and the random damage caused by a military bombardment.[89] Though published in late June, these lines had evidently been written several weeks earlier. By the later date Luria had traveled by bicycle to southern France, moving out of the way of the advancing German armies, en route to Spain and then Portugal. Luria would end up in the United States, where he changed his given name to Salvador E. in order to have a middle initial to give to immigration officials, and eventually shared with Delbrück and Alfred Hershey a Nobel Prize for work on phage.[90]

An idiosyncratic version of target theory continued to be offered by Jordan who, in the late 1930s and into the 1940s, saw in it the best exemplar of his envisioned "quantum biology," which would uncover the "directing centers" of organic life. His conception of the importance of the technique went far beyond biophysics, however, in that he linked his quantum-physicality approach to biology with philosophical and political concerns: his attempts to delegitimize materialism (organisms were governed by indeterministic controlling centers) and to legitimize authoritarianism (the sensitive target regions of the cell were analogous to dictatorial forms of government). His appropriation of target theory was yet another example of the way in which this reconceptualization of the organic could readily be made to carry cultural and social meanings, and in the context of the Third Reich, this was particularly volatile stuff.[91]

For Jordan, the 3MP was a superb confirmation of this model of the organism as directed by "steering centers" that were susceptible to indeterministic quantum effects. Thus in a 1937 paper in the first issue of the new journal *Radiologica*, Jordan cited the 3MP as demonstrating that "the appearance of a mutation is to be seen as determined by a single elementary quantum-physical event" and therefore showing the "macroscopic development of an organism [to be] dependent on the occurrence or nonoccurrence of a single elementary atomic-physical event."[92] Similarly, in an extensive review article in *Physikalische Zeitschrift*, Jordan concluded the first paragraph by citing the 3MP as demonstrating that "biological elementary processes are simultaneously physical elementary processes."[93] And in a 1941 book intended for a popular audience, Jordan presented the 3MP as an interdisciplinary confirmation of his own prior hypothesis, on quantum-theoretical grounds, that the gene was a single molecule:

> This conclusion (Jordan 1934), at first judged skeptically or decidedly negatively by critics, experienced after just one year a confirmation of a definitive sort by an investigation, whose authors already embodied a new fruitful collaboration of

the natural-scientific disciplines they represented: a geneticist, an experimental and a theoretical physicist presented, in a work by Timoféeff-Ressovsky, Delbrück, and Zimmer which appeared in 1935, a proof of the mono-molecular nature of the gene, based on an investigation of the production of mutations by x-rays.[94]

As noted previously, Delbrück was, from the outset of his move into biology, perturbed by Jordan's simultaneous (or slightly prior) incursion into the same field and by the philosophical and methodological errors he thought Jordan was making.[95] Timoféeff-Ressovsky and Zimmer, though valuing Jordan's technical-mathematical contributions, were at best lukewarm toward his philosophical enthusiasms.[96] For them, the institutionalization of a truly transdisciplinary biophysics was enough, and here the support of a leading physical theorist such as Jordan was consequential.

One indication of the perceived success of target theory was the planning for the transformation of the KWI for Brain Research after the war.[97] Timoféeff-Ressovsky, along with Zimmer, had disappeared into Russian custody, but plans to continue work at the institute were put forward under the aegis of the Academy of Sciences in Berlin and with the backing of Robert Rompe who, as noted above, had increasingly become part of the Berlin-Buch circle. Rompe was, moreover, a solid Communist and was now becoming one of the main authorities for education and research in the Soviet occupation zone. In 1947–48 Rompe and the academy sounded out Jordan—by then in Göttingen, and academically unemployed due to problems with denazification—as the possible leader of a new research center for biophysics, in which target-theoretical investigations were to have a central place. From the academy's perspective, it seems that continuity in the work of one of the major biophysical research centers of the Soviet Zone and nascent German Democratic Republic was a top priority, as was recruiting a prominent scientist from western Germany to lead it— even if that scientist had a questionable past. Jordan's position is less clear: he may have been genuinely seeking a chance to return to Berlin, where he had been appointed professor in 1943, or he may have been using an offer from an institution in the Soviet occupation zone to smooth the path of his rehabilitation in western Germany. In the end, Jordan ended up with a professorship in Hamburg, while the Berlin-Buch institute was reestablished under the leadership of physicist Walter Friedrich—who early in his career had worked briefly in Dessauer's institute—but with considerably more modest scope and methodological agenda.[98]

A contemporary snapshot of the state of the art can be found in *Das Trefferprinzip in der Biologie*, a textbook authored by Timoféeff-Ressovsky and Zimmer as the first volume of a projected series on biophysics and published in 1947 (though probably written before the end of the war). A brief historical overview reviews the work of (inter alia) Dessauer's group, Condon and Terrill, Holthusen, Holweck and Lacassagne, Glocker, Wyckoff, Rajewsky, Lea and his colleagues, and the 3MP collaboration itself.[99] The text proceeds to a discussion of the theoretical principles: one- and multiple-hit processes on simple and complex target zones; variation of number of hits necessary to produce and effect; and so on.[100] With chapter 5 began reviews of various experimental results on various biological objects and various forms of radiation, and consideration of the problem of energy transport in various solutions and tissues—a set of problems that was by then clouding the previously seemingly clear results of the theory.[101] Separate chapters provided target-theoretical analyses of radiogenic mutations; killing of bacteria, viruses and phage; and various other phenomena such as human light perception.[102]

Although in the orthodox narrative the 3MP serves primarily as a kind of signpost on the eventual road to the double helix, this biophysical text by two of its three authors devotes only 4 (out of 267) pages of text specifically to "the nature of the gene": here Timoféeff-Ressovsky and Zimmer stuck to the rather cautious formulation that the genes are "physicochemical units" (*physikalisch-chemische Einheiten*) that could, based on the work of Torbjörn Caspersson, probably be identified with nucleoproteins; an analogy between genes and viruses, and Wendell Stanley's identification of viruses as "giant nucleo-proteid molecules" provided further confirmation. In sketching future possibilities, the authors reveal some conspicuous absences in target-theory research as it existed in Germany up to that point: "Further research therefore raises the highly interesting problem of analyzing the details of the structure and the multiplication mechanism of these elementary biological structures." In this connection, experimental genetics had the advantage of "clear definition of easy verification of structural changes of the elementary units, that is, of mutations. An essential help in this domain will be offered by the modern biochemistry of hormones."[103] From this text, then, tracing the chemical identity of the gene appears as only a modest part of the target-theoretical agenda. The extant field was defined largely by its methods rather than its objects.

A similar picture of the state of target theory in the 1940s emerges from the volume on biophysics, edited by Boris Rajewsky and Michael Schön,

in the FIAT Review of German Science.[104] An editors' note indicated
that Jordan, an intended coauthor of the historical overview section of
the volume, was prevented by "compelling external circumstances" from
completing his section and that therefore "the work of the Berlin circle
around N. W. Timoféeff-Ressovsky" as well as Jordan's own contributions
did not receive as much discussion as originally intended.[105] Nevertheless,
Rajewsky's chapter still positioned the 3MP as an "especially fruitful and
successful" confirmation of the theoretical analyses and deployment of experimental techniques that had been developed by Dessauer and his colleagues, Crowther, Condon and Terrill, Holweck and Lacassagne, Curie,
Glocker, Jordan, Lea, and Rajewsky himself. The 3MP had also, Rajewsky
noted, led to further "more refined applications" by Timoféeff-Ressovsky
and Zimmer, as well as theoretical elaborations by Jordan, while Sommermeyer followed in the direction of research initiated by Glocker.[106] Hanns
Langendorff's survey of the literature on the cytological effects of radiation cited the 3MP, alongside papers by Dessauer, Blau and Altenburger,
Crowther, Curie, Holweck, Lacassagne, and Rajewsky, as laying the experimental and theoretical foundation for the field.[107] In a review of work
on genetic effects of short-wavelength and particle radiations written by
Hans Bauer of the KWI for Biology, Timoféeff-Ressovsky was the most
cited author. The 3MP paper, specifically, appeared as the source of the
"hypothesis that gene mutations represent changes in well-defined assemblages of atoms, and that the mutation process follows the reaction-kinetic
laws of monomolecular reactions," a hypothesis that had in the meanwhile
received "more precise quantitative extensions."[108]

Schön's article on energy migration in molecular complexes began with
a citation of the 3MP as demonstrating the connection between single ionizations (in target areas of molecular size) with gene mutations.[109] But the
article also reflected recent research on energy transfer that adumbrated
problems for the 3MP argument. So, too, did the recurring critiques about
the variability observed in organisms' response to radiation. As the target
theory became more elaborate, its original elegant simplicity—the possibility of reading off the result from a plotted dosage-response curve—was
lost. The theory became more and more unwieldy as applied to the problem
of the molecular structure of genes. Muller's worries from the mid-1930s
proved valid; researchers realized that simplifying assumptions of target
theory could not always account for the physiologically significant effects
of energy transfers across (relatively) large distances and those of complex
biochemical processes; and by 1950 Rajewsky conceded that "only a small

fraction of the primary processes in the biological action of ionizing radiation are explicable in terms of the classical target theory."[110] Muller specified that a more detailed examination of the diffusion of activated chemical species and of energy transfer within tissues was necessary.[111]

With this transformation of the field, the meaning of the 3MP for the generation of researchers after the 1940s—for those who were aware of it at all—shifted considerably. Although the molecular conception of the gene went from strength to strength, the methodology used by the 3MP was, as of about 1950, increasingly dubious. Meanwhile, other techniques and instruments that had never been integral to the target-theoretical enterprise were galvanizing the attention of various schools of the emerging new molecular biology. Structural biochemistry, X-ray crystallography, ultracentrifugation, electrophoresis, chemical mutagenesis, digital computing, bacterial genetics—these techniques and instruments and others provided compelling new insights into the submicroscopic genetic realm that radiation target theory could not hope to aspire to.[112] As signaled by the fates of the 3MP authors, Germany had gone through the decimation of its scientific talent due to racial and ideological purges, increasing self-imposed international isolation, and the personal and material destructions of the war and its aftermath, just as international exchange was becoming increasingly significant for the development of this new science.[113] Conversely, target theory as a whole hardly disappeared, but on the contrary continued to flourish as a central methodology in radiobiology, a field of growing importance in the atomic age.[114] But these later results were not implicit in any burial of the 3MP in 1935. Following upon the appearance of the first appearance of the paper in 1935, a vigorous international and interdisciplinary network of researchers in many locations developed, applied, and debated its target-theoretical approaches to genetics and related biological topics. As we have seen, the result was a broad spectrum of work ranging from the highly technical to the philosophically speculative. In the context of its original appearance, the 3MP was a vital part of this rich, multicentered scientific exchange on the theory and practice of radiation biophysics.

Notes

1. N. V. Timoféeff-Ressovsky, K. G. Zimmer, and M. Delbrück, "Über die Natur der Genmutation und der Genstruktur," *Nachrichten von der Gesellschaft der Wissenschaften zu Göttingen, mathematisch-physikalische Klasse, Fachgruppe*

VI: Biologie, 1 (1935): 189–245; Carolyn Harding interview of Max Delbrück, session 3 (July 20, 1978); transcript available at http://oralhistories.library.caltech .edu/16/01/OH_Delbruck_M.pdf (accessed March 10, 2008), 50–51; discussed in L. E. Kay, "Conceptual Models and Analytical Tools: The Biology of Physicist Max Delbrück," *Journal of the History of Biology* 18 (1985): 221; and in E. P. Fischer and C. Lipson, *Thinking about Science: Max Delbrück and the Origins of Molecular Biology* (New York: Norton, 1988), 94. References to specific pages in the 3MP will be given by page number as found in the translation in this volume, followed by the pagination in the original paper in parentheses. In his German- and French-language publications, Timoféeff-Ressovsky invariably wrote his middle initial as *W*; this volume, however, follows the American (Library of Congress) convention of transforming this to a *V*.

2. See, e.g., J. Cairns, G. S. Stent, and J. D. Watson, eds., *Phage and the Origins of Molecular Biology* (Cold Spring Harbor, NY: Cold Spring Harbor Press, 1966), hereafter *Phage and the Origins*; G. S. Stent, "That Was the Molecular Biology That Was," *Science* 160 (1968): 390–94; L. E. Kay, "The Secret of Life: Niels Bohr's Influence on the Biology Program of Max Delbrück," *Rivista di Storia della Scienza* 2 (1985): 487–510; Kay, "Conceptual Models."

3. S. E. Luria, *A Slot Machine, a Broken Test Tube: An Autobiography* (New York: Harper and Row, 1984), 20; Fischer and Lipson, *Thinking about Science*, 94–95.

4. E. Schrödinger, *What Is Life? The Physical Aspect of the Living Cell* (Cambridge: Cambridge University Press, 1944), reprinted in a combined edition as *What Is Life? The Physical Aspect of the Living Cell & Mind and Matter* (Cambridge: Cambridge University Press, 2000); D. Fleming, "Émigré Physicists and the Biological Revolution," *Perspectives in American History* 2 (1968): 176–213, reprinted in Fleming and B. Bailyn, eds., *The Intellectual Migration: Europe and America, 1930–1960* (Cambridge, MA: Harvard University Press, 1969), 152–89; R. C. Olby, *The Path to the Double Helix* (New York: Dover, 1994; first published 1974), 240–47; E. J. Yoxen, "Where Does Schrödinger's *What Is Life?* Belong in the History of Molecular Biology?" *History of Science* 17 (1979): 17–52; L. E. Kay, *The Molecular Vision of Life: Caltech, the Rockefeller Foundation, and the Rise of the New Biology*, Monographs on the History and Philosophy of Biology (New York: Oxford University Press, 1993), 132–33; E. F. Keller, *Refiguring Life: Metaphors of Twentieth-Century Biology*, Wellek Library Lectures (New York: Columbia University Press, 1995), 19–20; Michel Morange, *A History of Molecular Biology*, trans. M. Cobb (Cambridge, MA: Harvard University Press, 1998), 67–78; L. Ceccarelli, *Shaping Science with Rhetoric: The Cases of Dobzhansky, Schrödinger, and Wilson* (Chicago: University of Chicago Press, 2001), 82–110; U. Deichmann, "A Brief Review of the Early History of Genetics and Its Relationship to Physics and Chemistry," in *Max Delbrück and Cologne: An Early Chapter of German Molecular Biology*, ed. S. Wenkel and U. Deichmann (Singapore: World

Scientific, 2007), 3–18, especially 8. For historiographical and sometimes critical analyses of the recruitment of physicists into molecular biology, see, e.g., P. Abir-Am, "The Discourse of Physical Power and Biological Knowledge in the 1930s: A Reappraisal of the Rockefeller Foundation's 'Policy' in Molecular Biology," *Social Studies of Science* 12 (1982): 341–82; E. J. Yoxen, "Giving Life a New Meaning: The Rise of the Molecular Biology Establishment," in *Scientific Establishments and Hierarchies*, ed. N. Elias, H. Martins, and R. Whitley, Sociology of the Sciences, vol. 6 (Dordrecht: D. Reidel, 1982), 123–43; P. G. Abir-Am, "Themes, Genres and Orders of Legitimation in the Consolidation of New Scientific Disciplines: Deconstructing the Historiography of Molecular Biology," *History of Science* 23 (1985): 73–117; E. F. Keller, "Physics and the Emergence of Molecular Biology: A History of Cognitive and Political Synergy," *Journal of the History of Biology* 23 (1990): 389–409; P. G. Abir-Am, "A Historical Ethnography of a Scientific Anniversary in Molecular Biology: The First Protein X-Ray Photograph (1984, 1934)," *Social Epistemology* 6 (1992): 323–54; E. F. Keller, *Secrets of Life, Secrets of Death: Essays on Language, Gender, and Science* (New York: Routledge, 1992), 39–55; M. P. Murphy and L. A. J. O'Neill, eds., *What Is Life? The Next Fifty Years* (Cambridge: Cambridge University Press, 1995), especially the essays by S. J. Gould and S. A. Kaufmann; R. Doyle, *On Beyond Living: Rhetorical Transformations of the Life Sciences*, Writing Science (Stanford, CA: Stanford University Press, 1997), 25–38; P. G. Abir-Am, "The First American and French Commemorations in Molecular Biology: From Collective Memory to Comparative History," *Osiris*, 2nd ser., 14 (1999): 324–72; L. E. Kay, "Quanta of Life: Atomic Physics and the Reincarnation of Phage," *History and Philosophy of the Life Sciences* 14 (1992): 3–21; N. Rasmussen, "The Mid-Century Biophysics Bubble: Hiroshima and the Biological Revolution in America, Revisited," *History of Science* 35 (1997): 245–99.

5. Cairns et al., *Phage and the Origins*.

6. Olby, *Path*; E. A. Carlson, *The Gene: A Critical History* (Philadelphia: W. B. Saunders, 1966), 158–65.

7. For background on Delbrück's status in biology from the German perspective, see U. Deichmann, *Biologists under Hitler*, trans. T. Dunlap (Cambridge, MA: Harvard University Press, 1996), 314–17; Deichmann, "Brief Review."

8. The first use of the term *Treffertheorie* that I have found is in R. Glocker and A. Reuß, "Über die Wirkung von Röntgenstrahlen verschiedener Wellenlänge auf biologische Objekte I," *Strahlentherapie* 46 (1933): 142, although discussion of *Treffer* (hits) and *Treffbereiche* (targets) well predates this. For Glocker and Reuß, the term *Treffertheorie* covers a variety of theories that have in common the idea that the effect of radiation is due to the deposition of more or less discrete "hits," though among these there may be disagreement about the specific nature of the hits or their respective conception of the "target zone" (*Treffbereich*). Since "target theory" was the most common term in English, that is the term I will use in this chapter, though "hit theory" is both linguistically and conceptually a closer

rendering of the German term *Treffertheorie*, predicated as it was on a consensus about the occurrence of some kind of "hit" event, but not necessarily a consensus as to the nature of the relevant "target." For further discussion of this distinction, see especially the Summers chapter, this volume. Cf. F. Holweck, "Mesure des dimensions élémentaires des virus par la méthode d'ultramicrométrie statistique," *Comptes rendus hebdomadaires des séances de l'Académie des sciences* 209 (1939): 380–82.

9. 3MP 268 (238).

10. Kay, *Molecular Vision*, 133.

11. M. Demerec, "The Role of Genes in Evolution," *American Naturalist* 69 (1935): 125–38, quotations on 133, 131.

12. H. J. Muller and A. A Prokofyeva, "The Individual Gene in Relation to the Chromomere and the Chromosome," *Proceedings of the National Academy of Sciences* 21 (1935): 16; an earlier landmark article is H. J. Muller, "The Gene as the Basis of Life," in *Proceedings of the International Congress of Plant Sciences, Ithaca, New York, August 16–23, 1926*, vol. 1, ed. B. M. Duggar (Menasha, WI: George Banta, 1929), 897–921, reprinted in *Studies in Genetics: The Selected Papers of Hermann Joseph Muller* (Bloomington: Indiana University Press, 1962), 188–204. For discussion, see E. A. Carlson, *Genes, Radiation, and Society: The Life and Work of H. J. Muller* (Ithaca: Cornell University Press, 1981), 135–50, and 184–92 for Muller's visit to Berlin-Buch; also Carlson, *Gene*, 86–88; Carlson, "An Unacknowledged Founding of Molecular Biology: H. J. Muller's Contributions to Gene Theory, 1910–1936," *Journal of the History of Biology* 4 (1971): 149–70; E. F. Keller, *The Century of the Gene* (Cambridge, MA: Harvard University Press, 2000), 20.

13. See Carlson, *Gene*, 106–42; J. S. Fruton, *Molecules and Life: Historical Essays on the Interplay of Chemistry and Biology* (New York: Wiley-Interscience, 1972), 235–38; Fruton, *Proteins, Enzymes, Genes: The Interplay of Chemistry and Biology* (New Haven: Yale University Press, 1999), 427; M. R. Dietrich, "From Gene to Genetic Hierarchy: Richard Goldschmidt and the Problem of the Gene," in *The Concept of the Gene in Development and Evolution: Historical and Epistemological Perspectives*, ed. P. J. Beurton, R. Falk, and H.-J. Rheinberger, Cambridge Studies in Philosophy and Biology (Cambridge: Cambridge University Press, 2000), 91–114.

14. Some authors posit molecular biology as a post–World War II phenomenon, constituted largely by distinctive social, political, technological and cognitive features of the postwar era; e.g., Rasmussen, "Mid-Century Biophysics Bubble"; N. Rasmussen, *Picture Control: The Electron Microscope and the Transformation of Biology in America, 1940–1960*, Writing Science (Stanford, CA: Stanford University Press, 1997); S. de Chadarevian, *Designs for Life: Molecular Biology after World War II* (Cambridge: Cambridge University Press, 2002); Chadarevian and B. J. Strasser, "Molecular Biology in Postwar Europe: Towards a 'Glocal' Picture,"

Studies in History and Philosophy of Biology and Biomedical Sciences 33 (2002): 361–65, and the other articles in this special issue of the journal; B. J. Strasser, *La fabrique d'une nouvelle science: La biologie moléculaire à l'âge atomique (1945–1964)*, Bibliothèque d'histoire des sciences, vol. 8 (Florence: Leo S. Olschki, 2006). Other authors have posited various features of recent molecular biology that reach back into the prewar period, though of course with dramatic transformations along the way; representative examples from recent historical literature include P. G. Abir-Am, "From Multidisciplinary Collaboration to Transnational Objectivity: International Space as Constitutive of Molecular Biology, 1930–1970," in *Denationalizing Science: The Contexts of International Scientific Practice*, ed. E. Crawford, T. Shinn, and S. Sörlin, Sociology of the Sciences, vol. 16 (Dordrecht: Kluwer Academic, 1992), 153–86; Abir-Am, "The Politics of Macromolecules: Molecular Biologists, Biochemists, and Rhetoric," *Osiris*, 2nd ser., 7 (1992): 164–91; L. E. Kay, "Life as Technology: Representing, Intervening, and Molecularizing," *Rivista di Storia della Scienza*, series 2, 1 (1993): 85–103, reprinted in *The Philosophy and History of Molecular Biology*, ed. S. Sarkar, Boston Studies in the Philosophy of Science, vol. 183 (Dordrecht: Kluwer Academic, 1996), 87–100; Kay, *Molecular Vision*, 2–5; D. T. Zallen, "Redrawing the Boundaries of Molecular Biology: The Case of Photosynthesis," in *Philosophy and History*, ed. Sarkar, 45–65; P. G. Abir-Am, "The Molecular Transformation of Twentieth-Century Biology," in *Science in the Twentieth Century*, ed. J. Krige and D. Pestre (Amsterdam: Harwood, 1997), 495–524; L. E. Kay, "Problematizing Basic Research in Molecular Biology," in *Private Science: Biotechnology and the Rise of the Molecular Sciences*, ed. A. Thackray, Chemical Sciences in Society (Philadelphia: University of Pennsylvania Press, 1998), 20–38; Morange, *History of Molecular Biology*; P. G. Abir-Am, "Molecular Biology in the Context of British, French, and American Cultures," *International Social Science Journal* 168 (2001): 187–99. For lucid and concise reviews of this historiography, see A. N. H. Creager, "Building Biology across the Atlantic," *Journal of the History of Biology* 36 (2003): 579–89; J.-P. Gaudillière and H.-J. Rheinberger, "Life Stories," *Studies in History and Philosophy of Biological and Biomedical Sciences* 35 (2004): 753–64.

15. Kay, "Life as Technology," 90; Kay, *Molecular Vision*, 5.

16. Cf. Fleming, "Émigré Physicists," 161; Olby, *Path*, 237; Kay, "Life as Technology," 89; Kay, *Molecular Vision*, 4; Zallen, "Redrawing," 49.

17. On inter- or transdisciplinarity in later molecular biology, see (among many authors) Abir-Am, "From Multidisciplinary Collaboration"; Kay, "Life as Technology," especially 91; Kay, *Molecular Vision*, especially 5; Abir-Am, "Molecular Transformation"; Rasmussen, "Mid-Century Biophysics"; Abir-Am, "Molecular Biology"; Chadarevian, *Designs for Life*. On Delbrück as exponent of the "Copenhagen spirit," see, e.g., Fleming, "Émigré Physicists"; Olby, *Path*, 225–47; Kay, "Conceptual Models"; Fischer and Lipson, *Thinking about Science*; Keller, *Century of the Gene*, 22.

18. Historical studies stressing instrumentality in the history of molecular biology include Abir-Am, "Discourse of Physical Power"; Kay, "Life as Technology"; Kay, *Molecular Vision*; Robert E. Kohler, *Lords of the Fly:* Drosophila *Genetics and the Experimental Life* (Chicago: University of Chicago Press, 1994); Abir-Am, "Molecular Transformation"; Rasmussen, *Picture Control*; Kay, "Problematizing Basic Research"; N. Rasmussen, "Instruments, Scientists, Industrialists and the Specificity of 'Influence': The Case of RCA and Biological Electron Microscopy," in *The Invisible Industrialist: Manufactures and the Production of Scientific Knowledge*, ed. J.-P. Gaudillière and I. Löwy, *Science, Technology and Medicine in Modern History* (Basingstoke, UK: Macmillan, 1998), 173–208; L. E. Kay, *Who Wrote the Book of Life? A History of the Genetic Code*, Writing Science (Stanford, CA: Stanford University Press, 2000); Chadarevian, *Designs for Life*; S. de Chadarevian, "Reconstructing Life: Molecular Biology in Postwar Britain," *Studies in History and Philosophy of Biological and Biomedical Sciences* 33 (2002): 431–48; A. N. H. Creager, *The Life of a Virus: Tobacco Mosaic Virus as an Experimental Model, 1930–1965* (Chicago: University of Chicago Press, 2002); J.-P. Gaudillière and H.-J. Rheinberger, eds., *From Molecular Genetics to Genomics: The Mapping Cultures of Twentieth-Century Genetics*, Routledge Studies in the History of Science, Technology, and Medicine, vol. 20 (London: Routledge, 2004). On the introduction of X-rays into the biomedical context, see A. Hessenbruch, "Geschlechterverhältnis und rationalisierte Röntgenologie," in *Geschlechterverhältnisse in Medizin, Naturwissenschaft, und Technik*, ed. C. Meinel and M. Renneberg (Stuttgart: Verlag für Geschichte der Naturwissenschaft und der Technik, 1996), 148–58; M. Dommann, *Durchsicht, Einsicht, Vorsicht: Eine Geschichte der Röntgenstrahlen 1896–1963*, Interferenzen, vol. 5 (Zürich: Chronos, 2003). A useful historiographical overview of the theme of instrumentality is contained in Creager, *Life of a Virus*, chap. 8.

19. See Summers chapter, this volume, as well as N. V. Timoféeff-Ressovsky and K. G. Zimmer, *Das Trefferprinzip in der Biologie*, vol. 1 of *Biophysik* (Leipzig: S. Hirzel, 1947); B. Rajewsky and M. Schön, eds., *Biophysik*, 2 vols., Naturforschung und Medizin in Deutschland (FIAT Review of German Science), vols. 21–22 (Wiesbaden: Dieterich'sche Verlagsbuchhandlung, 1948); K. G. Zimmer, "The Target Theory" in Cairns et al., *Phage and the Origins*, 33–42; Zimmer, "That Was the Basic Radiobiology That Was: A Selected Bibliography and Some Comments," *Advances in Radiation Biology* 9 (1981): 411–67; R. H. Beyler, "'Consider a Cube Filled with Biological Material': Re-conceptualizing the Organic in German Biophysics, 1918–45," in *Biophysics 1918–1945: Fundamental Changes in Cellular Biology in the XXth Century*, ed. C. Galperin, S. F. Gilbert, and B. Hoppe, 39–46 (Turnhout, Belgium: Brepols, 1999).

20. On Dessauer, see Summers chapter, this volume, as well as F. Dessauer, "Lebenslauf," unpublished manuscript, Dessauer Papers, Kommission für Zeitgeschichte, Bonn; Walter Friedrich, oral history interview, May 15, 1963, Archive

for the History of Quantum Physics, American Institute of Physics, College Park, MD (cited with permission); B. Rajewsky, "Friedrich Dessauer zum Gedächtnis," in Friedrich Dessauer, *Quantenbiologie: Einführung in einen neuen Wissenszweig,* 2nd ed., xi–xix (Berlin: Springer, 1964); W. Pohlit, "Friedrich Dessauer 1881–1963," web page, http://www.physik.uni-frankfurt.de/paf/paf84.html, last modified April 16, 1999 (accessed August 16, 2007); H. Goenner, "Albert Einstein and Friedrich Dessauer: Political Views and Political Practice," *Physics in Perspective* 5 (2003): 21–66.

21. F. Dessauer, "Über einige Wirkungen von Strahlen, I," *Zeitschrift für Physik* 12 (1922): 38–47.

22. M. Blau and K. Altenburger, "Über einige Wirkungen von Strahlen, II," *Zeitschrift für Physik* 12 (1922): 315–29. Blau went on to do pioneering work in particle physics at the Radium Institute in Vienna; her career was disrupted by the Nazis' rise to power in Austria and, indeed, within her own institute. See P. L. Galison, "Marietta Blau: Between Nazis and Nuclei," *Physics Today,* 50, no. 11 (November 1997): 42–48; B. Strohmeier and R. Rosner, *Marietta Blau, Stars of Disintegration: Biography of a Pioneer of Particle Physics,* ed. P. F. Dvorak, Studies in Austrian Literature, Culture, and Thought (Riverside, CA: Ariadne Press, 2003); M. Rentetzi, *Trafficking Materials and Gendered Experimental Practices: Radium Research in Early 20th-Century Vienna* (New York: Columbia University Press, 2008), published also as an e-book through Gutenberg-e, http://www .gutenberg-e.org/rentetzi.

23. J. A. Crowther, "Some Considerations Relative to the Action of X-Rays on Tissue Cells," *Proceedings of the Royal Society of London B* 96 (1924): 207–11; Crowther, "The Action of X-Rays on *Colpidium colpoda,*" *Proceedings of the Royal Society of London B* 100 (1926): 390–404; for discussion of Crowther's work and influence, see Summers chapter, this volume.

24. E. U. Condon and H. M. Terrill, "Quantum Phenomena in the Biological Action of X-Rays," *Journal of Cancer Research* 11 (1927): 324–33.

25. D. E. Lea, R. B. Haines, and C. A. Coulson, "Mechanism of the Bactericidal Action of Radioactive Radiations," *Proceedings of the Royal Society of London B* 120 (1936): 47–76; Lea, Haines, and Coulson, "The Action of Radiations on Bacteria: III–γ-Rays on Growing and on Non-proliferating Bacteria," *Proceedings of the Royal Society of London B* 123 (1936): 1–21; D. E. Lea, "Determination of the Sizes of Viruses and Genes by Radiation Methods," *Nature* 146 (1940): 137–38.

26. C. Fabry, "Fernand Holweck (1890–1941)." *Cahiers de physique* 8 (1942): 1–16; Fabry, "Fernand Holweck," *Pensée* 4 (1944): 65–69; J. A. Thomas, "Le martyre de Fernand Holweck," *Pensée* 27 (1949): 21–28.

27. A. Lacassagne, "Action des rayons K [*sic*] de l'aluminium sur quelques microbes," *Comptes rendus hebdomadaires des séances de l'Académie des sciences* 186 (1928): 1316–17; F. Holweck, "Essai d'interprétation énergétique de l'action des rayons K de l'aluminium sur les microbes," *Comptes rendus hebdomadaires des sé-*

ances de l'Académie des sciences 186 (1928): 1318–19. *B. pyocyaneus* is also known as *Pseudomonas aeruginosa.*

28. F. Holweck, "Production de rayons X monochromatiques de grande longueur d'onde: Action quantique sur les microbes," *Comptes rendus hebdomadaires des séances de l'Académie des sciences* 188 (1929): 197–99; A. Lacassagne, "Action des rayons X de grande longueur d'onde sur les microbes: Établissement de statistiques précises de la mortalité de bactéries irradiées," *Comptes rendus hebdomadaires des séances de l'Académie des sciences* 188 (1929): 200–202; M. Curie, "Sur l'étude des courbes de probabilité relatives à l'action des rayons X sur les bacilles," *Comptes rendus hebdomadaires des séances de l'Académie des sciences* 188 (1929): 202–4.

29. F. Dessauer, "Problemstellung und Theorie," in *Zehn Jahre Forschung auf dem physikalisch-medizinischen Grenzgebiet*, ed. F. Dessauer (Leipzig: Georg Thieme, 1931), 179; Beyler, "Consider a Cube."

30. K. Sommermeyer, *Quantenphysik der Strahlenwirkung in Biologie und Medizin*, Probleme der Bioklimatologie, vol. 2 (Leipzig: Geest and Portig, 1952), 46; emphasis in original. We will return to Sommermeyer's role below.

31. H. Holthusen, "Über die Dessauersche Punktwärmehypothese," *Strahlentherapie* 19 (1925): 285–306; F. Pordes, "Zum biologischen Wirkungsmechanismus der Röntgenstrahlen," *Strahlentherapie* 19 (1925): 307–24.

32. Y. Nakashima, "Einige Versuche zum Grundvorgang der biologischen Strahlenwirkung," *Strahlentherapie* 24 (1926): 1–36; H. Holthusen, "Der Grundvorgang der biologischen Strahlenwirkung," *Strahlentherapie* 25 (1927): 157–73; F. Dessauer, "Über den Grundvorgang der biologischen Strahlenwirkung," *Strahlentherapie* 27 (1928): 364–81; H. Holthusen, "Bemerkungen zu obigen Ausführungen Dessauers über meine Arbeit 'Der Grundvorgang der biologischen Strahlenwirkung,'" *Strahlentherapie* 27 (1928): 382–85; R. Braun and H. Holthusen, "Einfluß der Quantengröße auf die biologische Wirkung verschiedener Röntgenstrahlenqualitäten," *Strahlentherapie* 34 (1929): 707–34.

33. Hessenbruch, "Geschlechterverhältnis," 155–56.

34. See Pohlit, "Friedrich Dessauer"; Goenner, "Einstein and Dessauer."

35. See Beyler, "Consider a Cube."

36. R. W. G. Wyckoff, "The Killing of Certain Bacteria by X-Rays," *Journal of Experimental Medicine* 52 (1930): 445. Wyckoff states that this result "seems to be in serious conflict" with the findings of Holweck and Lacassagne. Holweck, "Production," 199, however, deduces a "sensitive zone" (assumed to be spherical) of radius 0.43 microns within a bacterium of diameter 0.7 microns and length 4 microns. By my calculations, this also gives a sensitive zone that is approximately 5 percent of the total volume. See also R. W. G. Wyckoff, "The Killing of Colon Bacilli by X-Rays of Different Wavelengths," *Journal of Experimental Medicine* 52 (1930): 769–80; Wyckoff, "The Killing of Colon Bacilli by Ultraviolet Light," *Journal of General Physiology* 15 (1932): 351–61. By *B. coli* Wyckoff surely meant the bacterium now called *Escherichia coli*, not the protozoan *Balantidium coli*.

Scrutiny of these journals around 1930–32 reveals that this was a fairly common usage at that time. A recent World Health Organization guidebook notes that *B. coli* was an earlier, *E. coli* a later name for this bacterium; see L. Fewtrell and J. Bartram, *Water Quality: Guidelines, Standards, and Health* (London: IWA, 2001), 293.

37. J. W. Gowen and E. H. Gay, "Gene Number, Kind, and Size in *Drosophila*," *Genetics* 18 (1933): 30.

38. M. Demerec, "The Effect of X-Ray Dosage on Sterility and Number of Lethals in *Drosophila melanogaster*," *Proceedings of the National Academy of Sciences* 19 (1933): 1015–20.

39. E. Bunde, "In Memoriam Richard Glocker," in *Medizinische Physik in der klinischen Routine*, ed. H. J. Schopka, Medizinische Physik, vol. 9 (Heidelberg: Alfred Hüthig Verlag, 1978), 33–40.

40. R. Glocker, "Das Grundgesetz der physikalischen Wirkung von Röntgenstrahlen verschiedener Wellenlänge und seine Beziehung zum biologischen Effekt," *Strahlentherapie* 26 (1927): 147–55; Glocker, "Die Wirkung der Röntgenstrahlen auf die Zelle als physikalisches Problem," *Strahlentherapie* 33 (1929): 199–205.

41. R. Glocker, E. Hayer, and O. Jüngling, "Über die biologische Wirkung verschiedener Röntgenstrahlenqualitäten bei Dosierung in R-Einheiten," *Strahlentherapie* 32 (1929): 1–38; Glocker and Reuß, "Über die Wirkung"; H. Langendorff, M. Langendorff, and A. Reuß, "Über die Wirkung von Röntgenstrahlen verschiedener Wellenlänge auf biologische Objekte, II, IV," *Strahlentherapie* 46 (1933): 289–92, 655–62; R. Glocker, H. Langendorff, and A. Reuß, "Über die Wirkung von Röntgenstrahlen verschiedener Wellenlänge auf biologische Objekte, III," *Strahlentherapie* 46 (1933): 517–28.

42. R. Glocker, "Quantenphysik der biologischen Röntgenstrahlenwirkung," *Zeitschrift für Physik* 77 (1932): 656–57.

43. On Timoféeff-Ressovsky's career in Germany, see D. B. Paul and C. B. Krimbas, "Nikolai V. Timoféeff-Ressovsky," *Scientific American* 266 (February 1992): 86–92; Deichmann, *Biologists under Hitler*, 117–22, 219–27; T. Junker and E.-M. Engels, eds., *Die Entstehung der synthetischen Theorie: Beiträge zur Geschichte der Evolutionsbiologie in Deutschland 1930–1950*, Verhandlungen zur Geschichte und Theorie der Biologie, vol. 2 (Berlin: Verlag für Wissenschaft und Bildung, 1999), especially the essay by J. Haffer, "Beiträge zoologischer Systematiker und einiger Genetiker zur evolutionären Synthese in Deutschland (1937–1950)," 121–50; H. Bielka, *Geschichte der medizinisch-biologischen Institute Berlin-Buch*, 2nd ed. (Berlin: Springer, 2002), 61–67; H. Satzinger and A. Vogt, *Elena Aleksandrovna und Nikolaj Vladimirovic Timoféeff-Ressovsky (1898–1973; 1900–1981)*, preprint no. 112 (Berlin: Max-Planck-Institut für Wissenschaftsgeschichte, 2002); T. Junker, *Die zweite Darwinische Revolution: Geschichte des synthetischen Darwinismus in Deutschland 1924 bis 1950*, Acta Biohistorica, vol. 8 (Marburg: Basilisken-Presse, 2004), 91–107.

44. Zimmer, "That Was."

45. P. Jordan, "Die Quantenmechanik und die Grundprobleme der Biologie und Psychologie," *Die Naturwissenschaften* 20 (1932): 815–21; Jordan, "Quantenphysikalische Bemerkungen zur Biologie und Psychologie," *Erkenntnis* 4 (1934): 215–52; for background and discussion, see the Roll-Hansen and Sloan chapters, this volume, as well as F. Aaserud, *Redirecting Science: Niels Bohr, Philanthropy, and the Rise of Nuclear Physics* (Cambridge: Cambridge University Press, 1990); M. Norton Wise, "Pascual Jordan: Quantum Mechanics, Psychology, National Socialism," in *Science, Technology, and National Socialism*, ed. M. Renneberg and M. Walker (Cambridge: Cambridge University Press, 1994), 224–54; R. H. Beyler, "Targeting the Organism: The Scientific and Cultural Context of Pascual Jordan's Quantum Biology, 1932–1947," *Isis* 87 (1996): 248–73; Beyler, "Exporting the Quantum Revolution: Pascual Jordan's Biophysical Initiatives," in *Pascual Jordan (1902–1980): Mainzer Symposium zum 100. Geburtstag*, ed. Jürgen Ehlers et al., preprint no. 329 (Berlin: Max-Planck-Institut für Wissenschaftsgeschichte, 2007), 69–81.

46. On Delbrück's response to Jordan, see Sloan chapter, this volume, as well as Kay, "Secret of Life"; Aaserud, *Redirecting Science.*

47. See Zimmer, "Target Theory," 36–37; M. Delbrück, "A Physicist's Renewed Look at Biology: Twenty Years Later," *Science* 168 (1970): 1312; Carlson, *Genes, Radiation*, 188; Kay, "Conceptual Models"; P. Wurmbach, "Miassovo See–Berlin–Cold Spring Harbor," in *Wissenschaften in Berlin*, ed. T. Buddensieg et al. (Berlin: Mann, 1987), 3:91. Jordan, who was at that time at the University of Rostock, was also in correspondence with the Timoféeff-Ressovsky group; see Beyler, "Targeting the Organism," 264.

48. See the 3MP translation in this volume.

49. Muller, "Gene as the Basis"; H. J. Muller, "Physics in the Attack on Fundamental Problems of Genetics," *Scientific Monthly* 44 (1937): 210–14; see also Carlson, *Genes, Radiation.*

50. H. J. Muller, "An Analysis of the Process of Structural Change in Chromosomes of *Drosophila*," *Journal of Genetics* 40 (1940): 1–66; see also K. Mackenzie and H. J. Muller, "Mutation Effects of Ultra-Violet Light in *Drosophila*," *Proceedings of the Royal Society of London B* 129 (1940): 491–517; Muller, "Gene Mutations Caused by Radiation," in *Symposium on Radiobiology: The Basic Aspects of Radiation Effects on Living Systems*, ed. J. J. Nickson (New York: John Wiley and Sons, 1952), 296–332.

51. N. V. Timoféeff-Ressovsky and M. Delbrück, "Strahlengenetische Versuche über sichtbare Mutation und die Mutabilität einzelner Gene bei *Drosophila melanogaster*," *Zeitschrift für induktive Abstammungs- und Vererbungslehre* 71 (1936): 322–34.

52. H. Fricke and M. Demerec, "The Influence of Wave-Length on Genetic Effects of X-Rays," *Proceedings of the National Academy of Sciences* 23 (1937): 320–27; cf. Demerec, "Role of Genes."

53. W. V. Mayneord, "The Physical Basis of the Biological Effects of High Voltage Radiations," *Proceedings of the Royal Society of London A* 146 (1934): 867–79; B. Rajewsky, "Theorie der Strahlenwirkung und ihre Bedeutung für die Strahlentherapie," *Wissenschaftliche Woche zu Frankfurt am Main* 2 (1934): 75–90. The National Socialists drove Dessauer into exile—first to the University of Istanbul in Turkey, and eventually to Switzerland—primarily because of his activities as a leading Center Party politician, but also because of his partially Jewish ancestry. (Dessauer himself was a devout Catholic.)

54. N. V. Timoféeff-Ressovsky, *Experimentelle Mutationsforschung in der Vererbungslehre: Beeinflussung der Erbanlagen durch Strahlung und andere Faktoren*, Wissenschaftliche Forschungsberichte: Naturwissenschaftliche Reihe, vol. 42 (Dresden: Theodor Steinkopff, 1937), 124–34.

55. Timoféeff-Ressovsky, *Experimentelle Mutationsforschung*, 135–36.

56. N. V. Timoféeff-Ressovsky, "Eine biophysikalische Analyse des Mutationsvorganges," *Nova Acta Leopoldina*, n.s., 9 (1940): 230; Timoféeff-Ressovsky and K. G. Zimmer, "Über Zeitproportionalität und Temperaturabhängigkeit der spontanen Mutationsrate von *Drosophila*," *Zeitschrift für induktive Abstammungs- und Vererbungslehre* 79 (1941): 530. K. Pätau and N. V. Timoféeff-Ressovsky, "Statistische Prüfung des Unterschiedes der Temperaturkoeffizienten höher und normaler Mutationsraten nebst einem Beispiel für die Planung von Temperaturversuchen," *Zeitschrift für induktive Abstammungs- und Vererbungslehre* 81 (1943): 62–71. The latter article in part calls attention to a mathematical error in the 1941 article; K. Pätau and N. V. Timoféeff-Ressovsky, "Die Genauigkeit der Bestimmung spontaner und strahleninduzierter Mutationsraten nach der 'CLB'-Kreuzungsmethode bei *Drosophila melanogaster*," *Zeitschrift für induktive Abstammungs- und Vererbungslehre* 81 (1943): 181–90, confirms the prior result using new experiments and analysis. For an overview of Timoféeff-Ressovsky's wide-ranging and synthetical biological interests, see Paul and Krimbas, "Nikolai V. Timoféeff-Ressovsky"; Junker and Engels, *Die Entstehung der synthetischen Theorie*.

57. P. Jordan, "Biologische Strahlenwirkung und Physik der Gene," *Physikalische Zeitschrift* 39 (1938): 352–53; Jordan, "Über die Elementarprozesse der biologischen Strahlenwirkung," *Radiologica* 2 (1938): 26–28; discussed in Beyler, "Targeting the Organism."

58. N. V. Timoféeff-Ressovsky and K. G. Zimmer, "Neutronbestrahlungsversuche zur Mutationsauslösung an *Drosophila melanogaster*," *Die Naturwissenschaften* 26 (1938): 362–65; N. V. Timoféeff-Ressovsky, K. G. Zimmer, and F. A. Heyn, "Auslösung von Mutationen an *Drosophila melanogaster* durch schnelle Li+D-Neutronen," *Die Naturwissenschaften* 26 (1938): 625–26; K. G. Zimmer, "Dosimetrische und strahlenbiologische Versuche mit schnellen Neutronen, I," *Strahlentherapie* 63 (1938): 517–27; see also Beyler, "Targeting the Organism," 265; Bielka, *Geschichte der medizinisch-biologischen Institute*, 57. Timoféeff-Ressovsky's institute continued research on the biological effects of neutron radia-

tion into the war years, according to Karl Zimmer's postwar testimony to Soviet intelligence officers; see U. Hossfeld and M. Walker, "Hero or Villain? Stasi Archives Shed Light on Russian Scientist," *Nature* 411 (2001): 237. The Emergency Association for German Science (Notgemeinschaft der deutschen Wissenschaft), the original supporter of the 3MP research, was renamed the German Research Association (Deutsche Forschungsgemeinschaft) in October 1937; cf. Deichmann, *Biologists under Hitler*, 90.

    59. K. G. Zimmer and N. V. Timoféeff-Ressovsky, "Dosimetrische und strahlenbiologische Versuche mit schnellen Neutronen, II," *Strahlentherapie* 63 (1938): 534–35.

    60. Zimmer and Timoféeff-Ressovsky, "Dosimetrische," 535–36; see also K. G. Zimmer and N. V. Timoféeff-Ressovsky, "Über einige physikalische Vorgänge bei der Auslösung von Genmutationen durch Strahlung," *Zeitschrift für induktive Abstammungs- und Vererbungslehre* 80 (1942): 353–72; E. Wollman, F. Holweck, and S. Luria, "Effect of Radiations on Bacteriophage C_{16}," *Nature* 145 (1940): 935–36.

    61. M. Delbrück, "Radiation and the Hereditary Mechanism," *American Naturalist* 74 (1940): 360.

    62. K. G. Zimmer, "Zur Berücksichtigung der 'biologischen Variabilität' bei der Treffertheorie der biologischen Strahlenwirkung," *Biologisches Zentralblatt* 61 (1941): 208–20.

    63. R. Prigge and H. von Schelling, "Zur Analyse der Antigenwirkung," *Die Naturwissenschaften* 30 (1942): 661.

    64. K. G. Zimmer, "Zur treffertheoretischen Analyse der Antigenwirkung," *Die Naturwissenschaften* 30 (1942): 452–53; Richard Prigge to K. G. Zimmer, May 21, 1942, folder IIIA, Nachlaß Prigge, Archivzentrum, Universitätsbibliothek Frankfurt. For background and consequences, see R. Karlsch, "Boris Rajewsky und das Kaiser-Wilhelm-Institut für Biophysik in der Zeit des Nationalsozialismus," in *Gemeinschaftsforschung, Bevollmächtige und der Wissenstransfer: Die Rolle der Kaiser-Wilhelm-Gesellschaft im System kriegsrelevanter Forschung des Nationalsozialismus*, ed. H. Maier (Göttingen: Wallstein, 2007), 395–452.

    65. F. Möglich, R. Rompe, and N. V. Timoféeff-Ressovsky, "Bemerkungen zu physikalischen Modellvorstellungen über Energieausbreitungsmechanismen im Treffbereich bei strahlenbiologischen Vorgängen," *Die Naturwissenschaften* 30 (1942): 418; see also N. Riehl, N. V. Timoféeff-Ressovsky, and K. G. Zimmer, "Mechanismus der Wirkung ionisierender Strahlen auf biologische Elementareinheiten," *Die Naturwissenschaften* 29 (1941): 625–39; F. Möglich, R. Rompe, and N. V. Timoféeff-Ressovsky, "Über die Indeterminiertheit und die Verstärkererscheinungen in der Biologie," *Die Naturwissenschaften* 32 (1944): 316–24; F. Möglich, R. Rompe, N. V. Timoféeff-Ressovsky, and K. G. Zimmer, "Über Energiewanderungsvorgänge und ihre Bedeutung für einige biologische Prozesse," *Protoplasma* 38 (1943): 105–26; for discussion of the industrial physicists' participation in the Timoféeff-Ressovsky circle, see D. Hoffmann and M. Walker, "Friedrich Möglich:

A Scientist's Journey from Fascism to Communism," in *Science and Ideology: A Comparative History*, ed. M. Walker (London: Routledge, 2003), 236.

66. Deichmann, *Biologists under Hitler*, 116.

67. "Denkschrift zur Errichtung eines Kaiser-Wilhelm-Instituts für Biophysik zu Frankfurt am Main," draft memorandum, October 22, 1936, in Magistratsakten 8.283, Institut für Stadtgeschichte, Frankfurt.

68. Deichmann, *Biologists under Hitler*, 227–29.

69. B. Rajewsky and N. V. Timoféeff-Ressovsky, "Höhenstrahlung und die Mutationsrate von *Drosophila melanogaster*," *Zeitschrift für induktive Abstammungs- und Vererbungslehre* 77 (1939): 488–500.

70. The previous incarnation of the KWI for Metals Research in Berlin closed due to budgetary difficulties in 1933.

71. H. Langendorff and K. Sommermeyer, "Strahlenwirkung auf Drosophilaeier, I," *Fundamenta Radiologica* 4 (1939): 196–209; Langendorff and Sommermeyer, "Strahlenwirkung auf Drosophilaeier, II–V," *Strahlentherapie* 67 (1940): 110–18, 119–29; 68 (1940): 42–52, 656–68.

72. U. Dehlinger, "Über die Morphologie des Gens und den Mechanismus der Mutation," *Die Naturwissenschaften* 25 (1937): 138. Schrödinger, *What Is Life?*, famously ascribed a crystallike structure to the gene; significantly, however, Schrödinger stressed that the crystal must be "aperiodic" for information-theoretic reasons. This element of the argument is missing from Dehlinger's brief note.

73. K. Sommermeyer, "Über die Treffertheorie und ihre Anwendung in der Theorie der Genmutation," *Die Naturwissenschaften* 26 (1938): 154; Sommermeyer, "Bemerkung zur Theorie des strahlenbiologischen Sättigungseffektes," *Die Naturwissenschaften* 30 (1942): 104–5; Sommermeyer and U. Dehlinger, "Beiträge zur Diskussion eines Gen-Modells," *Physikalische Zeitschrift* 40 (1939): 67–70.

74. C. Packard, "The Biological Roentgen," *Radiology* 27 (1936): 191–95; M. Demerec, "Hereditary Effects of X-Ray Radiation," *Radiology* 30 (1938): 212–20; H. H. Plough and G. P. Child, "Autosomal Lethal Mutation Frequencies in *Drosophila*," *Proceedings of the National Academy of Sciences* 23 (1937): 435–40; A. Gulick, "What Are the Genes? I, The Genetic and Evolutionary Picture," *Quarterly Review of Biology* 13 (1938): 1–18; T. H. Goodspeed and F. M. Uber, "Radiation and Plant Cytogenetics," *Botanical Review* 5 (1939): 1–48; K. Wohl, "The Mechanism of Photosynthesis in Green Plants," *New Phytologist* 39 (1940): 33–64; S. Wright, "Genes as Physiological Agents: General Considerations," *American Naturalist* 79 (1945): 289–303. This list was arrived at by searching the JSTOR database, an extremely useful bibliographical reference but one that makes no claim to completeness, even among English-language journals. A conventional bibliographical tool for this sort of work is Science Citation Index, now included in the Web of Knowledge database; however, its coverage in this period is likewise problematic.

75. N. Giles, "The Effect of Fast Neutrons on the Chromosomes of *Tradescantia*," *Proceedings of the National Academy of Sciences* 26 (1940): 567–75; E. R. Dempster, "Dominant vs. Recessive Lethal Mutation," *Proceedings of the National*

Academy of Sciences 27 (1941): 249–50; A. Marshak, "Relative Effects of X-Rays and Neutrons on Chromosomes in Different Parts of the 'Resting Stage,'" *Proceedings of the National Academy of Sciences* 28 (1942): 29–35; U. Fano, "On the Interpretation of Radiation Experiments in Genetics," *Quarterly Review of Biology* 17 (1942): 244–52. On Fano, see below.

76. R. R. Gates, "The Species Concept in the Light of Cytology and Genetics," *American Naturalist* 72 (1938): 340–49; Karl Sax, "Chromosome Aberrations Induced by X-Rays," *Genetics* 23 (1938): 494–516; C. W. Emmons and A. Hollaender, "The Action of Ultraviolet Radiation on Dermatophytes: II, Mutations Induced in Cultures of Dermatophytes by Exposure of Spores to Monochromatic Ultraviolet Radiation," *American Journal of Botany* 26 (1939): 467–75; K. Sax and E. V. Enzmann, "The Effect of Temperature on X-Ray Induced Chromosome Aberrations," *Proceedings of the National Academy of Sciences* 25 (1939): 397–405; T. Dobzhansky and S. Wright, "Genetics of Natural Populations: V, Relations between Mutation Rate and Accumulation of Lethals in Populations of *Drosophila pseudoobscura*," *Genetics* 26 (1941): 23–51; B. P. Kaufmann, "Reversion from Roughest to Wild Type in *Drosophila melanogaster*," *Genetics* 27 (1942): 537–49; H. C. Fryer and J. W. Gowen, "An Analysis of Data on X-Ray-Induced Visible Gene Mutations in *Drosophila melanogaster*," *Genetics* 27 (1942): 212–27; C. Rick, "The Genetic Nature of X-Ray Induced Changes in Pollen," *Proceedings of the National Academy of Sciences* 28 (1942): 518–25.

77. T. Just, review of *Mutationsforschung [sic] in der Vererbungslehre* by N. V. Timoféeff-Ressovsky, *American Midland Naturalist* 18 (1937): 308.

78. Anonymous review of *Introduction to Biophysics* by O. Stuhlman, *Quarterly Review of Biology* 18 (1943): 393–94.

79. D. E. Lea, "A Theory of the Action of Radiations on Biological Materials Capable of Recovery," *British Journal of Radiology* 11 (1937): 489–97, 554–66.

80. D. E. Lea, "A Radiation Method for Determining the Number of Genes in the X-Chromosome of *Drosophila*," *Journal of Genetics* 39 (1940): 181–87; articles that don't cite the 3MP directly but cite Timoféeff-Ressovsky's 1937 book and associated relevant literature include Lea, "Determination"; Lea and K. M. Smith, "The Inactivation of Plant Viruses by Radiation," *Parasitology* 32 (1940): 405–16.

81. D. E. Lea and D. G. Catcheside, "The Bearing of Radiation Experiments on the Size of the Gene," *Journal of Genetics* 47 (1945): 41–50, quotation on 41.

82. D. E. Lea, *Actions of Radiations on Living Cells* (Cambridge: Cambridge University Press, 1947), 382; the term "classic" comes from F. L. Holmes, "Seymour Benzer and the Convergence of Molecular Biology with Chemical Genetics," in Beurton et al., *Concept of the Gene*, 123–24, in the context of a discussion of Lea's influence on the early career of Seymour Benzer.

83. N. V. Timoféeff-Ressovsky, "Le mécanisme des mutations et la structure du gène," in *Réunion internationale de physique-chimie-biologie: Congrès du Palais de la découverte, Paris, Octobre 1937*, vol. 8, *Biologie*, Actualités scientifiques

et industrielles, no. 725 (Paris: Hermann and Cie, 1938), 2:83–104; Timoféeff-Ressovsky, *Le mécanisme des mutations et la structure du gène*, Actualités scientifiques et industrielles, no. 812, ser. Génétique, ed. B. Ephrussi, no. 4 (Paris: Hermann and Cie, 1939). The latter publication is a somewhat expanded version of the former.

84. E. Wollman and A. Lacassagne, "Évaluation de la taille relative des bactériophages par leur radiosensibilité," *Comptes rendus des séances de la Société de biologie et de ses filiales* 131 (1939): 959–61; A. Lacassagne and E. Wollman, "Évaluation des dimensions des bactériophages au moyen des rayons X," *Annales de l'Institut Pasteur* 64 (1940): 4–39.

85. Fano's path was in certain respects similar to Luria's: he also left Italy due to the intensification of Fascist racial policies in 1938, in his case going directly to the United States, where for a few years he worked on radiation biology with Demerec in Cold Spring Harbor. See, e.g., U. Fano, "The Significance of the Hit Theory of Radiobiological Actions," *Journal of Applied Physics* 12 (1941): 347; Fano, "On the Interpretation"; M. Demerec and U. Fano, "Mechanism of the Origin of X-Ray Induced Notch Deficiencies in *Drosophila melanogaster*," *Proceedings of the National Academy of Sciences* 27 (1941): 24–31.

86. Luria, *Slot Machine*, 22–23.

87. S. E. Luria, "Actions des radiations sur le *Bacterium coli*," *Comptes rendus hebdomadaires des séances de l'Académie des sciences* 209 (1939): 604–6.

88. Wollman et al., "Effect of Radiations," 936; cf. F. Holweck, S. E. Luria, and E. Wollman, "Recherches sur le mode d'action des radiations sur les bactériophages," *Comptes rendus hebdomadaires des séances de l'Académie des sciences* 210 (1940): 639–42.

89. S. E. Luria, "Radiobiologie quantique," *Paris médical* 30, no. 26 (1940): 305–11.

90. Luria, *Slot Machine*, 26–32; for additional biographical background, see R. M. Burian, J. Gayon, and D. Zallen, "The Singular Fate of Genetics in the History of French Biology," *Journal of the History of Biology* 21 (1988): 387–88; Morange, *History of Molecular Biology*, 53–55; R. M. Burian and J. Gayon, "The French School of Genetics: From Physiological and Population Genetics to Regulative Molecular Genetics," *Annual Review of Genetics* 33 (1999): 325–27; P. G. Abir-Am, "The Rockefeller Foundation and Refugee Biologists: European and American Careers of Leading RF Grantees from England, France, Germany, and Italy," in *The "Unacceptables": American Foundations and Refugee Scholars between the Two Wars and After*, ed. G. Gemelli, Euroclio, no. 18 (Brussels: P.I.E.–Peter Lang, 2000), 231–32; R. E. Selya, "Salvador Luria's Unfinished Experiment: The Public Life of a Biologist in a Cold War Democracy" (PhD diss., Harvard University, 2002). Two of Luria's main collaborators in Paris were not so fortunate. Fernand Holweck died in 1941 while in German custody after his arrest for resistance activities. Eugène Wollman, together with his wife Elisabeth, died after their deporta-

tion to the Auschwitz extermination camp in December 1943. See Thomas, "Le martyre"; Luria, *Slot Machine*, 24; U. Deichmann, "Emigration, Isolation and the Slow Start of Molecular Biology in Germany," *Studies in History and Philosophy of Biological and Biomedical Sciences* 33 (2002): 464.

91. See Wise, "Pascual Jordan"; Beyler, "Targeting the Organism"; Beyler, "Evolution als Problem für Quantenphysiker," trans. R. Brömer, in *Evolutionsbiologie von Darwin bis Heute*, ed. R. Brömer, U. Hossfeld, and N. Rupke (Berlin: Verlag für Wissenschaft und Bildung, 1999), 137–60; Beyler, "Exporting the Quantum Revolution."

92. P. Jordan, "Über die Rolle atomphysikalischer Einzelprozesse im biologischen Geschehen," *Radiologica* 1 (1937): 24. In a follow-up article, Jordan again cited the 3MP and thanked Timoféeff-Ressovsky and Zimmer "for friendly advice" ("Über die Elementarprozesse," 22, 33).

93. Jordan, "Biologische Strahlenwirkung," 345; Jordan's reference to his own work is to "Quantenphysikalische Bemerkungen."

94. P. Jordan, *Physik und das Geheimnis des organischen Lebens* (Braunschweig: Vieweg and Sohn, 1941), 82.

95. See Sloan and Roll-Hansen chapters, this volume.

96. Timoféeff-Ressovsky and Zimmer, *Trefferprinzip*, 251–69.

97. The following paragraph summarizes information in Beyler, "Targeting the Organism," 270–71; Bielka, *Geschichte der medizinisch-biologischen Institute*, 61–67; Beyler, "Exporting the Quantum Revolution," 76–77.

98. A final irony of the story is that, after German reunification, the Berlin-Buch institute was transformed once again, receiving the new name of Max Delbrück Center for Molecular Medicine.

99. Timoféeff-Ressovsky and Zimmer, *Trefferprinzip*, 1–6. Almost certainly the book, or most of it anyway, must have been written a couple of years earlier, given the authors' detention after the end of the war. The latest works cited in its bibliography date from 1944. The anticipated subsequent volumes of the series evidently never appeared.

100. Timoféeff-Ressovsky and Zimmer, *Trefferprinzip*, 6–76.

101. Timoféeff-Ressovsky and Zimmer, *Trefferprinzip*, 76–142.

102. Timoféeff-Ressovsky and Zimmer, *Trefferprinzip*, 142–221.

103. Timoféeff-Ressovsky and Zimmer, *Trefferprinzip*, 247–51.

104. Rajewsky and Schön, *Biophysik*. The FIAT reviews were surveys of German science commissioned by the U.S.-British Field Intelligence Agency Technical following World War II; for background, see J. Gimbel, *Science, Technology, and Reparations: Exploitation and Plunder in Postwar Germany* (Stanford, CA: Stanford University Press, 1990).

105. Rajewsky and Schön, *Biophysik*, 1:1. The "external circumstances" are not specified; however, it is likely that they were connected to the difficulties in regaining an academic appointment that Jordan faced after the war, because of his prior

association with National Socialism; see Wise, "Pascual Jordan"; A. Schirrmacher, "Physik und Politik in der frühen Bundesrepublik Deutschland: Max Born, Werner Heisenberg, und Pascual Jordan als politische Grenzgänger," *Berichte zur Wissenschaftsgeschichte* 30 (2007): 13–31.

106. B. Rajewsky, "Grundlagen der Treffertheorie der biologischen Strahlenwirkung," in Rajewsky and Schön, *Biophysik*, 1:9–14, quotations on 12–13.

107. H. Langendorff, "Wirkungen auf Zellen und Zellkomplexe," in *Biophysik*, ed. Rajewsky and Schön, 1:52.

108. H. Bauer, "Genetische Wirkungen von Kurzwelligen und Korpuskularstrahlungen," in *Biophysik*, ed. Rajewsky and Schön, 1:73–74.

109. M. Schön, "Energiewanderung in Molekülkomplexen und Struktureinheiten," in Rajewsky and Schön, *Biophysik*, 1:25.

110. B. Rajewsky, "The Limits of the Target Theory of the Biological Action of Radiation," *British Journal of Radiology* 25 (1950): 552.

111. Muller, "Gene Mutations"; see also L. J. Stadler, "The Gene," *Science* 120 (1954): 812.

112. See, e.g., the references in notes 4 and 14.

113. See Fleming, "Émigré Physicists"; Abir-Am, "From Multidisciplinary Collaboration"; Abir-Am, "Molecular Transformation"; Chadarevian and Strasser, "Molecular Biology in Postwar Europe"; J.-P. Gaudillière, "Paris–New York Roundtrip: Transatlantic Crossings and the Reconstruction of the Biological Sciences in Post-War France," *Studies in History and Philosophy of Biological and Biomedical Sciences* 33 (2002): 389–417; Deichmann, "Emigration, Isolation"; Creager, "Building Biology"; Deichmann, "Brief Review"; B. J. Strasser, "Building Molecular Biology in Post-War Europe: Between the Atomic Age and the American Challenge," in *Max Delbrück*, ed. Wenkel and Deichmann, 58–65. The internment and political persecution of Timoféeff-Ressovsky and Zimmer in the Soviet Union is certainly not the least significant of these dislocations; see the references in note 43, as well as Karlsch, "Boris Rajewsky."

114. See, e.g., the survey and extensive bibliography in Zimmer, "That Was."

PART II

Philosophical Perspectives
on the Three-Man Paper

CHAPTER FIVE

Niels Bohr and Max Delbrück: Balancing Autonomy and Reductionism in Biology

Nils Roll-Hansen

The up-and-coming young physicist Max Delbrück was inspired by Niels Bohr to stake his scientific career on basic problems of biology. More precisely, it was Bohr's speculations on complementarity in the lecture "Light and Life" at the International Congress on Light Therapy in Copenhagen in August 1932 that fired Delbrück's imagination and launched the career that gave him a Nobel Prize for work on phage genetics. But what, more precisely, was Bohr's view of the relationship between biology and physics,[1] and what was the idea that drove Delbrück? I have previously argued that he saw biology as a potential "source of phenomena that demanded new fundamental advances in physics."[2] He was, in other words, nurturing the hope of promoting a new revolution in physics similar to the transition from classical mechanics to quantum and relativity theory that had taken place a few decades earlier, and this idea was not foreign to Bohr. But as we shall see, there were important differences between the two.

This chapter attempts to develop a more precise understanding of the relation between Bohr's philosophy of science and Delbrück's biological research program. I argue that Delbrück's biological research program was reductionist in the sense that it pursued physical and chemical explanations for biological phenomena. Bohr's primary concern was with fundamental features of empirical natural science that implied a radical difference between physics and biology, however successful reductionism might be. He found that Delbrück's reductionist program did not contradict his own philosophical view, but he did not involve himself in empirical biological research. Daniel McKaughan has argued, in apparent contradiction to this claim, that "Bohr and Delbrück shared an antireductionist outlook" and that Delbrück hoped to demonstrate a legitimate role for "teleological concepts" in biological science.[3] In the present volume McKaughan further develops his argument that the shared purpose of the two was to reveal the limits of physical and chemical explanation in biology.

However, the two interpretations may not be as irreconcilable as appears at first glance. A deep conviction that living organisms can never be fully explained in terms of mechanical, physical and chemical causes is not necessarily inconsistent with a belief that only such causes are legitimate in scientific explanations of biological phenomena. In his *Critique of Teleological Judgment*, Immanuel Kant described an antinomy, a paradoxical contradiction, for biology. On one hand, the nature of human understanding demands that all production of material things must be explained according to mechanical laws. On the other hand, some natural objects, namely living organisms—purposes of nature (*Naturzwecke*) as Kant termed them—appear to demand explanation according to a quite different kind of law, namely that of final causes.[4] Kant suggests that the solution to the paradox lies in the limited powers of human understanding. Living organisms are simply too complex for full explanation in terms of human science. Beyond the reach of a strict science using mechanical explanation, final causes and teleological explanations are our substitute means of creating order in the world. Within what can strictly be called the science of living organisms, there is only room for mechanical explanations. But beyond strict science, in our broader thinking about living things, final causes and teleological explanations are legitimate and unavoidable.[5]

Translated into the modern debate about reductionism, this interpretation of Kant suggests that, although living organisms will never be completely explained in terms of the mechanical causes of physics and chemistry, such explanations are nevertheless the only kind that can be pursued within the limitations of a strictly scientific biology. This can be taken to correspond well to the actual situation in modern biology. Teleological concepts and explanations have a heuristic role as shorthand expressions of complex mechanical systems and relationships, and they are essential in guiding the formulation of research problems. Still they have a small and secondary role in the resulting *scientific explanations*—if any at all. I suggest that much recent philosophical discussion of reductionism in biology has not been sufficiently sensitive to the difference between, on one hand, a reductionism that envisages complete reduction as an epistemic ideal, though it may never be reached or may even in principle be unreachable, and on the other hand, a reductionism that has given up or explicitly rejected such thinking. On the latter view the ideal of complete reduction, as it is used for instance in much present standard physicalist analysis of reductionism in biology and other science, is misleading. It tends to stimulate scientific dogmatism by covering up the fundamental dependence of science upon common sense, suggesting that scientific knowledge has a principally different source. Perhaps a modest entity realism that takes objects to exist independently of the observer and emphasizes the fundamental difference between the object and our idea ("Picture") of it is sufficient as a regulative idea for science.

Høffding's Philosophy of Science

The Danish philosopher Harald Høffding (1843–1931) was representative of the view of science held in the close-knit intellectual milieu of Copenhagen of the late nineteenth and early twentieth centuries. He was the product of a local neo-Kantian tradition, and he reached considerable international recognition as an authority on epistemology for other contemporary Danish natural scientists who affected world science, such as the geneticist Wilhelm Johannsen (1857–1927), and the physicist Niels Bohr (1885–1962).

In Høffding's view, all knowledge, including scientific, was historically dependent in a fundamental sense. A major theme in his philosophy was the "uncompletability of knowledge."[6] He had no doubt about Kant's great achievement in the analysis of causation, but nevertheless found his idea of causal *law* too dogmatic. Causality was best understood as a principle "leading human thought constantly to set itself new tasks, whether they can be solved or not." Whatever law is formulated, it will again and again take on new forms.[7] Høffding saw natural science as being fundamentally of the same nature as the humanities (*Aandsvidenskaben*). Science in general is a product of mental work, and thus it is determined by the nature of such work.[8] As I interpret Høffding, not only was scientific knowledge interminably changing, but also the knowledge of natural science would be fundamentally different from the naturally given objects that it describes.

Thus the very idea of a complete and final causal explanation does not make good sense for science. It has no clear and precise meaning.

Høffding combined this outlook with a strong and uncomplicated trust in the ongoing progress of scientific knowledge, most obviously exemplified by the natural sciences. Høffding saw himself basically as a realist in the sense that science is assumed to provide knowledge of a world that does not depend on the knowing subject for its existence. But he rejected the naïve realist conception of truth simply as correspondence of knowledge with an external object,[9] and characterized his own view as "constructive realism."[10]

Jan Faye has maintained that Bohr's philosophy of science is closely similar to that of Høffding. The two men were often in contact from Bohr's childhood on, and they exchanged views on philosophy at various times, not least in the period at the end of the 1920s when Bohr developed his ideas on complementarity.[11] While Bohr's philosophical texts are short, turgid, and ambiguous, Høffding's were the product of a professional philosopher of high international standing. His views thus form an obvious source for the interpretation of Bohr, but they have so far not been widely utilized.[12] Faye's interpretation of Bohr as an "objective anti-realist" is built on an extensive and careful analysis of Høffding. Faye recognizes that Høffding saw himself as a realist, a "constructive realist," but finds "objective anti-realist" to be a more suitable designation. Objective anti-realism, according to Faye, assumes, with other realisms, that the world exists independently of our minds. But it rejects the possibility of factual descriptions that have a stable meaning and that are true in an absolute sense. Thus the idea of a fixed and objective reality that can in principle be exactly described in a unique and true way does not make good scientific sense.[13] Høffding's "constructive realism" includes a "dynamic concept of truth" as opposed to the "static" concept of naïve realism.[14] The close similarity of Faye's "objective anti-realism" and Høffding's "constructive realism" indicates that calling Bohr "realist" or "antirealist" can refer to more or less identical interpretations.

Høffding was a personal friend of Niels Bohr's father, the physiologist Christian Bohr (1855–1911). The philosopher and the experimental biologist had a common interest in a much discussed problem of the time—how far can "mechanical" physicochemical explanations go in biology? Like most active contemporary scientists, they rejected traditional vitalism—the idea of a special substance or power of life—as well as an all-encompassing mechanism. But what did this middle position imply? If physical science and biology were autonomous spheres of knowledge, was it then possible to set a precise limit to reduction, to identify specific biological phenomena that were irreducible?

A colleague and friend of Christian Bohr, the British physiologist J. S. Haldane (1860–1936) thought such limits could be specified. Like Bohr he worked on respiration with a particular interest in the organic uptake of oxygen.[15] As late as 1923 he argued that the nature of oxygen changes fundamentally when it leaves hemoglobin in the capillaries and enters cell metabolism: "We can imagine no form of chemical combination that will now explain the behaviour of oxygen . . . . It is life and not matter that we have before us."[16] Høffding's obituary of Christian Bohr indicates that he had at some time been close to this view, but later backed away. For years his interest had been in "tracing the border" between the living and the inorganic, "to see if there was a determinate border, and if so where it was located." The physiology of breathing was particularly well suited for answering such questions, explained Høffding, and Christian Bohr had at one time been misunderstood as wanting to return to vitalism. As Høffding explained in 1911, his intention was to prove simply that the actions of the nervous system had to be taken into consideration.[17]

Høffding's own view of the relationship between biology and physical science is a historicized version of Kant's. For Høffding the historical changes in scientific concepts and theories go deeper than Kant recognized, as I have explained above. But with this reservation, the philosophical core of Høffding's view is well expressed in Kant's solution to the antinomy of biology: strictly scientific knowledge demands mechanical causal explanation, but the heuristic and speculative aspects of biology cannot do without teleology. Kant saw this as a consequence of limitations in the human intellect compared to that of an omniscient spirit. Trained as a theologian, Høffding nevertheless did not use such a God's-eye view as the frame of his analysis. David Favrholdt has pointed out how Bohr's explicit rejection of such a perspective was at the core of his insistence that "we are both spectators and participants" in our relation to natural phenomena, and that this becomes particularly evident with respect to the microphysical world.[18]

The Kantian view of the relationship between physics and biology has been designated in various ways, for example as "critical teleology"[19] and as "teleomechanical."[20] In a lecture to the Biological Society in Copenhagen in 1898, Høffding attributed this view, which he called "neo-vitalism," to Claude Bernard and Rudolf Virchow among others. They held that "only the methods of physics and chemistry can lead to understanding of the laws of life, but by no means thought that the innermost nature of life could be understood by these laws alone," which also appears as his own preferred position. This view is so different from classical vitalism, however, that Høffding emphasized that it is misleading to use the word "vitalism."[21]

In the concluding discussion of this talk, Høffding pointed to what he called the "analogy" between the "antinomy" of biology and the series of oppositions between the rational and the empirical, the formal and real, in science. There is no barrier that stops rational or formal knowledge from progressing indefinitely through deductions and definitions. But "empirical or real knowledge often is forced to stop at facts, which may well be described and analyzed, but cannot be defined or derived from other facts. A fact of this kind is life, and thus biology will perhaps forever belong to empirical knowledge."[22] Thus Høffding suggests that there are empirical facts that cannot be totally included in the formal system of physical science, therefore implying a fundamental autonomy of biology from physical science. However, "life" is a general commonsense concept with no precise scientific definition. The brute existence of "life" does not point to specific limits to the reduction of biology to physical science.

In 1898, Høffding was careful not to set specific limits to the future physical reduction of biology. He avoided pointing out particular biological phenomena that simply had to be taken at face value and could never to some extent be explained by physical theory. Thus he was more careful than Kant had been in the late eighteenth century when he claimed, for instance, that it was impossible to understand the growth of "a grass-blade from mechanical laws alone."[23] And he was more careful than Claude Bernard (1813–78), who claimed in the late nineteenth century that evolutionary changes in hereditary biological form were inaccessible to experimental method and thus outside the reach of a strictly scientific biological explanation. Such topics belonged to history of nature, according to Bernard.[24] Another example is the claim of the philosopher-biologist Joseph Henry Woodger (1894–1981), who in 1929 argued against the chromosome theory of heredity formulated by T. H. Morgan and collaborators. Woodger held that genes could not possibly be chemical molecules. Genes are supposed to propagate through division into two entities that both have properties identical to the original, but if one divides a molecule, the products will inevitably be different from the original entity.[25] It can be argued that neither Kant, Bernard, nor Woodger fully appreciated the role of radical conceptual and theoretical change in the historical development of science. They tried dogmatically to fix biological descriptions that had simply to be taken at face value as fundamentally without possibility of further explanation through causal physicochemical analysis, contrary to Høffding's idea of the "uncompletability" of scientific knowledge.

Philosophers should not try to intervene in the work of natural science, wrote Høffding in a discussion of modern physics in 1930, but consider "what light this work so far throws on the nature and conditions of knowledge."[26] Quantum theory has demonstrated, he argued, that old established concepts of causation are not valid for the entities of the subatomic world. Electrons and other parts of an atom do not obey the laws of classical mechanics. But this does not imply that the principle of causation loses its relevance, nor does it deny that novel concepts of causation can be created to satisfy the observed phenomena. Høffding rejected the radical empiricist conclusion of Bertrand Russell—the claim that quantum mechanics shows that in modern natural science there is no place for the category of "thing."[27] The fundamental role of the classical concept of "substance" had been undermined in recent centuries, for instance, through Kant's interpretation in terms of the *Ding-an-sich*, which by definition was not knowable, acknowledged Høffding. But he sympathized with those nineteenth- and twentieth-century philosophers who would not discard the concept of the "thing" as a fundamental category. Those most knowledgeable about natural science agree, argued Høffding, that "the popular conception of a 'thing' should be corrected, but the last trace of it cannot disappear," and that "correction, and continually new corrections, is not the same as exclusion."[28] He warned against the kind of a priori transcendental argument concerning the limits of empirical natural science to be found in Kant and still popular among Høffding's contemporary philosophers. Niels Bohr was well aware of Høffding's persistent message about radical historical change in scientific concepts, and he saw the recent revolution in physics as a confirmation of this view.[29]

"Light and Life"

The details of Max Delbrück's early years and his migration into biology have been discussed in the introduction to this volume. Of particular relevance to this chapter is his encounter with Niels Bohr's reflections on the relations of biology and physics, particularly as developed in the lecture Bohr delivered in August 1932 on the relation of the biological and physical.

Delbrück consistently referred to this lecture, entitled "Light and Life," as his great inspiration for a career in molecular biology.[30] What was it that fired the imagination of an ambitious young physicist? Bohr gave the opening lecture at the International Congress on Light Therapy to an illustrious audience with little knowledge of physics. Once again Bohr introduced his notion of complementarity, exemplified by the dual nature of light—as either a wave or a particle phenomenon—and reviewed the recent revolution of physics, which had supplanted classical mechanics with relativity theory and quantum mechanics, thereby changing our basic intuitions about the nature of matter. The salient question now, he said, is "whether some fundamental traits are still missing in the analysis of natural phenomena, before we can reach an understanding of life on the basis of physical experience." This can hardly be answered "without an examination of what we may understand by a physical explanation, still more penetrating than that to which the discovery of the quantum of action has already forced us."[31]

Pursuing this suggestion of a new and more penetrating analysis of basic concepts of physics, Bohr drew an analogy between the situation in atomic physics before quantum theory and the contemporary situation in physiology. The stability of atoms could not be explained by classical physics. Similarly, life appeared to be a phenomenon inaccessible to complete analysis in physicochemical terms, since an investigation would kill the organism long before the molecular structure had been charted in detail. Life seemed to reside in a molecular complexity inaccessible to available approaches of physics and chemistry. But perhaps a new revision of fundamental physical concepts could open new possibilities in explaining biological phenomena like "self-preservation" and "propagation" of individual organisms.[32]

Bohr immediately added, however, that there was no reason to believe that reduction of biology based on a new physics could ever be carried to completeness. This was implicit in the notion of complementarity as a condition arising from the psychological nature of scientific observation and explanation in general. Bohr ended the argument by pointing to the impossibility of a complete explanation of self-consciousness. "Without entering into metaphysical speculations, I may perhaps add that any analysis of the very concept of an explanation would, naturally, begin and end with a renunciation as to explaining our own conscious activity."[33]

Bohr's argument can be interpreted in terms of Høffding's view of natural science as fundamentally a "spiritual science" (*Aandsvidenskab*). In this perspective, the complementarity of biology has a common source with complementarity in atomic physics, but cannot be derived from it. It is significant that Bohr concluded by stressing that he does not intend to express "any kind of scepticism as to the future development of physical and biological sciences."[34] This can be understood as a warning against setting any specific limits to the explanation of biological phenomena in terms of physics and chemistry, as J. S. Haldane and Bohr's father had tended to do. Any such attempts are likely to be overturned by future conceptual advances not only in biology, but also in physics and chemistry. The recent revolution in physics was a striking example of such basic conceptual innovation.

But how well does this interpretation fit Bohr's apparent point about an experimental limit to reductionist explanation in biology—the claim that the organism will be dead long before its molecular structures and processes have been precisely investigated? This "thanatological principle"[35] is based on a thought experiment similar to the one supporting Heisenberg's uncertainty principle: since even the least possible intervention needed for a measurement will change the object, there is a definite limit to our knowledge about the state of the object determined by the physical quantum of action. But the relevance of such an argument in the case of biology is unclear. Knowledge about molecular structures and processes are not built on direct observation alone, but also on indirect inference from the results or products of interventions under various specified conditions. So Bohr's argument is perhaps best understood as a not very precise historical analogy between atomic physics and the science of living organisms. His own discussion suggests such a historical analogy rather than an epistemically limiting physiological fact: "On this view, the existence of life must be taken as a starting point in biology. . . . The asserted impossibility of a physical or chemical explanation of the function peculiar to life would in this sense be analogous to the insufficiency of the mechanical analysis for the understanding of the stability of atoms."[36]

Since quantum theory does explain the stability of atoms and molecules, the analogy must be between the situation in classical mechanics before quantum mechanics and the present situation in biology. But even with the most advanced physical theory—quantum mechanics—there are limits to precise causal explanation. The quantum of action still has to be taken as an elementary fact. There is no reason ever to expect everything to be explained. Science is essentially uncompletable (*uafsluttelig*), as Høffding claimed. In biology one can expect fundamental progress in causal-physical explanation, but there will nevertheless remain some unexplained basic facts of "life" representing the basic autonomy or irreducibility of biology to physics.

Bohr had a deep interest in biology and its relation to physical science. He also hoped that philosophical analysis of the general nature of relationships between these two sciences could suggest fruitful paths for experimental research in biology. Finn Aaserud has pointed out how Bohr's philosophizing on biology in the late 1920s and the 1930s was linked to his search for new topics and new sources of funds for his institute, as it was losing momentum after an early glorious period.[37] Bohr himself never had the intention to start experimental biological work. He was presumably well aware of his lack of knowledge in that field. But he was interested in stimulating young physicists like Max Delbrück to try their luck.

Causal "Mechanism" versus "Life"

As I indicated above, the force of the thanatological principle—the fact that detailed experimental investigation will sooner or later kill the organism—depends on its epistemological interpretation. On a radical empiricist view that ties scientific knowledge closely to direct observation, the possibilities of causal explanation may appear to stop where the experiment kills the organism. With an epistemology that also admits of more indirect inference to knowledge, and thus makes possible knowledge about entities that cannot be observed, the thanatological principle will have less impact. Bohr's realism, albeit "constructive," points in the latter direction, as does my interpretation of his use of the thanatological principle in a historical analogy.

Ernst Pascual Jordan (1902–80) was another brilliant young physicist who discussed the fundamental methodological problems of biology with Bohr and was inspired to make the quest for a new biology a main project of his life. His scientific and political career as a scientist with Nazi sympathies has been thoroughly described by Richard Beyler.[38] Jordan's attempt to create a new discipline called "quantum biology" was quite different from Delbrück's persistent hunting for the paradox in molecular biology. Jordan never invested much of his own time in experimental physiological research. He remained primarily a theoretical physicist with a vision of how biology would undergo a revolution due to the fundamental insights of the new physics. In the radical empiricist spirit of the Copenhagen interpretation, he took quantum mechanics rather dogmatically as a given basis for all natural science.[39] Thus Jordan's research program was radically different from that of Delbrück, who entertained the idea that experimental investigation of biological phenomena might precipitate a new revolution in physics. While Jordan made a particular philosophical interpretation of quantum mechanics his basis, Delbrück wanted to challenge the general validity and scope of that theory through experimental biology.

Finn Aaserud has described how Bohr's view of the relation between biology and physics matured in the course of his discussions leading up to Jordan's 1932 paper "Quantum Mechanics and the Basic Problems of Biology and Psychology." A manuscript for this paper, which Jordan sent to Bohr in the spring of 1931, became a focus in their discussions. Bohr's low-key but persistent criticism made little impact on Jordan. The editors of the journal *Die Naturwissenschaften* were a little reluctant to accept the paper, feeling they had already published enough philosophy, but in November 1932 it finally appeared.[40]

The amplifier (*Verstärker*) theory that Jordan proposed in this paper is built on the assumption that certain behaviors of individual atoms are essentially acausal, that is, they have no causal explanation. A favorite example is radioactive disintegration, which follows strict statistical laws on the aggregate level but is acausal on the level of the individual atom. Jordan suggested that living organisms contain amplifiers that translate the acausal behavior of individual atoms and molecules into behavior on the macroscopic level. Jordan's scheme had obvious similarities to old solutions of the problems with scientific determinism, the ancient atomists' suggestion about irregularities in movement of individual atoms, theological speculations about a "god of the gaps," and so forth. The explanatory vacuum could be filled by various kinds of entelechies and vitalisms, or stark political voluntarism for that matter. The acausal speculations were built on a dogmatic interpretation of quantum theory in physics that did not appeal to Bohr. It was not in tune with the kind of solution of the Kantian antinomy between mechanistic and teleological explanation in biology that he had absorbed from his Danish philosophical environment.

Jordan began his argument from David Hume's analysis of causation. Hume's great merit, on Jordan's interpretation, was to have conclusively traced knowledge back to observation. He had shown that the primary physical concepts are essentially given by immediate observation.[41] Jordan applied his Humean interpretation of causation in atomic physics to biology. He argued that if it made no sense scientifically to ask for the causes of the disintegration of a radium atom, it was probably no more meaning-ful to ask what caused a specific genetic mutation.[42] In the last section of the paper, Jordan defended his amplifier theory against the objection that pure chance, the simple absence of causality, explains nothing. In response Jordan proposed that the activities of an organism could be divided into two "zones," one being the macroscopic and observable causally deter-mined zone, where events are subject to mechanical and chemical laws, the other the microscopic and acausal "directing" zone: "Firstly [there is] the zone of macroscopic causality, where all reactions are accessible to observation and run according to mechanical and chemical laws. Secondly [there is] the zone of 'directing' reactions, which involve very minute quantities of substance down to atomic orders of magnitude, which (on the evidence of atomic physics) are not causally determined, and which on the other hand (on the evidence of physiology) act in a directly triggering manner on the reactions of the first zone."[43]

Jordan admitted that this did not explain, as Bohr had pointed out, the striking stability of organisms as expressed in their unitary purposive na-ture, that is, as "purposes of nature" (*Naturzwecke*) in Kant's terminology. To explain this feature of living organisms, Jordan proposed "an 'inner' zone of acausal reactions" that is the seat of the "unity" of the organism and is characterized by "a still higher degree of non-observability of its physical state, than we know from atomic physics."[44] As an example of what belonged to the inner zone, he mentioned the "state within a chro-mosome." According to Jordan, it is absolutely impossible, with physical and chemical methods, precisely "to determine the internal physical state of a chromosome" without destroying its vital functions.[45] And as I noted, the entities of the "inner zone," which direct the macroscopic behavior of organisms in their characteristic biological behavior, are unobservable in an even "higher degree" than the parts of atoms. There is a clear similar-ity of the directing "unity" situated in the inner zone to Hans Driesch's neovitalist concept of the entelechy.

Bohr had rejected Jordan's amplifier theory in a letter of June 1931. In the words of Finn Aaserud, Bohr had explained in his characteristic polite manner that he could not accept Jordan's "narrow analogy" between wave-particle dualism and the fundamental problems of biology and psychology with the "attempt to reduce psychic phenomena as well as biology, directly to quantum mechanics."[46] In his "Light and Life" lecture in August 1932, Bohr developed his own view, warning against the ideas of Jordan without mentioning his name. "For example," said Bohr, "when it has been suggested that the will might have as its field of activity the regulation of certain atomic processes within the organism, for which on the atomic theory only probability calculations may be set up, we are dealing with a view that is incompatible with the interpretation of the psycho-physical parallelism here indicated."[47]

Jordan's conclusion was that quantum theory set specific limits to progress toward a reductionist (mechanistic) biology. The explanation of biological phenomena through physical and chemical analysis, which had contributed so much to progress of biology during the nineteenth and early twentieth century, had now reached an ultimate limit, for example, in the analysis of chromosomes as the seat of heredity. Jordan's dogmatic antireductionism built on a fundamental understanding of nature derived from atomic physics. In this sense he can also be characterized as a physical reductionist with little sense for a fundamental autonomy of biological science.

Thus Bohr's claim for the autonomy of biology was of a different nature than Jordan's. It was not based on the ontological priority of physics, but on fundamental epistemic limitations common to physics, biology, and all other empirical sciences. Bohr's view was antireductionist in the sense that biological phenomena can never be completely explained in terms of physics and chemistry. And it was reductionist in the sense that search for physical and chemical causal mechanisms is the right method in experimental biology. Bohr emphasized the openness of the reductionist perspective; it was conceivable that the discovery of new biological phenomena could precipitate a new revolution in physics. This was the possibility that Delbrück picked up. But as I will argue in the following, Delbrück never quite grasped the content of Bohr's antireductionist metaphysics. He took it too literally as having specific scientific implications.

Delbrück's Research Program in Genetics

In contrast to Bohr's philosophizing, Delbrück's interest in biology was from the start focused on experimental science.[48] Bohr's philosophy of biology was a source of inspiration, but there are no attempts to develop Bohr's philosophy in Delbrück's early publications.

As described in Sloan's chapter in this volume, Pascual Jordan's lecture on biology and quantum mechanics in Berlin-Dahlem in November 1934 was severely criticized by biologists for professing obscure vitalism. As in the 1932 paper, Jordan claimed to build on ideas very similar to Bohr's. For Delbrück it became important to explain that Bohr was no vitalist like Jordan. To explain the difference, Delbrück wrote a short note to Max Hartmann, who had been particularly sharp in his criticism of Jordan.[49]

In this note to Hartmann, Delbrück rejected Jordan's use of the thanatological argument. Experimental physiology will not meet any definite limitations because its investigations tend to kill the organism. "On the contrary: For genetic *and* developmental mechanics *and* physiology *and* biochemistry *and* biophysics, it is characteristic and essential that they study processes in the *living* organism," maintained Delbrück.[50] His juxtaposition of disciplines suggested that none of them are epistemically fundamental and privileged. He put "biology *and* physics *and* chemistry" on the same footing, and stressed the irreducibility of chemistry to physics: "physics and chemistry are not causally reducible to one another." Delbrück appears to have taken autonomy of each of these disciplines for granted. At the same time he commented condescendingly on the intellectual sharpness of biologists in his letter to Bohr: they are used to reading longer publications hastily and do not quite grasp a subtle new idea, but "must always force it into the ready-made scheme of their concepts."

Delbrück's contribution to the "Three-Man-Paper" was his first serious publication in biology. As described in more detail in the introduction to this volume, the bulk of the paper, more than half the total number of pages, is a survey of the state of the art in genetics by Nikolai Timoféeff-Ressovsky. In his discussion, Timoféeff-Ressovsky depicts mutation studies as the preferred way to penetrate questions about the nature of the gene. The most quoted authority is Hermann Muller. As discussed in the introduction, Muller was the theoretician pushing for clarification of the gene concepts and for investigations of the physicochemical nature of the gene.[51] Muller's papers between 1927 and 1935 were a main source of inspiration for Timoféeff-Ressovsky's project of radiation genetics into which Delbrück was drawn.[52]

Delbrück's contribution was a physicist's analysis of the idea of the gene as molecule. The characteristic stability of the gene followed from the chemical stability of its molecular structure. Mutations were interpreted as a displacement (*Umlagerung*) of atoms within this structure. Delbrück's main tool was X-rays. It was the same tool that Muller had used for his studies of artificially induced mutations in the late 1920s. This work won for him in 1946 the second Nobel Prize awarded to a geneticist. The linking of transmission genetics to cytology had opened the path to a physicochemical understanding of the gene, explained Delbrück. The detailed analyses of radiation genetics in *Drosophila* had led to an estimate of the size of the gene that was "on the order of the largest, distinctively structured molecules known to us."[53] This was the starting point of his theorizing about the nature of the gene.

The introduction to Delbrück's portion of the paper reflects his stand in the ongoing debate on vitalism. He points out how the concept of observation in transmission genetics is quite different from that in atomic physics. The differences in characters that form the experimental basis of genetics refer to properties of whole organisms. In this sense the whole living organism is the elementary entity. One could draw the conclusion, wrote Delbrück, that genetics is completely autonomous, and "must not be mixed up with physico-chemical ideas."[54] But, as already mentioned, he believed that the chromosome theory showed the way to physical and chemical investigations of the gene. What is special to genes as molecular structures is that they are so few and far between, he explained. They act as individual molecules, not as ensembles. From this point of view, genes differ essentially from the ordinary "chemical definition" of molecules.[55]

Somewhat carried away by the idea of the organism as the elementary indivisible entity, Delbrück apparently made the exaggerated claim that there is "by definition" only one instance of the specific *Genmolekül* in each living being. One in each cell is probably what he meant. Even in a thought experiment, pondered Delbrück, it is not possible to conceive of the gene entering a uniform chemical reaction, unless we are able to isolate the same gene from a large number of genetically similar organisms. However, this is pure speculation (*Spielerei*) as long as "we do not want to draw any specific, directly testable claims from the presumption of its in principle possibility."[56] We are left to express the underlying idea in some other way, Delbrück concluded. He then went on to discuss the gene as a "well-defined assemblage of atoms" (*wohldefinierte Atomverband*), and proposed that a study of displacements in this structure can be a way to elucidate its whole architecture. The effect of electromagnetic radiation, especially X-rays, followed naturally as the specific topic of his part of the paper.[57]

It is striking how Delbrück's speculations were built on the idea of the gene as a stable molecular structure. The speculative biochemical approach to the understanding of this structure was stopped short, however, as he was unable, not surprisingly, to imagine the methods developed in following decades to multiply single DNA molecules into large ensembles of identical molecules, thereby making them accessible to ordinary chemical analysis. But we also see how the same underlying idea of the gene as a stable molecule guided his radiation experiments. The *Genmolekül* was the hard core of Delbrück's research program at this time.[58] A quantum-mechanical perspective was his special contribution. But he was well aware that he was now working in the field of biology, or more precisely in biochemistry, and that the work had to start from established biological and biochemical knowledge. There was no special shortcut for physicists.

There is no mention of Bohr or complementarity in the paper. When Delbrück sent a copy of the manuscript to Bohr, he explained in the accompanying letter that the paper contained "no kind of complementarity argument." To the contrary "*it turns out* that one can formulate a unified atomic-physical theory of mutation and molecular stability."[59] This could possibly be read as an indication that Delbrück held a restrained hope of more dramatic results that would have needed consideration in term of complementarity. The results so far did not contradict the traditional reductionist view that the structure of the gene could be investigated independently of its effects on the organism.

An obvious desideratum for Delbrück's research program on the physical nature of the gene was to find a suitable experimental system of small and chemically simple organisms that multiply rapidly. Just before he left Berlin for the United States on a Rockefeller fellowship in 1937, he wrote down some speculations on the recent startling discoveries about viruses. The crystallization of tobacco-mosaic virus by Wendell M. Stanley indicated that the virus as a whole had a well-defined three-dimensional structure, that is, it could be viewed as a kind of large molecule. A key question was then whether the virus acted simply as a chemical stimulus that made the living cell produce new virus entities of the same kind, or whether the multiplication was "an essentially autonomous accomplishment" of the virus itself. Delbrück in his 1937 note argued for the latter view. The virus did in an important sense replicate itself, like a gene: "We want to look upon the replication of viruses as a particular form of primitive replication of genes, the segregation of which from nourishment supplied by the host should in principle be possible." However, in the following sentence he stated that the "essentially autonomous" replication of the virus could be compatible with traditional reductionism: "In this sense, one should view replication not as complementary to atomic physics but as a particular trick of organic chemistry." Apparently Delbrück in 1937 pursued a reductionist program in genetics, with "complementarity" as a following shadow.[60]

As described in the introduction, Delbrück found the ideal system for his work in bacteriophage when he went to the California Institute of Technology in 1937 on a research fellowship, and his disciplinary transformation was completed when he was appointed professor of biology at Caltech in 1947. Delbrück's underlying hope in the phage work was that gene replication would reveal fundamentally new physical phenomena. By the mid-1940s this hope was fading, as it turned out that the replication process also involved some kind of recombination. It was not the simple autocatalytic process that Delbrück originally hypothesized. When he returned to Caltech in 1947, his first graduate student was set to work on the light response of a bacterium.[61] By 1950 he had turned to sensory physiology, focusing on photoreception in the microscopic fungus *Phycomyces*.[62] He was still inspired by the ideas in Bohr's 1932 lecture, which had also discussed the amazing ability of the eye to register single photons. Bohr had concluded that "the sensitiveness of the eye may even be said to have reached the limit imposed by the atomic character of the light effects."[63]

In his retrospective and programmatic lecture of 1949, "A Physicist Looks at Biology," Delbrück was still sympathetic to "Bohr's suggestion of a complementarity situation in biology, analogous to that in Physics." He had not given up the hope that "paradoxes" would "crop up" and point the way to new discoveries in physics.[64] But the tone of Delbrück's essay was scientistic and dogmatic compared to Bohr's philosophically guarded pronouncements. Delbrück maintained that "quantum mechanics is the final word as regards the behaviour of atoms," due to the "analysis by Bohr and Heisenberg of the possibilities of observation." This appears closer to Heisenberg's radical empiricism than to the "constructive realism" of Høffding, which Bohr adopted. For Delbrück, nevertheless, the "final word" of quantum mechanics represented "liberation" and not "renunciation" as far as new discoveries in physics goes: "The limitation in the applicability of present day physics may then prove to be, not the dead end of our search, but the open door to the admission of fresh views on the matter."[65] He had in mind "the fact that the individual organism represents an indissoluble unit, barring us, at least at present, from a reduction to the terms of molecular physics." Perhaps this barrier will be overcome, said Delbrück, "but a physicist is well prepared to find that it is essential."[66] I interpret Delbrück to say that as a theoretical physicist inspired by Bohr, he was well prepared to find a limit to the explanation of living organisms in terms of existing chemistry and physics, that is, to be confronted with "the paradox."

Delbrück's hope for a paradox in genetics faded through the 1940s and 1950s as the pioneering of molecular genetics proceeded under the auspices of the Phage Group, orchestrated by Delbrück himself. But it did not disappear completely. For instance, it briefly revived in the spring of 1953 when Delbrück learned about Crick and Watson's discovery of the strikingly simple double-helical structure of DNA. In a letter to Bohr he wrote: "Very remarkable things are happening in biology. I think Jim Watson has made a discovery which may rival that of Rutherford in 1911."[67] Bohr, a decade later, in his last lecture, "Light and Life Revisited," recalled how Delbrück had told him about the discovery of the DNA structure and said it "might lead to a revolution in microbiology comparable with the development of atomic physics, initiated by Rutherford's nuclear model of the atom."[68]

At the centennial celebration of the Carlsberg laboratories in Copenhagen in 1976, Delbrück devoted the main part of his lecture, "Light and Life III," to progress in understanding photochemical mechanisms of photosynthesis. He started with a brief discussion of Bohr's 1932 suggestion of complementarity in biology as "an uncertainty principle regarding life, analogous to Heisenberg's uncertainty principle in quantum mechanics." Delbrück acknowledged that he himself had pursued this suggestion in genetics. But the discovery of the structure of DNA had with "one blow" shown that the trick could be accomplished within established chemical theory. For genetics and related subjects, "the earlier epistemological question" was now a dead issue, "and it has not yet become a live issue in the area of psychobiology."[69] But as I have shown, Delbrück's hopes of meeting the paradox in genetics did not immediately disappear with the discovery of the DNA structure in 1953.

Delbrück versus Bohr

The methodological discussions between Bohr and Delbrück in 1954–58 are revealing for the philosophical differences between the two. As Delbrück's romantic and revolutionary, physics-inspired research program in molecular genetics was being overtaken by the progress of a more traditional and trivial mechanistic biochemistry in the 1950s, he sought new guidance in methodological discussions with Bohr. Delbrück evidently hoped that a deeper and more precise understanding of Bohr's philosophy would help him develop fruitful new ideas for experimental physiological research. Instead he soon felt frustrated by Bohr's vague philosophizing.

After a private discussion in the autumn of 1954, Delbrück wrote to Bohr explaining his frustration: "I would like to tell you some supplementary thoughts in connection with our Princeton conversations. I would like to make clearer the connection between three things. First, why I work on Phycomyces, second why I am not enthusiastic about discussing with you at the present stage of development the problems concerned with replication and crossing over and the like, and third, why I am really desirous that you, or Aage and you, should write up the thoughts about complementarity in physics in greater detail."[70]

Delbrück had two topics in mind as needing methodological guidance, namely light reception of the fungus *Phycomyces* and replication of DNA molecules. The replication of DNA was a crucial problem in the new molecular genetics that still had no satisfactory answer. Delbrück explained that in the *Phycomyces* research, he was pursuing a concrete application of complementarity to biology. He was not interested in complementarity as a general epistemic idea, but wanted to press the experimental analysis so far that the paradox appeared. This was, so to speak, Delbrück's working hypothesis.

> What I failed to stress was my suspicion, you might almost say hope, that when this analysis is carried sufficiently far, it will run into a paradoxical situation analogous to that into which classical physics ran in its attempt to analyze atomic phenomena. This, of course, has been my ulterior motive in biology from the beginning. What I have in mind is an application of the complementarity principle not in a form which is just vaguely analogous to the way it is used in physics, as having something to do with the shift in the dividing line between observer and object, but as something much more closely related to the physics situation, springing directly from the individuality and indivisibility of the quantum processes.[71]

He complained how Bohr "in the old days" stressed that "the quantum of action" might be expected in biology to "play a decisive role exactly in the same manner as in physics," but in recent years he only discussed "complementarity arguments for situations in which we have no hint as to how one could possibly measure the interaction or convince oneself otherwise of its finiteness." Delbrück's letter concluded by asking for "a double statement of the complementarity principle," first in terms of physics, as Bohr had done, for example, in his conversations with Einstein, "and once in a more general and abstract form, which should enable the student to see for himself whether indeed the same argument ought to have applications in other fields."[72]

Three years later Delbrück himself tried to respond to this challenge for a dual statement. In November 1957, Bohr delivered the Karl Taylor Compton Lectures at MIT on "Complementarity in Quantum Mechanics," and Delbrück was invited to comment on the fifth lecture where biology was a main topic. Delbrück proposed that a distinction be made between "complementarity of the first kind," which is described in purely physical terms and exemplified in atomic physics, and a generalized "complementarity of the second kind," which applied also to "an experimental situation in which the observations are not expressible in terms of classical physics." Delbrück explained that by pushing the chemical analysis, one might reach a point where it was possible to say, "Here is a manifestation of life that could not rationally be reduced to an interpretation in molecular terms."[73] In other words, he hoped for a phenomenon that was truly inexplicable in terms of existing physics and chemistry and that would precipitate a radical change in physics analogous to that of the 1910s and 1920s. This historical analogy, Delbrück stressed, could only be a motivation, and could not by itself change the traditional reductionist approach of molecular biology. But it could affect speculation and choice of working hypotheses. Delbrück developed at length how genetic maps and molecular sequences of DNA might turn out not to be "simple images of another" and thus provide a concrete example of "a true complementarity of the second kind."[74]

In a subsequent letter, Delbrück explained to Bohr that the only way he could find "the complementarity argument" to be relevant and fruitful in biology was "in some such manner as I tried to illustrate in the example on genetic mapping." But he also had "a feeling that this is not what you have in mind, and, therefore that I do not properly understand your point of view."[75] Bohr was, however, busy with other things and answered evasively half a year later through his secretary that he did not think there was a difference "in the conception of the basic problems, but rather in the attempt to find a formulation which allows obtain the best understanding between the different biological trends."[76]

The serious scientific discussion between the two appears to have reached a dead end, though their philosophical discourse continued on a playful level. Delbrück was asking for a general principle of complementarity that could fruitfully guide the formulation of working hypotheses in molecular biology. But this was something Bohr could not provide, and probably had never really meant to do. Bohr tried repeatedly to explain his philosophical point of view to Delbrück, who continued to ask for something more concrete and specific. By 1960, Bohr was also reluctant to use the suggestive term "complementarity" in discussions of biology. He used instead the term "complementary," and spoke of approaches entailed by the "immense complexity" of living organisms.[77]

In 1962, Delbrück invited Bohr to "elaborate on what you said 30 years ago, in the light of the new developments that have taken place."[78] In "Light and Life Revisited" Bohr stressed the impressive recent success of reductionism in biology and deflated his own earlier suggestion that "the very existence of life might be taken as a basic fact in biology in the same sense as the quantum of action has to be regarded as a fundamental element irreducible to classical physical concepts." He still thought it inevitable that "teleological terms will be used in complementing the terminology of molecular biology." But this did not "in itself imply any limitation in the application to biology of the well-established principles of atomic physics."[79]

Bohr gave this lecture in June 1962, at the opening of a new institute for genetics that Delbrück was setting up at the University of Cologne, and he died in November. As Delbrück commented the following year, there is no claim of "observational complementarity" in this text, "no real analogy to the complementarity argument in physics."[80] Bohr in his lecture had only talked about deeper understanding of the complexity of organisms, which pointed to "possibilities of mechanisms which hitherto have escaped notice."[81] There was no suggestion that it might be necessary to go beyond established physics and chemistry to find such mechanisms.

A decade later, in 1974–75, Delbrück discussed the relation between mind and matter in a lecture series at Caltech. A summarizing essay, "Mind from Matter?," was published in 1978,[82] and a book-length version posthumously in 1986. These texts are revealing for understanding Delbrück's basic philosophical outlook. Here he combined a straightforward and simple naturalistic story about origins of life and mind on Earth with a radical empiricist epistemology. Delbrück overcame troublesome metaphysical dualism, due to a misconceived "Cartesian cut" between subject and object, by insisting that physical law "refers very explicitly to situations actually or potentially experienced by an observer and to nothing else." Physical science is "objective" in the sense of "being 'reproducible' for each observer, and the 'same' for different observers."[83] As Gunther Stent wrote in his introduction to the book-length version of the

essay, Delbrück here "resorts to complementarity to develop an argument
to show the inadequacy of all versions of realism."[84] Perhaps it was radical empiricist leanings that led Delbrück to take Bohr's complementarity
message of 1932 too literally, and made him persist in complaining about
the lack of concrete guidance in searching for the paradox.

In other words, Delbrück lacked the philosophical training to follow
Bohr's understanding of Kant's antinomy of teleological judgment. Delbrück did not quite grasp the difference between physics and biology implied by the historicized version of Kant that Bohr had absorbed from
Høffding. For Bohr the idea of complementarity in biology was not derived
from physics. Rather the ideas of complementarity in the two disciplines
had a common basis in general principles of epistemology. In the 1932
"Light and Life" lecture, Bohr had emphasized that the ultimate basis for
his ideas on complementarity, in physics as well as in biology, was psychological phenomena like consciousness, purposeful action, and freedom of
the will, which appear so strikingly incompatible with the scientific ideals
of causal explanation. It was his firm belief that "any analysis of the very
concept of an explanation would, naturally, begin and end with a renunciation as to explaining our own conscious activity."[85]

Delbrück's reductionist research program, his hunting for the "paradox," was tempered by Bohr's insistence on autonomy for the different
disciplines of natural science. Respect for the specific knowledge of biology was also reinforced by Delbrück's own experimental physiological
work. But in the end his methodological outlook was also influenced by
the widespread expectation in contemporary theoretical physics that a
theoretical unification of natural science was close at hand—to the extent
that such a unity of science was at all possible. And for Delbrück, as for
most other physicists, it was a matter of course that this fundamental theory would be a product of physics. Bohr's conception of an autonomous
biological science had a deeper philosophical grounding. I have tried to
show how it was inspired by Høffding's constructive realism, which did not
accept physics as the one fundamental science of nature.

Interpreting Bohr's Philosophy of Biology

It is tempting to speculate that a main obstacle to understanding Bohr's
philosophy of biology has been a prejudice favoring the ontology of physics as the basis for all natural science. Such a preference is implicit in a

kind of physicalism that has long dominated philosophy of science. David Papineau, for instance, takes the progressive reduction of biology to support a general physicalist ontology, meaning that everything is "physically constituted."[86] Without it being stated explicitly, "physical" apparently refers primarily to the science of physics. Such a privileged epistemic position for physics is more explicitly stated, for instance, in Alexander Rosenberg's analysis of the relationship between physics and biology. He holds that while physical theory can be interpreted in realistic terms, biology has an instrumental character.[87] The physical realism of Rosenberg is formulated in direct opposition to John Dupré's pluralistic realism,[88] which gives the theoretical entities of biology a real existence on par with those of physics.

I have argued that Bohr's view of biology should be interpreted in the spirit of Høffding's "constructive realism." Høffding's third way, rejecting traditional vitalism, as well as an all-inclusive mechanism, is built on the Kantian idea that a true explanation in natural science must be causal.[89] But Høffding rejected Kant's dogmatic a priori interpretation of fundamental physical theory by holding that all concepts of natural science are subject to historical change. Høffding's historicizing of Kant undermined the status of physics as the fundamental science of nature. It thus paved the way for a pluralistic realism as the basis for the autonomy of the different branches of natural science, for example, of biology in relation to physics.

Presupposing Høffding's ideas on the "uncompletability of knowledge," and the continuing change in basic scientific concepts, it makes little sense to look for precise biological phenomena that are absolutely irreducible and have forever to be taken at face value. Kant's grass-blade example suggests a transcendental argument of this kind. And some of Bohr's contemporaries, like J. S. Haldane, E. S. Russell, J. H. Woodger, and Hans Driesch, were still trying to pinpoint biological phenomena that were in principle irreducible.[90] In his thanatological argument, Bohr pointed to the fragility of life and the impossibility of total investigation of structure on the atomic level without causing the death of the organism, and he suggested that "the existence of life itself" should be taken at face value as a "basic postulate of biology, not susceptible to further analysis."[91] This has sometimes been read as an attempt to set a precise limit to reductionism. But read in context, I would argue, it is better understood as a warning against the dogmatic adherence to existing physics and chemistry, for which classical vitalism, with its dogmatic denial, presents a mirror image.

Bohr, like Høffding, sought a third way. And Bohr illustrated its promise of radical progress by pointing to the recent revolution in physics.

The Bohr and Delbrück program has often been characterized as searching for "other laws of physics." But as recently argued, this characterization is misleading because the expression was first used by Schrödinger, who held a quite traditional reductionist view compared to both Bohr and Delbrück.[92] The new laws or principles appealed to by Schrödinger were not something alien to existing physical theory. As Schrödinger wrote in his famous book *What Is Life?*, the other laws were "nothing else than the principle of quantum theory over again."[93] His idea built directly on Delbrück's analysis of the gene in the Three-Man Paper of 1935, but lacked the revolutionary character of his speculations on biological complementarity: "From Delbrück's general picture of the hereditary substance it emerges that living matter, while not eluding the 'laws of physics' as established up to date, is likely to involve 'other laws of physics' hitherto unknown, which however, once they have been revealed, will form just as integral a part of this science as the former."[94]

In his 1945 review of Schrödinger's book, Delbrück was direct and dismissive in his evaluation of its scientific novelty: "There is nothing new in this exposition, to which the largest part of the book is devoted, and biological readers will be inclined to skip it." When he described the gene as an "aperiodic crystal" instead of using the well-established term "complicated molecule," Schrödinger was simply introducing a fancy term with no new explanatory value, wrote Delbrück. He did not mention Schrödinger's metaphor "code-script."[95] But it is doubtful that Schrödinger himself had any notion of its potential. He elaborated only very briefly, presenting "code-script" as an analogy for a blueprint of the organism rather than a language for describing its hereditary properties.[96] Delbrück nevertheless welcomed the book as an inspiration for physicists and biologists alike to think about the topic.

In the interpretation of Daniel McKaughan, Schrödinger's "other laws" represented simply "an enlargement or expansion of an essentially mechanistic physics."[97] Andrew Domondon points out that to Schrödinger "the paradox of life was how individual traits could be passed down from one generation to another with such high fidelity, given the amount of thermodynamic noise at the molecular level."[98] The general answer was in terms of the stability of chemical molecules, as Delbrück had developed this point for macromolecules. But something was needed to bridge the gap between molecular and biological stability, and this he called "other laws of physics." Schrödinger was not interested in, and did not mention, "complementarity." He had little sympathy for Bohr's philosophical speculations in this direction. In retrospect, one may also say that Schrödinger's view was close to what later turned out to be the answer—an explanation of heredity in terms of established biochemistry without any important consequences for fundamental physical theory.

I suggested above that physicalist prejudice has prevented an adequate understanding of Bohr's philosophy of biology. Let me try to explain the nature of this obstacle.

Paul Hoyningen-Huene, for instance, has criticized "Bohr's antireductionist argument" for being circular.[99] He reconstructs the analogical argument in Bohr's papers of 1929–33 in the following way:

> *Premise* 1: An explanation of the stability of atoms by means of (classical) mechanics is impossible given the relation of complementarity which governs such explanation.
>
> *Premise* 2: The explanation of characteristic biological functions by physics is governed by the same relation of complementarity.
>
> *Conclusion*: The explanation of characteristic biological functions by physics is thus impossible.[100]

Through an incisive and illuminating analysis, Hoyningen-Huene shows how this argument fails because it presupposes the irreducibility that it pretends to prove. The argument lacks the universal premise that would allow the conclusion to be deduced, namely that "Whenever A and B are complementary, it is impossible to explain B by A."[101] However, if my interpretation of Bohr is right, he is not trying to present a logical argument deriving antireductionism in biology from complementarity in atomic physics. To the contrary, such a derivation is impossible according to the epistemological principles of Høffding's "constructive realism." By interpreting Bohr's intention in terms of the above argument, Hoyningen-Huene is attributing to him a physicalist ontology, giving physics priority in describing the nature of the world. Bohr's general idea of complementarity was built on a radical uncompletability and the existence of continuing conceptual change in science. Complementarity in atomic physics was one expression, and Bohr suggested in 1932 that a similar situation could emerge in the experimental investigation of biological phenomena. Complementarity in atomic physics and physiology are, in other words, parallel phenomena that have a common philosophical ground. On Bohr's view there is no direct argument from specific complementarities in one area to parallel phenomena in another area. However, this was just what Delbrück asked Bohr for in the 1950s. Hoyningen-Huene's analysis thus illuminates the difference between Bohr and Delbrück. He mistakenly assumes Bohr to be an ontological reductionist taking physics as our primary guide to fundamental entities of the world.[102] Bohr's reductionism was methodological and not ontological. But as we have seen, Delbrück's speculations on the relation of mind to matter ended up with a standard physicalist and ontological reductionism.

A similarly misleading physicalist assumption turns up when Bohr is interpreted to hold that scientific explanation in physiology is necessarily teleological. Jan Faye appears to make a claim of this kind when he writes that for Bohr "teleological considerations are necessary for the characterization of living organisms," and that "the finalistic mode of description is complementary to the physical mode of description in virtue of the fact that the observational conditions required for each, taken individually, are incompatible."[103] I have suggested that the fundamental difference between biology and physics lies in the descriptive concepts, but does not depend on biological terms being teleological. Different descriptions of biological phenomena can be just as causal in the nonteleological sense, and still fundamentally complementary, as are the classical descriptions in the examples from atomic physics. Faye, on the contrary, interprets Bohr and Høffding to hold that irreducibility follows from the untranslatability of teleological into physicalist language: "Høffding and Bohr argued that life itself cannot be described in terms of causal statements for the simple reason that the teleological language relating to the life functions cannot be 'translated' into physicalist statements because it referred to properties which 'supervene' on the member organs of the organism and are not reducible to them."[104] I have argued that such physicalist assumptions are not consistent with Kant's solution to the antinomy of teleological judgment on which Høffding built, and through him, Bohr. According to Kant, teleological explanations of living organisms have only a secondary scientific status. It is mechanistic causal explanations that provide strictly scientific knowledge. Teleological explanation is a substitute in an area with phenomena too complex to handle for the limited human mind. Characteristically, modern experimental biology keeps teleological concepts at arms length. They are important heuristically and in general reflections on philosophical foundations, but are mostly avoided in strictly scientific explanations.

In my interpretation of Bohr, the fundamental difference between physics and biology arises from the interaction of the subjective preconditions, inherent in all science, with the differences in the objects under investigation. It is the reference of descriptive, nonteleological terms to specific biological phenomena (things and events) that provides the immediate basis for autonomy. Their role in describing the phenomena cannot be *completely* substituted by physical and chemical terms. The qualification—not *completely*—is essential, because partial substitution is going on all the time, following the guidance of a methodological reductionism.

Concluding Remarks

Methodological reductionism is an appropriate characterization of Bohr's position as I have interpreted it. Progress in scientific knowledge about living organisms has to build on mechanistic causal methods and explanations. Delbrück similarly approached biology in the spirit of methodological reductionism. But his particular motivation for entering experimental molecular biology was the hope of discovering paradoxes leading to a new revolution in physics. Bohr had vaguely suggested that such a revolution might be achieved, in accordance with a historicized version of Kant's solution to the antinomy of teleological judgment. But the epistemological position that Bohr had learned from Høffding was never well understood by Delbrück. His view was, in the end, closer to the physical reductionism of Schrödinger, who had no understanding of Bohr's philosophizing about complementarity.

On this background, Delbrück's persistent hunt for "the paradox" can hardly be interpreted as an attempt to pin down specific phenomena in the living world that are "not reducible," as McKaughan seems to have argued.[105] To the contrary, the hope of Delbrück was that diligent pursuit of standard mechanistic physiology would reveal paradoxes analogous to those in atomic physics. These new paradoxes would then stimulate a new revision of basic physical concepts to allow a further extension of causal mechanist explanation of living organisms.

Bohr was no doubt a philosophically convinced antireductionist. The impossibility of a *complete* reduction of biology to physical science was an epistemic principle derived from general philosophical reflection, in particular the nature of consciousness. As a general principle, it was confirmed by experience of fundamental conceptual change in science, as in the recent revolution of atomic physics. But it was not a thesis to be directly supported or tested by concrete experimental results or discoveries.

Perhaps one can argue that in important ways Bohr, under the inspiration of Høffding's "constructive realism," anticipated the radical historicizing of science that Feyerabend and Kuhn promoted in the 1960s. And that he also saw, through a glass darkly, how this was compatible with a realist interpretation of scientific theorizing and could support a fundamental autonomy of biology from physics.

Notes

1. A useful historical overview of Niels Bohr's ideas on the relationship between biology and physics is found in F. Aaserud, *Redirecting Science: Niels Bohr, Philanthropy, and the Rise of Nuclear Physics* (Cambridge: Cambridge University Press, 1990). See especially the section "Interest in Biology, 1929 to 1936," pp. 68–100.

2. N. Roll-Hansen, "The Application of Complementarity to Biology: From Niels Bohr to Max Delbrück," *Historical Studies in the Physical and Biological Sciences* 30 (2000): 417–42.

3. D. McKaughan, "The Influence of Niels Bohr on Max Delbrück: Revisiting the Hopes Inspired by 'Light and Life,'" *Isis* 96 (2005): 528–29.

4. Immanuel Kant, *Kritik der Urteilskraft* (Berlin: Philosophische Bibliothek, 1926), paragraph 70. All references will be to this edition, with cross-references, where relevant, to the well-known English translation by W. S. Pluhar (Indianapolis: Hackett, 1987). Translations from the German are my own.

5. Ibid., paragraphs 71–78.

6. Edgar Rubin, "Erkendelsens Uafsluttelighed som et Grundmotiv hos Høffding" [The inconclusiveness of knowledge as a basic motive of Høffding], in *Harald Høffding in Memoriam* (Copenhagen: Gyldendal, 1932), pp. 5–15.

7. Harald Høffding, *Bemerkninger om Erkendelsesteoriens nuvaerende Stilling* (Copenhagen: A. F. Høst & Søn, 1930), pp. 3, 8.

8. Harald Høffding, *Begrebet Analogi* (Copenhagen: A. F. Høst & Søn, 1923), p. 109. "Hvad vi kaller Naturvidenskab, er jo dog en Frugt af aandeligt Arbejde og dens Metoder saavel som de Resultater der naaes ved Hjælp av disse Metoder, maa være bestemt af det aandelige Arbejdes Væsen." (Natural science is after all a fruit of mental work, and its methods as well as the results that are reached must be determined by the nature of mental work.)

9. Harald Høffding, *Psykologi* (Copenhagen: P. G. Philipsens Forlag, 1882), p. 266. "Naar det viser sig umulig at anvende den populære Definition af Sandhed som Erkendelsens Overenstemmelse med Genstanden, da selve Genstanden forudsætter vor Erkendelse, saa maa vi søge Sandhedkriteriet *inden for* Bevissthedens Verden og ikke *uden for*." (Since it turns out to be impossible to apply the

common definition of truth as the accordance of cognition with its object, as the object itself presupposes our cognition, we have to look for the criterion of truth *inside* of the world of consciousness and not *outside*.)

10. Jan Faye, *Niels Bohr: His Heritage and Legacy: An Anti-realist View of Quantum Mechanics* (Dordrecht: Kluwer, 1991), p. 214.

11. Faye, *Niels Bohr*, pp. 52–76.

12. Most of Høffding's works have been published only in Danish, but there are a number of translations into French, English, Dutch, etc. With respect to philosophy of science, the correspondence with Emile Meyerson is important.

13. Faye, *Niels Bohr*, p. 199. Objective antirealism, according to Høffding, maintains that "we cannot sustain a notion of our descriptive language as one having a content which makes it possible for us to speak about a fixed and objective reality."

14. Ibid., p. 214.

15. Aaserud, *Redirecting Science*, pp. 69–77.

16. J. S. Haldane, *Mechanism, Life and Personality* (New York: Dutton, 1923), p. 104. For a further discussion see N. Roll-Hansen, "E. S. Russell and J. H. Woodger: The Failure of Two Twentieth-Century Opponents of Mechanistic Biology," *Journal of the History of Biology* 17 (1984): 399–428.

17. Harald Høffding, "Mindetale over Christian Bohr," *Tilskueren*, March 1911, pp. 209–12.

18. David Favrholdt, "General Introduction: Complementarity beyond Physics," in Favrholdt, ed., *Niels Bohr: Collected Works*, vol. 10 (Amsterdam: Elsevier, 1999): pp. xxxviii–xliii. Hereafter "Complementarity beyond Physics."

19. Roll-Hansen, "Bohr to Delbrück."

20. McKaughan, "Influence of Niels Bohr."

21. Harald Høffding, "Om Vitalisme," in Høffding, *Mindre Arbejder* (Copenhagen: Nordisk Forlag, 1899), pp. 46–48.

22. Ibid., pp. 48–49.

23. Kant, *Kritik der Urteilskraft*, paragraph 77; Pluhar, p. 294; my translation.

24. See discussion in Nils Roll-Hansen, "Critical Teleology: Immanuel Kant and Claude Bernard on the Limitations of Experimental Biology," *Journal of the History of Biology* 9 (1976): 59–91.

25. See Roll-Hansen, "Russell and Woodger," pp. 425–26.

26. Høffding, *Bemerkninger*, p. 15.

27. Bertrand Russell, *An Outline of Philosophy* (London: Unwin, 1979; first published 1927), pp. 80–84.

28. Høffding, *Bemerkninger*, pp. 22–23. According to Faye (*Niels Bohr*, pp. 63, 72–76), Høffding discussed the contents of this paper with Bohr before he presented it to the Royal Danish Academy on January 30, 1930.

29. Faye, *Niels Bohr*, pp. 69–70, referencing a speech given by Bohr in August 1932.

30. Max Delbrück interview with Carolyn Harding, July 14–September 11, 1978, California Institute of Technology Archives, http://oralhistories.library.caltech .edu/16/01/OH_Delbruck_M.pdf, session 3, 1978, pp. 39–41. Hereafter cited as Harding interview. All quotations used with permission of the Archives of the California Institute of Technology.

31. Niels Bohr, "Light and Life" (part 2), *Nature* 133 (1933), p. 458b, reprinted in *Niels Bohr: Collected Works*, vol. 10, edited by D. Favrholdt (Amsterdam: Elsevier, 1999), 33. All references to the English text will be to this edition, which differs in important ways from the later reprinting in Niels Bohr, *Essays 1932–1957 on Atomic Physics and Human Knowledge*, vol. 2 of *The Philosophical Writings of Niels Bohr* (Woodbridge, CT: Ox Bow Press, 1987), pp. 4–12.

32. Ibid., p. 458b (*Collected Works*, 34).

33. Ibid., p. 459a (35).

34. Ibid., p.459b (35).

35. The term "thanatological principle" is used by David Favrholdt (1999), "Introduction, Part 1: Complementarity in Biology and Related Fields," in *Niels Bohr: Collected Works*, vol. 10, pp. 12, 19. The expression is central to Joseph Needham's 1936 criticism of Bohr's speculations on the irreducibility of biology (J. Needham, *Order and Life*, [Cambridge, MA: MIT Press, 1968], pp. 27–33; first edition, Yale University Press, 1936).

36. Bohr, "Light and Life," p. 458a (34).

37. Aaserud, *Redirecting Science*.

38. R. Beyler, "From Positivism to Organicism: Pascual Jordan's Interpretations of Modern Physics in Cultural Context" (PhD dissertation, Department of the History of Science, Harvard University, 1994). Some of the main points are found in R. Beyler, "Targeting the Organism: The Scientific and Cultural Context of Pascual Jordan's Quantum Biology, 1932–1947," *Isis* 87 (1996): 248–73.

39. J. Heilbron, "The Earliest Missionaries of the Copenhagen Spirit," *Revue d'histoire des sciences* 38 (1985): 194–230.

40. Aaserud, *Redirecting Science*, pp. 82–94; Pascual Jordan, "Die Quantenmechanik und die Grundprobleme der Biologie und Psychologie," *Die Naturwissenschaften* 20 (1932): 815–21. Additional comments on this paper and Jordan's subsequent 1934 follow-up are found in Sloan chapter, this volume.

41. Jordan, "Die Quantenmechanik," p. 815. "Die Definitionsunterlagen der primärsten physikalischen Begriffe müssen durch unmittelbar *Beobachtbares* gegeben sein."

42. Ibid., p. 818.

43. Ibid., p. 820.

44. Ibid. As the explanation, he states that "dürfte vielmehr die 'innere' Zone der nichtkausalen Reaktionen, welche als der Sitz der die *Einheit* eines Organismus konstituierenden Rektionsfähighkeiten anzusehen ist, durch einen *noch höheren Grad der Nichtbeobarkeit* ihrer physikalischen Zustände ausgereichnet sein, als wir aus der Atomphysik kennen."

45. Ibid., 821.

46. Aaserud, *Redirecting Science*, p. 85.

47. Bohr, "Light and Life," p. 459a (*Collected Works*, 35).

48. An extensive description and analysis of Delbrück's research program from its start around 1932 until the late 1940s has been given by Lily Kay, "Conceptual Models and Analytical Tools: The Biology of Physicist Max Delbrück," *Journal of the History of Biology* 18 (1985): 207–46. The intellectual context of biological and physical theorizing as well as the role of the Rockefeller Foundation in supporting Delbrück's research as part of the embryonic discipline of "molecular biology" is treated there in detail. A more penetrating discussion of Delbrück's own thinking is found in Robert Olby, *The Path to the Double Helix* (New York: Dover Publications 1994; first edition 1974), pp. 232–40.

49. Delbrück's note to Hartmann is found enclosed with a letter from Delbrück to Bohr, November 30, 1934, as translated in D. Favrholdt, ed., "Selected Correspondence," in *Niels Bohr: Collected Works*, vol. 10 (Elsevier: Amsterdam, 1999), pp. 466–69. Hereafter "Selected Correspondence."

50. Ibid.

51. Nils Roll-Hansen, "*Drosophila* Genetics: A Reductionist Research Program," *Journal of the History of Biology* 9 (1978): 159–210.

52. Hermann Muller, "Artificial Transmutations of the Gene," *Science* 66 (1927): 84–87. For further discussion of his gene concept see the introduction to this volume. On this radiation research see Summers and Beyler chapters, this volume. See also E. P. Fischer and C. Lipson, *Thinking about Science: Max Delbrück and the Origins of Molecular Biology* (New York: Norton, 1988), pp. 88–90.

53. N. Timoféeff-Ressovsky, K. G. Zimmer, and M. Delbrück, "Über die Natur der Genmutation und der Genstruktur," *Nachrichten von der Gesellschaft der Wissenschaften zu Göttingen, mathematisch-physikalische Klasse, Fachgruppe VI: Biologie*, 1 (1935): 189–245; 3MP 256 (225). References to specific pages in this paper hereafter will be given by page number in the translation in this volume, followed by the pagination in the original paper in parentheses.

54. 3MP 255 (224).

55. 3MP 256 (225).

56. Ibid.

57. 3MP 256–58 (225–27).

58. "Hard core" in the sense of Imre Lakatos, "Falsification and the Methodology of Scientific Research Programmes," in Imre Lakatos and Alan Musgrave, eds., *Criticism and the Growth of Knowledge* (Cambridge: Cambridge University Press, 1970), pp. 91–195.

59. Delbrück to Bohr, April 5, 1935, in Bohr, "Selected Correspondence," p. 472, emphasis mine. The whole sentence in the original German reads: "Die Arbeit enthält *keinerlei Komplementaritäts*-argumente. In gegenteil, es zeigt sich, dass man eine einheitliche atomphysikalische Theorie der Mutation und der Stabilität der Moleküle austellen kann. Das liegt daran, dass man über die nähere Wirkungsweise

der Gene in der Entwicklung gar nichts zu wissen braucht; die Merkmalsdifferenzen, die als Folge einer Mutation resultieren, sind nur die *Indikatoren* dessen, was im Gen passiert" (ibid., p. 468).

60. Max Delbrück, "A Physicist's Renewed Look at Biology: Twenty Years Later," *Science* 168 (1970): 1314–15. This is the lecture that Delbrück delivered when he received the Nobel Prize in December 1969. The quotation is from an added note (appendix 1) in the English translation, "Preliminary Write-up on the Topic 'Riddle of Life' (Berlin, August 1937)."

61. Fischer and Lipson, *Thinking about Science*, p. 217.

62. Olby, *Path to the Double Helix*, p. 240. On Delbrück's early interest in photobiology, see Sloan chapter, this volume.

63. Bohr, "Light and Life," p. 457b (*Collected Works*, 33).

64. Delbrück, "Renewed Look at Biology," 1314–15.

65. Max Delbrück, "A Physicist Looks at Biology," reprinted in J. Cairns, G. S. Stent, and J. D. Watson, eds., *Phage and the Origins of Molecular Biology* (Cold Spring Harbor, NY: Cold Spring Harbor Press, 1966), pp. 19–20.

66. Ibid., pp. 17–18.

67. Delbrück to Bohr, 14 April 1953, in Bohr, "Selected Correspondence," p. 474.

68. Niels Bohr, "Light and Life Revisited" (1962), in Niels Bohr, *Essays 1958–1962 on Atomic Physics and Human Knowledge*, vol. 3 of *The Philosophical Writings of Niels Bohr* (Woodbridge, CT: Ox Bow Press 1987), p. 27.

69. Max Delbrück, "Light and Life III," *Carlsberg Research Communications* 14 (1976), p. 301.

70. Delbrück to Bohr, December 1, 1954, in Bohr, "Selected Correspondence," pp. 475–76.

71. Ibid.

72. Ibid.

73. Max Delbrück, "Atomic Physics in 1910 and Molecular Biology in 1957," unpublished manuscript, Delbrück Papers, California Institute of Technology DP 35:2, pp. 6–7. Cited with permission of the Archives of the California Institute of Technology and the Delbrück family.

74. Ibid., p. 19. As it turned out, this was a good guess, but the presently accepted explanation is in terms of complexity of molecular-level structures and processes rather than complementarity.

75. Delbrück to Bohr, December 23, 1957, Niels Bohr Scientific Correspondence, Archives for the History of Quantum Physics MF 28:2. Cited with permission of the Niels Bohr Archives, Copenhagen. (Not in "Selected Correspondence.")

76. S. Hellmann to Delbrück, June 18, 1958, Niels Bohr Archives, Copenhagen. Cited with permission.

77. "Quantum Physics and Biology," in Niels Bohr, *Causality and Complementarity*, edited by Jan Faye and Henry Folse, in *The Philosophical Writings of Niels Bohr*, vol. 4 (Woodbridge, CT: Ox Bow Press, 1998), pp. 184–85.

78. Delbrück to Bohr, April 15, 1962, in Bohr, "Selected Correspondence," p. 488.

79. Bohr, "Light and Life Revisited," p. 26.

80. Max Delbrück, "Biophysics," in *Commemoration of the Fiftieth Anniversary of Niels Bohr's First Papers on Atomic Constitution, Held in Copenhagen on 8–15 July 1963*, Session on Cosmos and Life, July 12, 1963 (duplicated typescript; Copenhagen: Institute of Theoretical Physics, 1963), p. 45.

81. Bohr, "Light and Life Revisited," p. 27.

82. Max Delbrück, "Mind from Matter," *American Scholar* 46 (1978): 339–53.

83. Ibid., p. 351.

84. Max Delbrück, *Mind from Matter? An Essay on Evolutionary Epistemology*, ed. Gunther Stent et al. (Oxford: Blackwell Scientific Publications, 1986), p. 14.

85. Bohr, "Light and Life," p. 459a (*Collected Works*, 35).

86. David Papineau, "The Rise of Physicalism," in Carl Gillett and Barry Loewer, eds., *Physicalism and Its Discontents* (Cambridge: Cambridge University Press, 2001), p. 3.

87. Alexander Rosenberg, *Instrumental Biology, or The Disunity of Science* (Chicago: University of Chicago Press, 1994).

88. John Dupré, *The Disorder of Things: Metaphysical Foundations of the Disunity of Science* (Cambridge, MA: Harvard University Press, 1993).

89. See Kant's discussion of the antinomy of teleological judgment. Kant, *Kritik der Urteilskraft*, paragraphs 68–78.

90. Roll-Hansen, "Russell and Woodger."

91. Bohr, *Essays 1932–1957*, p. 31.

92. McKaughan, "Influence of Niels Bohr"; Andrew T. Domondon, "Bringing Physics to Bear on the Phenomenon of Life: The Divergent Positions of Bohr, Delbrück and Schrödinger," *Studies in History and Philosophy of Biological and Biomedical Sciences* 37 (2006): 433–58. Olby, *Path to the Double Helix*, p. 245, makes the same point. See also discussion of Schrödinger in the introduction to this volume.

93. Erwin Schrödinger, *What Is Life? The Physical Aspect of the Living Cell, with Mind and Matter and Autobiographical Sketches*, reprint of first edition of 1944 with forward by R. Penrose (Cambridge: Cambridge University Press, 2000), p. 81.

94. Ibid., p. 68.

95. Max Delbrück, "What Is Life? And What Is Truth?" *Quarterly Review of Biology* 20 (1945): 370–72.

96. Schrödinger, *What Is Life?*, pp. 20–21.

97. McKaughan, "Influence of Niels Bohr," p. 524.

98. Domondon, "Bohr, Delbrück and Schrödinger," p. 446.

99. Paul Hoyningen-Huene, "Niels Bohr's Argument for the Irreducibility of Biology to Physics," in Jan Faye and Henry Folse, eds., *Niels Bohr and Contemporary Philosophy* (Dordrecht: Kluwer, 1993), pp. 252–53.

100. Ibid., pp. 240–41.
101. Ibid., p. 251.
102. Ibid., p. 239.
103. Faye, *Niels Bohr*, p. 160.
104. Ibid., p. 163.
105. McKaughan, "Influence of Niels Bohr," p. 529.

CHAPTER SIX

Was Delbrück a Reductionist?

Daniel J. McKaughan

Although the "Three-Man Paper" (hereafter 3MP) is now widely acknowledged as a key text in the foundations of molecular biology, it is still subject to serious misunderstanding.[1] Many biologists became aware of the article only after the Nobel Prize–winning physicist Erwin Schrödinger drew attention to the model of genetic mutation set forth in the 3MP as part of a discussion of the potential significance of quantum physics for biology in *What Is Life?* (1944). Schrödinger's popularization, shaped as it was by his own speculations about physics and biology, has continued to color the understanding of the 3MP among biologists, historians, and philosophers—particularly in the English-speaking world.[2] In the years to follow, molecular biology saw dramatic progress, marked by achievements such as the elucidation of the double-helical structure of DNA in 1953. Many took this progress as a vindication of mechanistic and reductionist approaches to biology, making it all too easy to forget about earlier, complex, and varied debates concerning the role of purposive, vitalistic, or organicist concepts in the biological sciences. All too frequently, readers of the 3MP have overlooked or misunderstood the subtle and complex set of philosophical ideas that form an important part of the background and motivation for Delbrück's collaboration with Timoféeff-Ressovsky and Zimmer and his subsequent research program. Guiding ideas such as these, if they appear at all in published scientific research, are often submerged, surfacing only in occasional remarks. They are but the tip of the iceberg, which shows itself more freely in correspondence or informal public addresses and which, once flipped, can expose a strikingly different side.

The argument of this chapter is that the key to understanding the larger issues raised by the 3MP and Delbrück's lifelong project is to situate his biological speculations with respect to what I shall call the "teleomechanical" framework for relating physics and biology espoused by Niels Bohr, who after all had inspired Delbrück's transition from physics to biology. Contrary to the claim of some scholars that Delbrück (much less Bohr) was motivated by an in-principle reductionism, I argue (1) that Bohr and Delbrück shared a core expectation that at least some aspects of life would turn out not to be reducible to physicochemical terms, but (2) that there were also some important differences in the way they eventually came to think about biological complementarity.[3] Although Delbrück's biological research was motivated by the dream of providing an empirical justification for Bohr's conceptual framework of biological complementarity, Delbrück regarded the question of whether advances in physics and chemistry would be sufficient for understanding all aspects of life as an open empirical question. Moreover, it turns out that whereas Bohr explicitly built the appeal to purposive concepts into his entire approach to the life sciences, Delbrück did not seem to envision a key role for teleology and apparently failed to grasp how central it was for Bohr.[4] We can see why some interpreters have mistakenly regarded Delbrück as a reductionist when we understand the methodology and philosophical attitudes that drove what I call Delbrück's "empirical antireductionism" as a research program.

Delbrück's Place in Reductionist Narratives of the History of Molecular Biology

The central goal of the 3MP is to propose "a both physically and genetically well-founded model of the general nature of gene mutation, from which we will then be able to draw certain conclusions about the nature of the gene" in light of the "very precise statements about the *mechanism of the effect of X-rays*" that the authors found themselves in a position to make.[5] According to the theoretical picture proposed in the 3MP, genes are conceived as relatively stable and fairly autonomous molecular structures. The authors remain neutral about whether their relative independence is to be explained by the fact that genes are individual molecules, or if they are instead fairly autonomous parts of a larger, perhaps periodic, structural unit such as the chromosome. Mutations are viewed as changes in the structural conformation of genes: "According to our hypothesis, a

particular mutation is a particular rearrangement of a particular molecule."[6]
Through an "external infusion of energy or a fluctuation of thermal energy," ionization, excitation of electrons, vibratory motions, or otherwise
dissipated energy can lead to mutations, which consist in "a rearrangement of atoms into a new equilibrium configuration within a larger assemblage of atoms."[7]

There are, however, aspects of the 3MP that are rather more philosophical in character and might strike the reader as puzzling or as of questionable relevance to its stated technical aims. Why, for example, does
Delbrück begin his contribution to part 3 of the 3MP, "Atomic-Physics
Model of Gene Mutation," with the "question of whether it is appropriate
to introduce speculations drawn from atomic physics into the conceptual
system of genetics."[8] Why would Delbrück here take up questions about
the potential autonomy of genetics and cell biology from chemistry and
physics, observing that, in comparison to chemistry, which "does not in
any way exhibit such an independence from physics," genetics is a "self-
contained," autonomous science, which takes for granted the "individual
organism" as a starting point for inquiry?[9] Why review considerations that
could lead one to "assert that genetics is autonomous and that it should
not be muddled with physical-chemical notions" and what is this busi-
ness about part-whole relations and heuristic schemes? Why do Delbrück
and his coauthors feel the need to explore the implications of the target
theory for dissolving the cell into genes as "the 'ultimate units of life,' " the
"overall functioning of the cell," or to speculate that "the 'starting points'
of the developmental sequences are not attributed to individual genes,
but rather to operations of the cell, or even to intercellular processes"?[10]
Topics such as these provide readers just a glimpse of the regular, rich, and
lengthy conversations between these authors dating from Delbrück's years
in Berlin (1932–37), conversations that touched on topics far beyond the
scope of the 3MP.[11]

In a standard history of "the molecular revolution in biology," Robert
Olby writes: "[F]rom the 1930s onwards the research programme of molecular biology was always considered to be reductionist in character, i.e.
biological function was to be accounted for in terms of structure going
right down to the molecular level."[12] Should the 3MP be understood in
this way? One general and fairly standard answer to the questions posed
in the previous paragraph, an answer that I am here calling into question,
is that Delbrück was a reductionist who (1) repudiated the idea that biology is autonomous from physics and chemistry and who (2) thought that

by proposing the molecular model of the gene he was paving the way for a
research program that he believed would eventually yield explanations of
cellular function and replication that appeal only to concepts drawn from
physics and chemistry. As an example of a scholar who specifically attri-
butes a reductionist agenda to Delbrück beginning in the 3MP, and who
argues for this claim clearly and in some detail, consider John Fuerst:

> The origins of the informational school and the Phage Group are inextrica-
> bly bound to the scientific and philosophical career of Max Delbrück, and to
> Delbrück's theoretical work on mutagenesis with N. W. Timoféeff-Ressovsky
> and K. G. Zimmer, culminating in the so-called "Three-Man-Work" or Green
> Pamphlet . . . . Delbrück wrote the theoretical section of the Green Pamphlet
> (which otherwise reported experiments on the X-ray mutagenesis of *Drosoph-
> ila*), and with his contribution, entitled "Physical-atomic Model of Gene Muta-
> tion" (which became known as the quantum mechanical model of the gene), he
> rephrased the experimental conclusions in the language of quantum mechan-
> ics . . . . Certainly Delbrück seemed to have no compunction himself at this
> stage about mixing up genetics with physico-chemical concepts, even though
> he was aware of the problems confronting any attempt at complete physico-
> chemical understanding . . . . I would claim that at the stage of the Green Pam-
> phlet, Delbrück does not seem to have accepted any inherent "insufficiency" or
> limitations to physico-chemical explanations in genetics—in the sense of Niels
> Bohr's "asserted impossibility of a physical or chemical explanation of the func-
> tion peculiar to life" analogous to the "insufficiency of the mechanical analysis
> for the understanding of the stability of atoms."[13]

The first step in Fuerst's argument is to question the degree to which Delbrück sympathized with, or shared, Bohr's basic outlook on the relation
between physics and biology. "Certainly there is little doubt that Bohr's
thoughts about biology and the relationship between physics and biology
were extremely important to Delbrück's scientific development, but not in
the sense of Delbrück becoming Bohr's disciple. Rather, Bohr's ideas first
aroused Delbrück's interest in biology (resulting in his move to Berlin and
collaboration with biologists), and served as a challenge for Delbrück to
solve the problem Bohr had posed—to prove that Bohr's 'limitations' did
not exist."[14] Fuerst's reading of Delbrück as a thoroughgoing reductionist
characterizes Delbrück's views both at the early stages of his career, during his years of research on phage, and into the late stages of his career.
"Whether based on physics or based on chemistry, the whole ethos of the
Phage Group, of the 'informational school,' was in fact a reductionist one,

the 'spiritual hallmark' of which was to be 'the quest for the physical basis of the gene.' "[15]

Fuerst, then, argues that Delbrück remained true to "his background of reductionism" over the course of his life. Despite remarks in Delbrück's own 1949 address that might suggest otherwise, Fuerst maintains that

> Delbrück dearly *desires* a reductionist approach to succeed, but fears that it may not—as exemplified by his statement that biology is depressing for the physicist because "insofar as physical explanations of seemingly physical phenomena go . . . the analysis seems to have stalled around in a semidescriptive manner without noticeably progressing towards a radical physical explanation." A reductionist driven to the wall of non-reductionism by epistemological frustration, Delbrück nevertheless does not forsake reductionist beliefs on the nature of adequate explanation.[16]

I shall be arguing that this view of Delbrück's attitudes toward reductionism is at odds with Delbrück's own self-understanding. Delbrück's hope throughout was rather to confront the scientific world with an empirical situation—a "paradox"—that was *not* analyzable in physicochemical terms. In this way, he would legitimize a "new intellectual approach to biology," by showing that the situation could only be resolved by adopting the biological complementarity envisioned by Bohr. If Delbrück experienced frustration, it was rather that the reductionist project proved so successful in areas like molecular genetics that he was driven to shift the focus of his research—precisely around the time when most other scientists thought molecular genetics was really getting exciting—from viral replication to the study of how the fungus *Phycomyces* responds to the stimuli of light and how light affects its growth, in the hope that he might still uncover a situation that required appeal to biological complementarity in sensory physiology. As we shall see, the commitment that Delbrück refused to forsake was rather the goal of vindicating "Bohr's suggestion of a complementary situation in biology, analogous to that in physics," which Delbrück described as "prime motive" for his own interest in biology.[17]

Now, what people mean by "reductionism" is sometimes less than maximally clear.[18] My own reason for describing Bohr and Delbrück as antireductionists derives from the fact that they employ the language of "irreducibility" in their own self-description rather than out of a desire to map their positions onto the range of positions on offer in, say, the contemporary philosophy of mind. Fuerst points out that Kenneth Schaffner, author of one of the strongest previous attempts to challenge the central place given to reductionism in standard accounts of the history of molecular biology, relies on a "rather strict logical empiricist conception of reduction."[19] This point is well taken. It would be a hollow victory for the nonreductionist reading if it turned out, say, that while Delbrück was not a proponent of the formal Nagelian model of intertheoretic reduction, he was still a reductionist in the straightforward sense of holding "a belief in [the] explanation of biological phenomena wholly in physical and chemical terms."[20] One might wonder, though, whether Fuerst defines reductionism rather too broadly as "a belief in the possibility of a mechanistic, analytical explanation of biological phenomena in terms of physics and chemistry."[21] If we take the idea of believing merely that *such a thing is, for all we know, possible in the long run of inquiry* at all seriously here, Fuerst's definition runs the risk of triviality in the other direction. If someone believes that it is *possible* that biochemical compounds or microorganisms from outer space are the source of life on earth, shall we count them among the advocates of panspermia? Still, if we understand "possibility" more in the sense of "expectation" and add the caveat that the person also takes some such belief as a "major organizing principle"[22] of their research, I am happy to oblige Fuerst in these ground rules for the discussion. Surely on any plausible definition of reduction, if discipline X employs concepts and terms that cannot be analyzed solely in terms of the concepts and terms of discipline Y, then discipline X is not reducible to Y.[23] And yet, as we shall see, in precisely that sense, Delbrück thought that biology is autonomous of and irreducible to physics and chemistry. He thought that those latter two sciences would not be able to account for at least some aspects of biological phenomena.[24]

It is not hard to find examples of card-carrying reductionists in the history of molecular biology. Jacques Loeb's book *The Mechanistic Conception of Life* (1912) expresses the "hope that ultimately life, i.e. the sum of all life phenomena, can be unequivocally explained in physico-chemico terms."[25] As Jacques Monod recalls in an interview with Horace Judson, biologists of the school of Jacques Loeb "were philosophical reductionists and were convinced that the living world could be reduced in principle to physics and chemistry, and nothing more."[26] Olby points to Francis Crick as another example: "F. H. C. Crick stated that the ultimate aim of the modern movement in biology was to explain '*all* biology in terms of physics and chemistry.' He added: 'Thus eventually one may hope to have the whole of biology "explained" in terms of the level below it, and so on right down to the atomic level.'"[27] Crick was quite explicit about his own reductionist motives, but he also sensed that this wasn't what drove Delbrück:

"I went into [molecular biology], for example, to try to show that you can explain all these phenomena . . . just by the laws of ordinary physics and chemistry. But then you ought to consider Max Delbrück."[28] As we shall see, Bohr and Delbrück represented a theoretical perspective, and expressed sentiments and expectations, of a rather different character.

Bohr's Philosophy of Biology: Kantian Teleomechanical Complementarity

A clear picture of Delbrück's approach to biology requires an understanding of the philosophical speculation that inspired it. In chapter 5, Nils Roll-Hansen has developed in detail Bohr's philosophy of biology and its origins in Harald Høffding's historicized neo-Kantianism. After explaining my own reading of Bohr, in the next two sections I will attempt to elucidate Delbrück's complex reactions to Bohr's philosophical framework as this can be pursued into Delbrück's work beyond the 3MP itself.[29]

The exploratory remarks in part 3 of the 3MP assess the "pros and cons" of principled considerations that might lead biologists to argue for the autonomy of genetics from physics and chemistry. Since the 3MP itself defends the idea that genes are atomic structures, perhaps it is not surprising that many scientists and historians would read the paper as a contribution to the reductionist project, overlooking passages that treat such questions with a bit more delicacy: "While genetics always deals provisionally with the isolation of *processes that do indeed have, physically-chemically, an unambiguous character*, they appear as only parts of processes when viewed biologically, and their *relation to the whole life process* remains problematic, unless their *coordination* is viewed as arising on the basis of some *heuristic scheme* in which the life process is postulated in principle as physical-chemical machinery."[30] On the face of it, talk of employing a heuristic scheme that construes life processes in entirely physicochemical terms might indeed sound reductionist. But the authors are also making a point about two ways of viewing biological processes, contrasting a perspective that analyses the organism into parts and processes viewed in isolation with a viewpoint that, at least heuristically, sees these processes as a coordinated and unified system of events that are crucial to the functions of the organism as a whole.

We can cast light on such passages—and the motivation for discussing the relationship between biology and physics at all in a paper on X-ray-induced mutagenesis—by contextualizing the 3MP against the backdrop of Niels Bohr's distinctive way of envisioning the relationship between biology and physics. Bohr's 1932 "Light and Life" address to the International Congress on Light Therapy in Copenhagen—an event that served as a key inspiration for Delbrück's migration from physics to biology—explicitly speculates about the philosophical significance of developments in quantum theory for the life sciences.[31]

Recent historical work lends strong support to the suggestion that Bohr's perspective can be usefully understood as falling within a broadly Kantian tradition. By the late-nineteenth century this tradition was drawing on Kant's Third Critique to stake out a position that advocated the a priori employment of teleological concepts as a starting point for biological research, at least as a heuristic device, in debates between vitalists and mechanists.[32] Bohr's interest was to develop a conceptual framework in which this routine deployment of both purposive and mechanistic concepts makes sense.

I use the label "teleomechanical complementarity" to refer to what I take to be Bohr's view of the life sciences.[33] To briefly characterize Bohr's mature view: his basic thought is that "mechanistic and vitalistic arguments are used in a typically complementary manner."[34] Bohr had formally introduced the concept of complementarity[35] in a 1927 lecture at Como to deal with very specific problems arising in quantum mechanics as a way of defining the conditions under which particular phenomena appear. The empirical phenomena presented by light in various experimental contexts apparently require mutually exclusive descriptions, which are jointly necessary for our understanding of the evidence. Unlike Pascual Jordan, Bohr was not suggesting that the complementarity needed to understand biological phenomena was in any way a consequence of the complementarity arising in quantum mechanics.[36] As he put it later in 1954, "With these remarks it is in no way meant to imply that, in atomic physics, we possess a clue to the explanation of life."[37] Rather, Bohr's interest in the 1932 lecture was in exploring a provocative *analogy* between light and life: "the analogy considered is the typical relation of complementarity . . . [;] the concept of purpose, which is foreign to mechanical analysis, finds a certain field of application in problems where regard must be taken of the nature of life."[38] Bohr's work in atomic physics had legitimized the idea that in special circumstances, perhaps owing to conditions necessary for experimental arrangements, a complete understanding of the phenomena in some domain might require mutually exclusive yet jointly necessary descriptions. Bohr was raised in a household where questions about whether teleological explanations and concepts employed in biology could be reduced to or accounted for solely in terms of the concepts of physics and chemistry were familiar topics of discussion. Given the vigorous interest in these topics exhibited by his physiologist father, Christian Bohr, and philosophical mentors like Harald Høffding, it is not at all surprising to find Bohr speculating about whether the concepts employed in biological modes of description might stand in a similar relation of complementarity.[39]

Perhaps the age-old puzzle about how to reconcile the apparent purposiveness of some aspects of life with the understanding provided by a mechanistic analysis of organisms into their physical parts was epistemological in character. While Bohr did, to be sure, distance his own more modest employment of terms like "vitalistic" or "finalistic" from "the old idea of a mystic life force" or the obscurantism with which they are often associated, he clearly saw the need for "two *scientific* approaches which only together exhaust the possibilities of increasing our knowledge. In this sense, mechanistic and vitalistic viewpoints may be considered as complementary."[40]

On the teleomechanical reading, Bohr is proposing a framework for relating physics and biology that takes teleological and mechanistic descriptions to be complementary in the precise sense that these dual modes of description are mutually exclusive yet jointly necessary for a complete understanding of life: "Although science will of course strive for ever more detailed knowledge of the physical mechanism underlying the functions of organisms, a description of life corresponding to the ideal of mechanism will only constitute one line of approach."[41] Teleological approaches, which had so often been marginalized or associated with "the old idea of a mystic life force," can now be recognized as "equally indispensable" for biology.[42]

One important component of Bohr's view, and the aspect on which Delbrück focused a good deal of attention, was the possibility that life might not be reducible to physicochemical terms, that a purely mechanistic description would not exhaust and indeed would leave out purposive aspects of organisms that are crucial to our understanding of them. We are now in a position to appreciate a clear sense in which Bohr took purposive concepts to be *irreducible* to mechanistic ones. He thought that purposive concepts could not be analyzed solely in terms of more basic concepts of physics and chemistry and must simply be assumed as a starting point for the description of organisms when seen as whole individuals.

Already in 1929 Bohr had expressed his expectation that further inquiry

will confront us with "a fundamental limit to the analysis of the phenomena of life in terms of physical concepts" and that such descriptions will
need to be supplemented with teleological ones.[43] Bohr looked back on
the 1932 discussion of the "asserted impossibility of a physical or chemical
explanation of the function peculiar to life"[44] in "Light and Life Revisited"
(1962), where he recalls: "Stressing the difficulties in keeping the organisms alive under conditions which aim at a full atomic account I therefore
suggested that the very existence of life might be taken as a basic fact in
biology in the same sense as the quantum of action has to be regarded in
atomic physics as a fundamental element *irreducible* to classical physical
concepts."[45] Unlike Schrödinger, Bohr did not say that, while the concepts
of classical mechanics proved inadequate as a basis for understanding life
in exclusively physicochemical terms, new concepts employed by quantum
theory will suffice.[46] As he put it in 1951:

> These remarks, however, do in no way imply that, even if quantum theory is
> indispensable for dealing with biological phenomena, it should in itself *suffice*
> for an explanation of life. On the contrary, the point which I want to emphasize
> is the wider implication of the lesson atomic physics has taught us about our po-
> sition as observers of natural phenomena and, in particular, about the rational
> use of words like *cause* and *purpose*.
>
> Although science will of course strive for ever more detailed knowledge of
> the physical mechanism underlying the functions of organism, *a description of
> life corresponding to the ideal of mechanicism will only constitute one line of ap-
> proach*. In fact, we must recognize that experimental conditions demanded for
> an exhaustive description conforming with this ideal would involve a control
> of the organism to an extent which would preclude the display of life. In actual
> biological research, a *vitalistic approach is equally indispensable*, since the pri-
> mary object must often be the studies of the reaction of the organism as a whole
> for the purpose of upholding life, a point of which we are not least reminded in
> medical research. We are here *neither speaking of any crude attempt of tracing an
> analogy to life in simple machinery, nor of the old idea of a mystic life force, but of
> two scientific approaches which only together exhaust the possibilities of increas-
> ing our knowledge*. In this sense, *mechanistic and vitalistic* viewpoints may be
> considered as *complementary*, and in the harmonious balance between their ap-
> plications we find the basis for the practical and rational use of the word life.[47]

These joint themes of the complementarity of teleology and mechanism
and of the irreducibility of teleological terms to mechanistic ones are also

persistent themes in Bohr's lectures from 1948 through 1962.[48] Bohr's view is that "we must recognize that the practical approach in biological research is characterized by the complementary way in which arguments, based on the full resources of physical and chemical science and *concepts referring to the integrity of the organism transcending the scope of these sciences* are employed."[49] Notice that the complementarity that Bohr envisions is not simply between a scientific description of the world and, say, what Wilfrid Sellars called the "manifest image" or our day-to-day way of thinking about the world. The complementarity between teleology and mechanism that Bohr envisions falls *within science*, where talk of organisms as a whole and purposive concepts enter not just to put people at ease about our place in the universe but, Bohr thinks, to do real scientific work.

> In *biological research*, references to features of *wholeness and purposeful reactions of organisms are used* together with the increasingly detailed information on structure and regulatory processes that has resulted in such great progress not least in medicine. We have here to do with a practical approach to a field where the means of expression used for the description of its various aspects refer to mutually exclusive conditions of observation. In this connection, it must be realized that the attitudes termed *mechanistic and finalistic* are not contradictory points of view, but rather exhibit a *complementary relationship* which is connected with our position as observers of nature.[50]

While it should be clear from the preceding discussion of Bohr's philosophy of biology that such a framework could be enlisted to support claims about the autonomy of biology from physics, for Bohr the more fundamental contrast is between teleology and mechanism. He took physiology (which pursues causal explanation in physical and chemical terms) and psychology (which employs functional concepts as primitive), for example, as examples of such complementary points of view.[51]

Bohr thought that, when we turn to the study of life, purely mechanistic modes of description face limitations. Neither he nor Delbrück refrained from speculating about specific domains in which biologists might hope to expose these limitations in areas such as (1) the replication of genes, viruses, or cells; (2) the self-regulatory activity of organisms or the responses of systems involving gas exchange in respiration or the fixation of oxygen by blood such as those that Christian Bohr had worked on, or in sensory perception and the study of how the fungus *Phycomyces* responds to the stimuli of light and how light affects its growth, issues that later became the focus of Delbrück's research interests; or (3) consciousness, and the meaning of mentalistic terms such as "thoughts," "feelings," and the goal-directed behavior manifested by conscious, free beings.

The underlining in Delbrück's copy of Bohr's "Discussion with Einstein on Epistemological Problems in Atomic Physics" from the Library of Living Philosophers volume on Einstein highlights Bohr's remark that "besides the complementary features conspicuous in psychology and already touched upon, examples of such relationships can also be traced in biology, especially as regards the comparison between mechanistic and vitalistic viewpoints."[52] In a section entitled "Limitations of Atomic Physics as Applied to Biology," from his own series of lectures in 1944, Delbrück writes:

> Physics has a tradition of long standing in biology of trying to lay down the law. So does chemistry . . . . To men like Helmholtz it seemed clear that both the substrate and processes are essentially the same in the organic as in the inorganic world, and that this precluded the existence of specific vital forces. This led to the concept of life as essentially a mechanical problem. It seemed as if the phenomena observed in living material should be deducible from the laws of Newton's mechanics as movements of particles due to forces originating in these particles. This is the mechanistic view of life, in contra-distinction to the postulate of specific vital actions. Such a view, I believe, is still in the back of the minds of many biologists of today and is the driving force of most biochemists. There is, of course, a disquieting feature in it that it leads to an impossibility of explaining cognition. How can we, who on this view are just a jumble of molecules, obtain a notion of these molecules? There are, of course, branches of biology which flourish without direct recourse to this ideal. The most notable is genetics, which in its pure form operates with "hereditary factors" and "phenotypic characters" in a perfectly logical system, without ever having to bother with the processes by which the "characters" originate from the "factors." The root of this science lies in the existence of natural units of observation, the individual living organisms which in genetics play much the same role as atoms and molecules in chemistry.
>
> However, though geneticists may be content to stop here, chemists and physicists will not . . . . Is this process simply a trick of organic chemistry which the organic chemists have not yet run across in their test tubes or is it a phenomenon of an entirely different nature, depending on the "complex organization of the living cell"? The answer to this question is at present totally unknown. The

old-line mechanists, of course, would consider it a sacrilegious thought that the reproduction of genes might not be a purely mechanical process, but the modern physicists may well be inclined to take a more liberal view.[53]

After rehearsing why he thinks that Bohr's complementarity argument "puts the relation between biology and physics on a new footing," Delbrück offers a list of biological phenomena that should be investigated.

This being the situation, it becomes a matter of great interest to the physicist, and perhaps to the biologist too, to find out just how far atomic physics does carry us in the understanding of the phenomena of the living cell, and it will be the aim of these lectures to trace the endeavors of atomic physics in biology in various directions in the following order:

1. External and internal metabolism
2. Photosynthesis
3. Genes and gene actions
4. Genes and gene mutations by X-rays
5. Bacterial viruses[54]

It is important to see that while Bohr's framework takes it that some aspects of life can be understood only by assuming purposive concepts as a starting point, neither Bohr nor Delbrück pretended to be setting out principled considerations that could demonstrate a priori where, if at all, mechanistic approaches would encounter these limits. As we have seen, both of them did venture some guesses about the sort of phenomena that might prove to resist analysis solely in mechanical terms (the replication of genes or viruses, various self-regulative processes of cells or organisms, consciousness, etc.), but their idea was that detailed empirical investigation would expose such limits somewhere. As Delbrück put it, "the individual organism presents an indissoluble unit, barring us, at least at present, from a reduction to the terms of molecular physics. It may well turn out that the bar is not really an essential one, but a physicist is well prepared to find that it is essential."[55] Vague, ambiguous, and persistent though Bohr and Delbrück were on these issues, dogmatic they were not. That is why, time and again, they could acknowledge things like "in fact, Bohr's suggestion turned out to be wrong," while holding out that consciousness and free will, for example, might still turn out to be phenomena that are not be reducible to physicochemical terms.[56]

While, as I shall argue, Delbrück was the one who set out to vindicate Bohr's philosophical framework as a research program, both of them altered their expectations about where such limits might be encountered in light of the impressive progress made toward the goal of causally explaining phenomena in these various domains in terms of the material constituents involved. Still, even late in life, when many biologists took dramatic advances in molecular biology and genetics to indicate that the tide had come in on behalf of reductionism, Bohr continued to maintain that "as long as the word 'life' is retained for practical or epistemological reasons, the dual approach in biology will surely persist."[57] Until the end of his life Bohr was committed to the idea that "surely, as long as for practical or epistemological reasons one speaks of life, such teleological terms will be used in complementing the terminology of molecular biology. This circumstance, however, does not in itself imply any limitation in the application to biology of the well-established principles of atomic physics."[58] Similarly, although the reductionist program was successful to such an extent that Delbrück was forced to switch topics for investigation at various points during his career, he remained steadfast in this central ambition. Even in 1978, Delbrück resolutely maintained that psychology and molecular biology "certainly are in a complementary relation which nobody can still formulate very well. They haven't been pursued to that bitter end where you *have* to make some kind of new dialectical approach."[59]

Delbrück's Empirical Antireductionism and Key Differences in the Ways That Delbrück and Bohr Understand "Biological Complementarity"

With a clearer understanding of Bohr's own philosophy of biology and his analogical extension of the term "complementarity" into the life sciences, we are now in a position to examine the nature of Bohr's influence on Delbrück. We saw above that Fuerst regarded Delbrück as a frustrated reductionist who set out to undermine Bohr's claims about the limitations of physicochemical analysis. The relationship between Bohr and Delbrück is indeed more complicated than has often been appreciated. However, a closer analysis of Delbrück's writings and correspondence reveals a picture quite different from that of Fuerst. In brief, Delbrück set out to vindicate "Bohr's subtle complementarity argument" empirically. Delbrück's lifelong project was to push experimental research to the point that a "paradox" (some biological phenomena that was not analyzable in physicochemical terms) would be encountered that could only be resolved as Bohr anticipated, by appeal to complementary *mechanical* and *teleological* descriptions.[60]

To what extent did Delbrück understand the subtle philosophical influences that influenced Bohr's views? Although Delbrück might have failed to appreciate some of the nuances and philosophical underpinnings of Bohr's more Kantian position, Delbrück clearly understood Bohr's basic stance. Biological phenomena might require the employment of descriptions that are mutually exclusive yet jointly necessary for understanding life processes, making biology autonomous from physics and chemistry in the clear sense that biology legitimately employs concepts that are not reducible to physical and chemical concepts. Here is how Delbrück expressed the idea in 1949:

> It may turn out that certain features of the living cell, including perhaps even replication, stand in a *mutually exclusive relationship* to the strict application of quantum mechanics, and that a new conceptual language has to be developed to embrace this situation. The limitation in the applicability of present day physics may then prove to be, not the dead end of our search, but the open door to the admission of fresh views of the matter. Just as we find features of the atom, its stability, for instance, which are *not reducible* to mechanics, we may find features of the living cell which are *not reducible* to atomic physics but whose appearance stands in a *complementary relationship* to those of atomic physics.
>
> This idea, which is due to Bohr, puts the relation between physics and biology on a new footing. Instead of aiming from the molecular physics end at the whole of the phenomena exhibited by the living cell, we now expect to find natural limits to this approach, and thereby implicitly new virgin territories on which laws may hold which involve new concepts and which are only loosely related to those of physics, by virtue of the fact that they *apply to phenomena whose appearance is conditioned on* not *making observations of the type needed for a consistent interpretation in terms of atomic physics.*[61]

Or again, here is how Delbrück still put it years later in 1978:

> [Bohr] talked about [the potential significance of complementarity] a lot, especially in relation to biology, in discussing the relation between life on the one hand and physics and chemistry on the other—whether there wasn't an experimental mutual exclusion, so that you could look at a living organism either as

a living organism or as a jumble of molecules; you could do either, you could make observations that tell you where the molecules are, *or* you could make observations that tell you how the animal behaves, but there might well exist a mutually exclusive feature, analogous to the one found in atomic physics.[62]

As we shall see, however, there is at least one key difference between Bohr and Delbrück that is important for understanding the relationship between their views on biological complementarity. Whereas Bohr explicitly built the appeal to purposive concepts into his entire approach to the life sciences, Delbrück did not seem to envision a key role for teleology and apparently failed to grasp how central it was for Bohr. Instead, the thread that Delbrück himself picks up on and endorses is the antireductionist element in Bohr.

I have elsewhere documented the claim that Delbrück tells us, in more than one place, that he took the main themes of "Light and Life" to be (1) irreducibility and (2) complementarity.[63] I take the textual and archival evidence for this to be indisputable. For example, in a letter to Bohr on November 30, 1934, Delbrück attempted to articulate a concise statement of the position: "It is *not* asserted that the laws of atomic theory can explain *specific* life phenomena. On the contrary."[64] The position he took himself to share with Bohr was instead: "*Assertion*: The assumptions having to do with the causal order of biological phenomena may in part stand in formal contradiction to the laws of physics and chemistry, because experiments on living organisms are *certainly* complementary to experiments establishing physical and chemical processes with *atomic* precision."[65] As a young scientist in transition, Delbrück was in need of concrete research projects, and if Bohr was right that life is not reducible to physics, the rigorous pursuit of a *reductionist* research program, pushed to its limits, should bring to light precisely where those limits are. I shall refer to this research program as "empirical antireductionism." Notice that the attitude with which this project is undertaken is not that of a dogmatic adherence to a thesis, but rather a kind of stance that takes on a conditional form, "If these limits exist, then we should find them by doing such and such kinds of research," coupled with a hopeful, though at times frustrated, expectation that such limits might eventually be found. The strategy and the sentiment that accompanied this project are nicely captured in a remark Gunther Stent made in a letter to Delbrück from May 20, 1968: "Good Guys dream like holists and work like reductionists."[66] Explaining his strategy in private correspondence with Bohr in 1954, Delbrück wrote: "I talked about this system as something analogous to a gadget of physics, and explained at some length why it seemed more hopeful to me to analyze this gadget in great detail, rather than the many other biological gadgets which have been the subject of conventional research for many years. What I failed to stress was my suspicion, you might almost say hope, that *when this analysis is carried sufficiently far*, it will run into a paradoxical situation analogous to that into which classical physics ran in its attempts to analyze atomic phenomena. *This, of course, has been my ulterior motive in biology from the beginning.*"[67] We saw that the crucial move in Fuerst's argument that Delbrück was a reductionist was the premise that Delbrück set out to "prove that Bohr's 'limitations' did not exist"[68] by looking for results that would stand in "direct contradiction to Bohr's application of complementarity to biology."[69] However, not only is there overwhelming evidence that Delbrück was deeply sympathetic to Bohr's outlook insofar as he understood it, the empirical antireductionist reading of Delbrück also explains why Fuerst's reading might appear plausible, on the face of it. Rather than attempt to develop a productive research program that employed purposive concepts as a starting point, Delbrück's strategy was to adopt a reductionist methodology with the hope of finding a "paradox" or clear empirical limitation in the mechanistic description of life that would demonstrate that some biological phenomena are *not* reducible to physicochemical terms. Delbrück's attitude to the reductionist program was this: "It looks sane until the paradoxes crop up and come into sharper focus. In biology we are not yet at the point where we are presented with clear paradoxes and this will not happen until the analysis and behavior of living cells has been carried into far greater detail. This analysis should be done on the living cell's own terms and the theories should be formulated without fear of contradicting molecular physics."[70]

Suppose that the reduction of genetics to molecular biology had been the sort of thing that Delbrück had been looking for. It would then be very difficult to explain why Delbrück lost interest in and indeed shifted the focus of his research away from molecular genetics just around the time people were making significant progress precisely along those lines. Yet, as Stent reports, in late 1949 and in the early 1950s, precisely when those working in structural chemistry and molecular genetics were hot on the trail of potential breakthroughs, Delbrück began to "lose interest" in phage research and molecular genetics.[71] Certainly by the mid-1950s, apparently because Watson and Crick's elucidation of the double-helical structure of DNA seemed to present a decisive reduction of the processes of replication to physicochemical explanations without uncovering a paradoxical situation that required appeal to "complementarity," Delbrück had shifted his research interests to sensory perception and the study of how the fungus *Phycomyces* responds to the stimuli of light and how light affects its growth.[72] This marks a significant difference between the sort of thing that Schrödinger predicted physicists would find by studying biology and Delbrück's own motivations.[73] Yet such a move is wholly intelligible if we understand Delbrück to be after empirical situations that expose the limits of mechanical explanation, thereby demonstrating a rightful place for the complementary use of teleological concepts taken as irreducibly primitive starting points in biology. Advances in the field on the reductionist side of biological inquiry forced Delbrück to switch topics, but not yet to lose faith in biological complementarity. *Phycomyces* represents a natural choice: in Delbrück's opinion understanding processes such as the ones exhibited by those microorganisms is "a prerequisite for a truly successful venture into neurobiology."[74] Sensory physiology is a step closer to the study of consciousness, Bohr's mainstay example of biological complementarity, but it is still simple and interesting enough to hold the possibility of leading to "decisive insights in the field."[75] As Delbrück says in "A Physicist's Renewed Look at Biology: Twenty Years Later," "we are not yet ready to tackle [neurobiology] in a decisive way."[76] Delbrück's faith in finding biological complementarity would be severely tested over the years and, as I see it, sometime between 1963 and 1976 Delbrück came to believe that "we now understand that organisms can be viewed very successfully as molecular systems."[77] In "Light and Life III" Delbrück reminisces once again about how Bohr's lecture motivated him "to learn biology and eventually to become a biologist":[78] "In this lecture, entitled 'Light and Life,' Bohr proposed the bold idea that *life might not be reducible* to atomic physics. He suggested that there might be a *complementarity* relation between life and atomic physics analogous to the complementarity encountered with the wave and particle aspects of atomic physics."[79] Once again, Delbrück takes the main theme of both "Light and Life" and "Light and Life Revisited" to be the "reducibility question," just as he had consistently throughout his career.[80] Delbrück now takes the "explosive progress in biochemistry and especially in molecular genetics" to have forced deep retreats with respect to Bohr's "earlier epistemological question." As Delbrück puts it, "This question is by now a dead issue in the area of ordinary biochemistry and physiology, and it has not yet become a live issue in the area of psychobiology."[81] Delbrück's earlier hopes had not panned out, but he still seems to cling to the possibility that complementarity might someday be required in psychology.

Sharing a common faith does not imply complete agreement at the level of doctrine, however. Delbrück's research program was fixated on the idea of biological complementarity, but the constant challenge of pushing this idea forward as part of an experimental program led him to modify Bohr's proposal in certain ways. By 1957 we see Delbrück still quite committed to investigating the idea of biological complementarity, even while he begins to push it in a form that departs significantly from Bohr. In "Atomic Physics in 1910 and Molecular Biology in 1957," a lecture delivered at MIT in November 1957, Delbrück writes:

> The possible relevance of the complementarity argument to biology has fascinated me for over 25 years, but it is a question which is difficult to discuss in concrete terms. Even within physics the complementarity argument occupies a strange position . . . . Its relevance to general problems of epistemology is apparent to many . . . . The formulations [of the complementarity argument] lean closely on Bohr's recent lectures . . . . The reader is referred to Professor Bohr's lectures, which are to be published by the MIT press. It appears to me that in biology we are still not in a position to be very definite about the complementarity argument and the examples discussed in this lecture are to be understood as *illustrations* of the manner in which this argument might apply, not as hypotheses.[82]

Clearly, Delbrück continues to see himself as heavily indebted to Bohr's speculations.

However, by 1957 Delbrück was inclined to look for biological complementarity in cases where biologists were trying to relate physical parameters (e.g., those that describe the structure of the nucleotide sequence) with data that he was inclined to think were not best interpreted as yielding a picturizable model (e.g., recombination frequency data). In his efforts to cash out Bohr's speculations in terms of situations that could be investigated and tested experimentally, Delbrück came to favor the example of genetic mapping as an illustration of how the notion of complementarity might be needed in biology. In genetic mapping, others were interpreting recombination frequency data as an indication of physical distance between genes on a linear sequence of DNA, whereas Delbrück was inclined to think that there was a much more complex and indirect relationship between the data and the physical structure of the DNA molecule. Recalling the history of atomic theory in the opening decades of the twentieth century, he was alert to the possibility that such recombination data might provide something more like the kind of indirect information about atomic structure that can be gleaned from the complex vibration and emission patterns that show up in spectroscopic data. Whether observation of the conditions under which recombination occurs required an experimental arrangement that would be mutually exclusive with those required for a specification of the physical structures involved, he regarded as a serious, but open, empirical question.

Whereas mutual exclusion of experimental arrangements might still be used as an argument for irreducibility, the essential role given to teleological concepts in Bohr's biological complementarity drops out of Delbrück's proposal in "Atomic Physics in 1910 and Molecular Biology in 1957":

> It should be noted that the generalized theory that we have in mind is a mechanistic one, in contradistinction to vitalist theories which endow living matter with special properties, or finalist theories, which employ the concept of purpose as an irreducible one. The theory we have in mind transcends physics in that it introduces genetic recombination as a phenomenon not reducible to physics in the usual sense, not because living matter has special properties, but because the conditions under which recombination occurs may be incompatible with conditions that might define the physical state. Whether the two sets of conditions are, in fact, incompatible is an open question. Presumably it will not long remain open.
>
> The point of view we are taking does not close the door to further experimental approaches. On the contrary, it emphasizes the need for further *efforts designed to ascertain whether the two experimental approaches are indeed mutually exclusive or not*. In addition, it points up the need to ascertain experimentally whether genetic mapping and chemical mapping are related in a simple manner. Without this point of view such a simple relation might be taken for granted with the result that important experiments might not be pushed with sufficient vigor.[83]

Later in 1957, Delbrück wrote to Bohr about the genetic-mapping example: "The more I think about the possible relevance of the complementarity argument for biology the more I seem to be driven to the conclusion that it can be fruitful only if it comes in some such manner as I tried to illustrate in the example of genetic mapping. At the same time I have a feeling that this is not what you have in mind and, therefore, that I do not properly understand your point of view."[84]

Remarks by Niels's son Aage Bohr in the discussion period following Delbrück's presentation on biological complementarity in 1963 reinforce the contrast that I have drawn between Delbrück's empirical antireductionism and Bohr's teleomechanical complementarity. Aage had worked with Delbrück at Caltech for a short period in the late 1940s and observes that whereas Delbrück has "always taken the view that this complementarity really applies to the very *concrete experimental* facts, namely that if we really study in great detail the organisms and the simplest ones, the bacteria, we will come at a certain moment to a real paradox," his own understanding of the relevance of complementarity to biology, in contrast, is that "one is faced with the situation that to describe an organism *one must use* many times *words that do not belong to physics and chemistry*."[85]

In spite of these differences with Bohr, even in 1957 Delbrück was still quite clearly trying to find a specific

> point that could be pinned down, where we could say "Here is a manifestation of life that could not rationally be reduced to an interpretation in molecular terms" . . . . To me this is an intriguing thought, and in fact the mere possibility of its being so can be one's principle motivation for an interest in biology. In practice it does not alter one's approach. The only avenue of progress in molecular biology today, as in atomic physics then [in 1910, prior to the discovery of the atomic nucleus], is to develop better techniques and to do more ingenious experiments, to drive molecular biology along its traditional and possibly naïve path. The difference between those who believe that this will all be plain sailing, and those who believe in the advent of an intellectual impasse to be resolved by a great revelation is like that between those who do not, and those who do, believe in a life hereafter. Their motivation is different, but their deeds are not.[86]

Reductionism and antireductionism were for Delbrück not differences in methodological approach, but rather differences in attitudes or expectations about what the results of inquiry might show. Delbrück leaves no doubt as to his own attitude.

The sought-after limitations of a purely physicochemical approach proved more difficult to identify than either Bohr or Delbrück anticipated. Delbrück concedes that Bohr's complementarity argument wasn't needed in some areas, but still fully expects that it, or at least his modified version of it, will be needed elsewhere. As the field of molecular biology advanced, each of them revised their expectations about where we might encounter such limits without giving up the essential antireductionist stance. Delbrück's 1975 Sevilla Lecture is one of many places where his antireductionist attitude is expressed, and it also illustrates my reading of the relationship between Delbrück's research methodology and the philosophical motivations that drove his program:

> How much further will this reductionist approach carry us? Ever since the radical breakthroughs in <u>molecular genetics</u> the mood has been to assume that other areas of <u>molecular biology</u> will be able to solve their problems in similarly complete fashion without encountering serious conceptual problems. However, it is plain as day that an impasse <u>will</u> be encountered, and some new kind of complementarity arguments will need to be introduced, when attempting to integrate the analysis of consciousness with neurobiology. Will there be other impasses at lower levels? We will not know the answer to this question until these lower levels have been fully reduced to the molecular level in at least a few cases, just as little as we could know whether classical physics would be adequate for understanding atoms *until the reductionist approach has been thoroughly tested and found to be inadequate.*[87]

In "Virology Revisited" (the opening lecture at an international symposium on the "Molecular Basis of Host Virus Interaction" at Banaras Hindu University, October 17, 1976), Delbrück recalls his work with Timoféeff-Ressovsky and Zimmer on the 3MP as one of his "first attempts at focusing on the problem of reductionism in biology." Looking back on the 3MP some forty-six years later, despite all the ways in which success on the reductionist side of the project of molecular biology had severely tested the hopes that Bohr's vision had inspired, Delbrück still describes the focus of his research as centered on those seemingly paradoxical and puzzling "accomplishments of life equally challenging to the physicist intrigued by the problem of reductionism."[88]

> I became fascinated by Niels Bohr's idea that the novel epistemological situation encountered in atomic physics might have wider applications, especially for biology (Bohr, 1933). Bohr put out the startling notion that life *might not be reducible* to atomic physics just as the atom is not reducible to classical mechanics. I believe that at that time I was the only one of the people working with Bohr who took this notion seriously. I decided to look into it and to see whether it had any merit, and to do so meant that I had to learn biology. During this period of five years in Berlin as assistant to Lise Meitner, I had the good fortune to meet Timoféeff-Ressovsky. Timoféeff-Ressovsky was engaged in an experimental study of the mutagenic effects of ionizing radiations. Discussions of his results gave one the feeling that the stability of the genes was not

really miraculous from the point of view of quantum mechanics (as the stability of atoms had been from the point of view of classical physics). We said so (Timoféeff-Ressovsky, et al., 1935) in a somewhat lengthy paper published in 1935 in a very inaccessible journal (Proceedings of the Göttingen Academy of Science). This paper, after a dormancy of 10 years, had an explosive aftereffect when Schroedinger quoted from it extensively in his little book, *What is Life?* (Schroedinger, 1945).[89]

Conclusion

I have challenged the idea that Delbrück was a reductionist and the suggestion that the 3MP should be read as an attempt to inaugurate a physicochemical reductionism that formed the "spiritual hallmark" of the phage side of the history of molecular biology, as claimed by Fuerst and others. A reevaluation of the evidence does not bear these claims out, and indeed, when the dust clears, the evidence supports an understanding of key figures such as Bohr and Delbrück that differs considerably from this traditional reading. We have seen that Bohr drew on a set of epistemological concepts and distinctions from the Kantian tradition that enabled Delbrück to pursue research that was, at the methodological level, often indistinguishable from that of reductionists, while being animated by a very different set of ideas, attitudes, and aspirations.

In his contribution to this volume, Roll-Hansen observes, in my view correctly, that Delbrück's research methodology was to seek to explain biological phenomena in mechanistic terms insofar as this would prove possible to do. While I would not wish to overshadow the extent of our agreement on many matters in this part of the history of molecular biology, there do seem to be some remaining disagreements between us that deserve comment.

First, Roll-Hansen reads Bohr as appropriating Kant in such a way that "only [physical and chemical] causes are legitimate in scientific explanations of biological phenomena." On this view, "within what can strictly be called the science of living organisms, there is only room for mechanical explanations," while purposive or teleological concepts have, at best, "a heuristic role as shorthand expressions of complex mechanical systems" as a kind of "substitute in an area with phenomena too complex to handle for the limited human mind" and hence do not enter into properly scientific explanations. However, we saw above that there is a strong case to be made that when Bohr, at least, speaks of complementary descriptions, he sees these as "two scientific approaches" that are "equally indispensable." This is what makes some aspects of biology autonomous from physics and chemistry. This is also part of why he suggests, in "Light and Life" and in subsequent writings, that we may need to rethink our understanding of what is involved in giving (and what we can demand of) a scientific explanation and why his examples consistently contrast complementary descriptions employed in what he takes to be legitimate sciences such as physiology (which pursues causal explanation in physical and chemical terms) and psychology (which employs functional concepts as primitive).

Second, while Roll-Hansen and I agree that Delbrück was hoping to make a breakthrough discovery in science, we seem to disagree about *which* science. Roll-Hansen claims that Delbrück undertook biological research in the hope that he could "precipitate a new revolution in *physics*," that would lead to "a radical change in *physics* analogous to that of the 1910s and 1920s" (emphasis mine). While no doubt that would be an exciting thing for a young scientist to contribute to, in my view Delbrück was hoping to contribute to biology rather than primarily to physics. I have elsewhere argued that the idea that biology might lead to the discovery of "other laws of physics"—which was indeed once the standard way of describing Delbrück's role in the history of molecular biology—has its source in the writings of Erwin Schrödinger, but that there is very little evidence to suggest that Delbrück shared this view and good reason to think that this traditional reading overlooks important differences between Schrödinger, Bohr, and Delbrück.[90]

Roll-Hansen sees this second point of disagreement as connected to a third, and perhaps our greatest, point of difference: our understanding of Delbrück's underlying motivations or expectations for pursing the research that he did, which I have argued were not reductionist. In his own concluding remarks in this volume, Roll-Hansen asserts: "Delbrück's persistent hunt for 'the paradox' can hardly be interpreted as an attempt to pin down specific phenomena in the living world that are '*not reducible*,' as McKaughan seems to have argued. To the contrary, the hope of Delbrück was that . . . [t]hese new paradoxes would stimulate a new revision of basic physical concepts to allow a further extension of causal mechanist explanation of living organisms."

I have argued that Bohr's views in biology are explicitly holistic and antireductionist in character. Moreover, while Delbrück may not have understood all of the subtleties of Bohr's position, there is strong textual and archival evidence supporting both of the following claims. First, Delbrück took Bohr to be suggesting that life might not be reducible to physics and chemistry and would instead require appeal to some form of biological complementarity that was difficult to spell out in precise terms. Second, Delbrück himself set out to explore the issue empirically with the hope of encountering experimental limitations to the reductionist project. As I see it, any adequate appraisal of Delbrück's thought must take these facts into account.

The language of irreducibility is not just an interpretive spin that I have put on the texts. That is just what they say. That is the way that both Bohr and Delbrück describe their own views. I do not see any other way to read remarks such as "with respect to biology, Bohr proposed in his 1932 lecture 'Light and Life' that life might not be reducible to atomic physics,"[91] or discussion of the years of concerted attempts to identify and pin down "a phenomenon not reducible to physics" and of "efforts designed to ascertain whether the two experimental approaches are indeed mutually exclusive or not," or the litany of other passages submitted for consideration above.[92]

It is, of course, important to remember that the context in which Bohr, Schrödinger, and Delbrück used the language of reducibility or irreducibility in biology was very different from our own. The proper context in which to locate Bohr's reflection on these issues is within ongoing debates between mechanists and neovitalists in continental biological thought of the late nineteenth and early twentieth centuries. Both Roll-Hansen and I have argued that we get a much better handle on what Bohr was thinking by attending to these Kantian resonances in his thought. There is no shortage of remarks in Bohr where he is clearly talking about a situation where the *relata* that stand in a mutually exclusive but jointly necessary relation of complementarity are mechanistic descriptions on one hand and a description he variously refers to as teleological, finalistic, or vitalistic on the other. For his part, Delbrück was certainly familiar with the reductionist sentiments that expressed themselves with increasing vigor as the discipline of molecular biology came to maturity, but the driving motivation for his program was the hope of vindicating Bohr's antireductionist speculations empirically.

It is also worth emphasizing that my interests in this chapter are historical; I have resisted getting drawn into questions about how, say, various concepts of "reduction" that are current in the contemporary literature on philosophy of science and philosophy of mind (e.g., the notions of Nagelian intertheoretic reduction, ontological or epistemological reduction, supervenience without reduction, and the like) might map onto these debates from an earlier era. But if Delbrück is allowed to speak in his own voice—and who could be better for the historian to ask?—the answer to the question posed in the title of this chapter must be negative. Indeed, Horace Judson did ask Delbrück directly about the attempt to provide a complete picture of the phenomena, including our own existence. Judson asked, "All science was inevitably reductionist at that level?" "Not reductionist," Delbrück replied, distancing himself from reductionism: "I think science, as it runs now, leads a deprived existence. This comes out very clearly when you try to do psychobiology. D'you want to look at anxiety as a hormonal phenomenon or do you want to look at it as an existential phenomenon? And if you look at it as an existential phenomenon then how can you incorporate it into your picture of science?"[93] By themselves, studies of the former sort, Delbrück continues, "would miss many aspects which you would like to understand."[94]

Notes

1. N. V. Timoféeff-Ressovsky, K. G. Zimmer, and M. Delbrück, "Über die Natur der Genmutation und der Genstruktur," *Nachrichten von der Gesellschaft der Wissenschaften zu Göttingen, mathematische-physikalische Klasse, Fachgruppe VI: Biologie*, 1 (1935): 189–245. References to pages in this paper will hereafter be given by page number as found in the translation, followed by the pagination in the original paper in parentheses.

2. This chapter deepens and extends arguments concerning the place of Bohr, Delbrück, and Schrödinger in the history of molecular biology set forth in Daniel J. McKaughan, "The Influence of Niels Bohr on Max Delbrück: Revisiting the Hopes Inspired by 'Light and Life,'" *Isis* 96 (2005): 507–29. See also Sloan and Fogel's introduction, this volume.

3. I would like gratefully to thank and acknowledge the California Institute of Technology and the Maurice A. Biot Archives Fund for the support and assistance provided during my research trip to the Max Delbrück collection at the Caltech Archives in June 2006. Part of the research for this project was also made possible by a Graduate Research Fellowship from the National Science Foundation.

4. See McKaughan, "Influence of Niels Bohr."

5. 3MP 262 (232).

6. 3MP 261 (231).

7. 3MP 265 (234).

8. 3MP 254 (223).

9. 3MP 254–55 (223–24).

10. 3MP 270 (240–41).

11. William Hayes, "Max Ludwig Henning Delbrück, 4 September 1906–10 March 1981," *Biographical Memoirs of Fellows of the Royal Society* 28 (1982), 58–90, especially 65–66. See also Sloan chapter, this volume. In his chapter for this volume, Roll-Hansen points out that, in a letter to Bohr on April 4, 1935, Delbrück explains that the 3MP contained "no kind of complementarity argument." However, the fact that Delbrück feels the need to explain why this doesn't figure explicitly in the paper reflects their previous discussions of this issue. Remarks such as those concerning the "heuristic scheme" indicate Delbrück's awareness of (and active interest in) the issue.

12. Robert C. Olby, "The Molecular Revolution in Biology," in Robert C. Olby, G. N. Cantor, and J. R. R. Christie, eds., *Companion to the History of Modern Science* (New York: Routledge, 1996), 504.

13. John A. Fuerst, "The Role of Reductionism in the Development of Molecular Biology: Peripheral or Central?" *Social Studies of Science* 12 (1982): 261–262.

14. Ibid., 262.

15. Ibid., 266.

16. Ibid., 265, emphasis in original.

17. Max Delbrück, "A Physicist Looks at Biology," *Transactions of the Connecticut Academy of Arts and Sciences* 38 (December 1949): 173–190. Reprinted in *Phage and the Origins of Molecular Biology*, ed. John Cairns, Gunther S. Stent, and James Watson (Cold Spring Harbor, NY: Cold Spring Harbor Press, 1966), 22.

18. For discussion of contemporary perspectives on what it means to be a reductionist in biology, which distinguishes between various sorts of ontological, epistemological, and methodological positions, see Ingo Brigandt and Alan Love, "Reductionism in Biology" in *The Stanford Encyclopedia of Philosophy (Fall 2008 Edition)*, ed. E. N. Zalta, http://plato.stanford.edu /entries/reduction-biology/.

19. Fuerst, "Role of Reductionism," 242. See Kenneth F. Schaffner, "The Peripherality of Reductionism in the Development of Molecular Biology," *Journal of the History of Biology* 7 (1974): 111–39.

20. Fuerst, "Role of Reductionism," 241.

21. Ibid., 242.

22. Ibid.

23. See Alexander Rosenberg, *The Structure of Biological Science* (Cambridge: Cambridge University Press, 1985).

24. I have offered extensive textual support for these claims in "Influence of Niels Bohr." The argument in the present article is self-contained, but rather than simply repeat the same evidence here, I will substantiate these claims with some alternative sources.

25. Jacques Loeb, *The Mechanistic Conception of Life: Biological Essays* (Chicago: University of Chicago Press, 1912), 3. Loeb's remarks were first given as a lecture in 1910, published separately in 1911 and then in the book in 1912.

26. Horace Freeland Judson, *The Eighth Day of Creation: Makers of the Revolution in Biology* (New York: Simon and Schuster, 1979), 201.

27. Olby, "Molecular Revolution in Biology," 509.

28. Judson, *Eighth Day of Creation*, 613.

29. See McKaughan "Influence of Niels Bohr" for further development of this reading.

30. 3MP 255–56 (224–25), emphasis mine.

31. Niels Bohr, "Light and Life," *Nature* 131 (1933): 421–23 and 133 (1933): 457–59, reprinted in *Niels Bohr: Collected Works*, vol. 10, edited by D. Favrholdt, (Amsterdam: Elsevier, 1999), pp. 29-35. All references to the English text will be to this edition, which differs in important ways from the later reprinting in Niels Bohr, *Essays 1932–1957 on Atomic Physics and Human Knowledge*, vol. 2 of *The Philosophical Writings of Niels Bohr* (Woodbridge, CT: Ox Bow Press, 1987), 4–12. The originally published German version can be found in Niels Bohr, "Licht und Leben," *Die Naturwissenschaften* 21 (1933): 245–50. The published English and German versions were translated with some "formal alterations" from the Danish version published in the congress report (*Naturens Verden* 17, 49).

32. See, for example, McKaughan, "Influence of Niels Bohr," and Roll-Hansen chapter, this volume. See also Henry J. Folse, *The Philosophy of Niels Bohr: The Framework of Complementarity* (Amsterdam: Elsevier and North-Holland, 1985); Folse, "Complementarity and the Description of Nature in Biological Science," *Biology and Philosophy* 5 (1990): 221–24; Jan Faye, "The Bohr-Høffding Relationship Reconsidered," *Studies in History and Philosophy of Science* 19 (1988): 321–46; Faye, *Niels Bohr: His Heritage and Legacy: An Anti-realist View of Quantum Mechanics* (Dordrecht: Kluwer Academic, 1991); Faye, "Once More: Bohr-Høffding," *Danish Yearbook of Philosophy* 29 (1994): 106–13; David Favrholdt, *Niels Bohr's Philosophical Background* (Copenhagen: Munksgaard, 1992); Jan Faye and Henry J. Folse, eds., *Niels Bohr and Contemporary Philosophy* (Dordrecht: Kluwer Academic, 1994); Catherine Chevalley, "Niels Bohr's Words and the Atlantis of Kantianism" (1994), in Faye and Folse, *Niels Bohr*, 33–55; and Nils Roll-Hansen, "The Application of Complementarity to Biology: From Niels Bohr to Max Delbrück," *Historical Studies in the Physical and Biological Sciences* 30 (2000): 417–42. For more general discussion of vitalist and mechanist currents in nineteenth-century German biology, see Timothy Lenoir, *The Strategy of Life: Teleology and Mechanics in Nineteenth-Century German Biology* (Chicago: University of Chicago Press, 1982).

33. Although it can stand as a reading of Bohr's texts independently of any putative Kantian influences, the label "teleomechanical" is intended to call attention to the fact that Bohr does not envision a simple union of teleology and mechanism. Instead, they stand in something like the complex relationship that Kant envisions in the *Critique of Judgment* (1790).

34. Niels Bohr, "On the Notions of Causality and Complementarity" (1948) in Bohr, *Causality and Complementarity: Supplementary Papers: The Philosophical*

Writings of Niels Bohr, vol. 4, ed. Jan Faye and Henry J. Folse (Woodbridge, CT: Ox Bow Press, 1998), 147. This reading of Bohr is developed and defended in detail in McKaughan, "Influence of Niels Bohr."

35. Two descriptions are *complementary* if and only if they require mutually exclusive experimental arrangements but are jointly necessary for our understanding of the phenomena. See Don Howard, "Who Invented the Copenhagen Interpretation? A Study in Mythology," *Philosophy of Science*, special issue, 71 (2004): 669–82; Dugald Murdoch, *Niels Bohr's Philosophy of Physics* (Cambridge: Cambridge University Press, 1987); Mara Beller, "The Birth of Bohr's Complementarity: The Context and the Dialogues," *Studies in History and Philosophy of Science* 23 (1992): 147–80.

36. For more discussion of Jordan's views, see the Beyler and Sloan chapters, this volume.

37. Niels Bohr, "Seventh International Congress of Radiology" (1954) in *Causality and Complementarity*, 163.

38. Bohr, "Light and Life," 458b (*Collected Works*, p. 34).

39. See also McKaughan, "Influence of Niels Bohr"; Roll-Hansen chapter, this volume.

40. Niels Bohr, "Medical Research and Natural Philosophy" (1951), in *Causality and Complementarity*, 153, emphasis mine.

41. Bohr, "Light and Life," 458b (*Collected Works*, p. 34).

42. Ibid.

43. Niels Bohr, "Introductory Survey" (1929) in *Atomic Theory and the Description of Nature*, vol. 1 of *The Philosophical Writings of Niels Bohr* (Woodbridge, CT: Ox Bow Press, 1987), 22.

44. Bohr, "Light and Life," 458a (*Collected Works*, p. 34), emphasis mine.

45. "Light and Life Revisited" (1962) in Niels Bohr, *Essays 1958–1962 on Atomic Physics and Human Knowledge*, vol. 3 of *The Philosophical Writings of Niels Bohr* (Woodbridge, CT: Ox Bow Press 1987), 26.

46. Erwin Schrödinger, *What Is Life? The Physical Aspect of the Living Cell, With Mind and Matter & Autobiographical Sketches* (Cambridge: Cambridge University Press, 1992). For discussion of Schrödinger's views concerning the relationship between physics and biology, see McKaughan, "Influence of Niels Bohr." As Schrödinger's "Warum" lecture indicates, his interest in these issues reaches back very early, at least into the early 1930s. Erwin Schrödinger, "Warum sind die Atome so Klein?" *Forschungen und Fortschritte* 9 (1933): 125–26.

47. Niels Bohr, "Medical Research and Natural Philosophy" (1951), in *Causality and Complementarity*, 153, emphasis mine.

48. See, for example, these lectures by Bohr: "On the Notions of Causality and Complementarity" (1948), 147; "Medical Research and Natural Philosophy" (1951), 153; "Seventh International Congress of Radiology" (1954), 163; and "Physical Science and Man's Position" (1956), 176, all in *Causality and Complementarity*; "Unity of Knowledge" (1954), "Atoms and Human Knowledge" (1955), and

"Physical Science and the Problem of Life" (1957), all in *Essays 1932–1957*; "The Connection between the Sciences" (1960), 21; and "Light and Life Revisited" (1962), 26, both in *Essays 1958–1962*.

49. Bohr, "Physical Science and Man's Position," 176, emphasis mine.

50. Bohr, "Atoms and Human Knowledge," 92, emphasis mine.

51. Bohr, "Light and Life," 458b (*Collected Works*, p. 34).

52. Niels Bohr, "Discussion with Einstein on Epistemological Problems in Atomic Physics" in *Albert Einstein: Philosopher Scientist*, Library of Living Philosophers, vol. 7, ed. Paul Schilpp (Evanston, IL: Library of Living Philosophers, 1949), 256. Delbrück's copy is in Delbrück Papers, California Institute of Technology (DP 38:8).

53. Max Delbrück, "Problems of Modern Biology in Relation to Atomic Physics: A Series of Lectures," given at the Vanderbilt University School of Medicine, April and May 1944, 2–3, Delbrück Papers, California Institute of Technology, DP 35.1. Available at http://www.vanderbilt.edu/delbruck/documents/Problems_of_Modern_Biology-Full.pdf. Delbrück points to Helmholtz again in taking up the history of mechanism in his "Physicist Looks at Biology" (1949), reprinted in *Phage and the Origins of Molecular Biology*, ed. J. Cairns, G. Stent, and J. D. Watson (Cold Spring Harbor, NY: Cold Spring Harbor Press, 1966), 16–17.

54. Delbrück, "Problems of Modern Biology," 8–9.

55. Delbrück, "Physicist Looks at Biology," 17.

56. Delbrück, *Mind from Matter? An Essay on Evolutionary Epistemology*, ed. G. S. Stent et al. (Palo Alto, CA: Blackwell Scientific, 1986), 236.

57. Bohr, "Connection between the Sciences," 21.

58. Bohr, "Light and Life Revisited," 26.

59. Interview with Max Delbrück by Carolyn Harding, July 14–September 11, 1978, Oral History Project, California Institute of Technology Archives, Pasadena, CA, available at http://oralhistories.library.caltech.edu/16/01/OH_Delbruck_M.pdf, 95–96. Italics are in the text of the interview. Hereafter cited as Harding interview. All quotations from this used with permission of the Archives of the California Institute of Technology.

60. McKaughan, "Influence of Niels Bohr," 521.

61. Delbrück, "Physicist Looks at Biology," in *Phage and the Origins*, 20–21, emphasis mine. See also Max Delbrück, "A Physicist's Renewed Look at Biology: Twenty Years Later," *Science* 168 (1970): 1312–15.

62. Harding interview, 40–41.

63. McKaughan, "Influence of Niels Bohr."

64. Bohr to Jordan, November 30, 1934, in Niels Bohr, "Selected Correspondence," trans. by D. Favrholdt, in *Niels Bohr: Collected Works*, vol. 10, ed. D. Favrholdt (Amsterdam: Elsevier, 1999), 468. Hereafter cited as "Selected Correspondence." For more details regarding this document, see Sloan chapter, this volume.

65. "Selected Correspondence," p. 468.

66. Gunther Stent to Max Delbrück, May 20, 1968, in DP 20:21.

67. Max Delbrück to Niels Bohr, December 1, 1954, DP 3:29, as in "Selected Correspondence," p. 476, emphasis mine.

68. Fuerst, "Role of Reductionism," 262.

69. Ibid., 263.

70. Delbrück, "Physicist Looks at Biology," in *Phage and the Origins*, 22.

71. Gunther S. Stent, "Introduction: Waiting for the Paradox," in *Phage and the Origins*, 7.

72. See, for example, Lily E. Kay, "The Secret of Life: Niels Bohr's Influence on the Biology Program of Max Delbrück," *Rivista di Storia della Scienza* 2 (1985): 503; E. P. Fischer and C. Lipson, *Thinking about Science: Max Delbrück and the Origins of Molecular Biology* (New York: Norton, 1988), 204. Note that, although sensory physiology now becomes a more explicit focus of Delbrück's research, his interest in the topic was not new as developed in the Sloan chapter, this volume. His concern with photobiology goes back to his earliest biophysical investigations and pre-dates his interests in genetics.

73. McKaughan, "Influence of Niels Bohr."

74. Delbrück, "Physicist's Renewed Look at Biology," 1312b.

75. Ibid., 1312b.

76. Ibid., 1312a.

77. Max Delbrück, "Light and Life III," lecture given at the Centennial of the Carlsberg Laboratory, Copenhagen, September 27, 1976, in *Carlsberg Research Communications* 41 (1976): 300.

78. Ibid., 300.

79. Ibid., 299, emphasis mine.

80. Ibid., 300.

81. Ibid., 301.

82. Max Delbrück, "Atomic Physics in 1910 and Molecular Biology in 1957," November 1957, DP 35:2.

83. Ibid., emphasis mine.

84. Max Delbrück to Niels Bohr, 23 December 1957, DP 3:29. (Not in "Selected Correspondence.")

85. Max Delbrück, "Biophysics," in *Commemoration of the Fiftieth Anniversary of Niels Bohr's First Papers on Atomic Constitution, Held in Copenhagen on 8–15 July 1963* (Copenhagen: Institute for Theoretical Physics, University of Copenhagen, 1963), 64, emphasis mine. Delbrück also perceived a difference between his own views and the positions of Niels and Aage. See, for example, Delbrück to Bohr, December 1, 1954, in Bohr, "Selected Correspondence," 475–78.

86. Delbrück, "Physics in 1910," DP 35:2.

87. Delbrück, "Sevilla Lecture" (1975), DP 35:14. Delbrück wrote this lecture in English and planned to deliver the talk. However, this trip was cancelled because

of a detached-retina operation. The paper was translated into Spanish by Arturo Eslava and Mariebel Alvarez and read by Corde-Oledo. The underlining is in the original; italics are mine.

88. Max Delbrück, "Virology Revisited" (the opening lecture at an international symposium, "Molecular Basis of Host Virus Interaction," Banaras Hindu University, October 17, 1976), later published in *Molecular Basis of Host-Virus Interaction*, IUB Symposium no. 78, ed. M. Chakravorty (Princeton, NJ: Science Press, 1979), 1–11; DP 37:9.

89. Ibid., 1.

90. McKaughan, "Influence of Niels Bohr."

91. Delbrück, *Mind from Matter?*, 236.

92. Delbrück, "Physics in 1910," DP 35:2.

93. Judson, *Eighth Day of Creation*, 615.

94. Ibid.

PART III

The Three-Man Paper

Translator's Preface

Brandon Fogel

The 3MP consists of four parts, one written by each of the authors individually and a fourth attributed to all three. Each part has a specific purpose; in the first, Timoféeff-Ressovsky presents the state of current mutation research; in the second, Zimmer elaborates the target theory and its experimental basis; in the third, Delbrück uses then-current theoretical physics to construct a model of mutation; and in the fourth, general reflections are offered on the consequences of that model for the conception of the gene and on how the model might be subjected to further test. Here I will present a brief summary of each of the 3MP's four parts, including some technical notes on the translation.

I would like to thank Phillip Sloan for his extensive assistance in the preparation of this translation.

Foreword

A brief, unattributed foreword emphasizes that the theoretical components of the paper are drawn directly from mutation research. The authors argue that, since the research concerns the genes directly, not their indirect effects by way of developmental processes, nettlesome questions about how genes actually take part in development can be avoided. They note that the chief value of their presentation lies in the application of the concepts of physics to those of mutation genetics.

Part 1, by Timoféeff-Ressovsky: Experimental Research

Timoféeff-Ressovsky's part is the largest of the four, composing roughly 50 percent of the paper. It summarizes the current state of experimental research on genes and mutation, much of which was performed by Timoféeff-Ressovsky himself, either on his own or in collaboration with Zimmer, and some of which was reported for the first time in the 3MP. The presentation is geared toward clarifying and justifying the use of radiation in mutation experiments and identifying a few key conclusions to be used in the later sections.

The brief introduction to Timoféeff-Ressovsky's part is worth a careful look. He notes that the conception of the gene has evolved from the formal to the concrete; genes are now understood not merely as abstract differentiating characteristics, but as parts of a "spatially constant and determinately ordered material system" (233 [191]). However, in an important warning that is reiterated in part 4, he argues that we cannot infer much about the nature of genes merely from their effects on developmental processes (the phenomena of gene expression); genes are only part of a larger developmental system, of which cells are the more natural fundamental units. Instead, a more direct manipulation of the genes is required to learn about their structure, and irradiation with X-rays provides just the needed tool.

Timoféeff-Ressovsky takes pains to justify the radiation methodology. He argues that experimentally triggered mutations are not qualitatively different from those that arise spontaneously, and irradiation is preferable to other triggers because it can be dosed carefully and because its mode of action is well understood physically. Most importantly, X-rays and gamma rays easily penetrate to the cell nucleus, making their action on the genetic material much more direct than other triggers, such as chemical exposure.

The data presented by Timoféeff-Ressovsky pertain almost exclusively to irradiation experiments with the fruit fly *Drosophila melanogaster*, and most of it is aimed at establishing that the mutation-triggering effect of irradiation is proportional to the dose, or total energy delivered, and is largely independent of the method of administration. Throughout the wavelength range of X-rays and gamma rays, the effect does not depend on the specific wavelength used, on the intensity of the radiation, or on whether it is administered continuously or in concentrated bursts. There is a mild temperature dependence, which is explained by the fact that all biological processes in the cold-blooded fruit fly are affected by temperature. In fact, this temperature dependence, measured in terms of a quantity referred to as the van't Hoff factor, becomes a critical tool in part 3 for Delbrück's comparison of his model to experiment. Most importantly, the proportionality of the mutation rate to the dose suggests that radiation does not act directly on genes, but rather indirectly through ionization of the surrounding material, a fact that features prominently in the rest of the paper.[1]

Finally, Timoféeff-Ressovsky examines mutation rates for specific mutations, especially those having to do with eye color in *Drosophila*. These mutations have well-defined rates, and most are reversible. From the fact that the rates of mutations differ, he concludes that the rate of any given mutation is determined by the physical structure of the alleles involved.

For the reader who does not wish to wade through the entirety of the lengthy and dense part 1, Timoféeff-Ressovsky provides a helpful summary of the main points in section III.

Part 2, by Zimmer: The Target Theory

Zimmer's relatively short part is intended to elucidate the target theory,[2] a set of experimental methods and theoretical techniques for deducing facts about the physical characteristics of genes from irradiation experiments, discussed extensively in the chapters by Summers and Beyler in this volume. The action of radiation on different forms of matter was reasonably well understood at the time; the goal of the target theory was to use that understanding to learn about the physical nature of genetic material.

Zimmer identifies three ways that bombardment with X-rays can be understood to achieve a "target-strike" on a gene: (1) a quantum of radiation might hit the gene directly, (2) the gene might be hit by a secondary electron that has been dislodged from a neighboring atom by a quantum of radiation, or (3) a pair of ions, created by a careening secondary electron, might interact with the gene. Each of the three possible kinds of target-strike leads to different empirical predictions for dependence on the various characteristics of the incoming radiation. He considers each in turn.

If mutation occurred through direct hit by radiation, then the mutation rate should drop as the wavelength decreases, since the number of incoming quanta is inversely proportional to the wavelength when the total energy is held constant; because the mutation rate is independent of the wavelength, this is ruled out. Direct interaction with a secondary electron should also lead to a dependence on wavelength, although Zimmer's argument for this is somewhat elliptical and obscured by less-than-transparent calculations. The connection can be made intuitively clear by recognizing that high-energy secondary electrons are generally produced only through direct hit by a high-energy photon, not through impact by other electrons; thus, there should be a close correlation between the number of secondary electrons and the number of incoming quanta, and this is ruled out for the same reason as the first case. Somewhat oddly, in the course of this disconfirmation of the secondary-electron hypothesis, Zimmer slips in the only size estimate to be found in the 3MP; the "target zone" is given a radius of approximately 1 micron.

The only viable option is thus the third, interaction with ions produced by secondary electrons, although Zimmer leaves for Delbrück the task of explaining why this should produce target-strikes in proportion to the total incoming energy.

Part 3, by Delbrück: Physical Model of Mutation

In the third part, conceptually the most interesting of the four, Delbrück applies theoretical physics to the experimental matters discussed in the preceding parts. He concludes that a gene is a large molecule and that a mutation consists of a stable rearrangement within the molecule, prompted either by random thermal fluctuation or an external influx of energy.

Delbrück's part consists of 4 parts: an introduction, a presentation of the model of mutation, a comparison of the model to experiment, and concluding remarks. The introduction begins with a justification of the use of physical concepts in genetics, and this is conceptually the most difficult portion of the 3MP. First, he characterizes the opposite point of view, that genetics is "autonomous" from physics. Chemistry and physics are unified because they share a common notion of measurement, in that all measurements are reducible to intervals of space and time. But measurements in genetics concern character traits, which are definable only in the context of individual organisms and thus cannot be reduced to the fundamental concepts of space and time. His counterargument is that experiment has shown that genes are spatially localized and can be tracked and studied as physical objects, even if the mechanics of their expression is unknown. These experiments indicate that genes are about the size of large molecules, a good reason for equating the two. But an even more important reason for doing so is the peculiar stability of genes, the fact that they maintain their identity over lengthy periods of time, a characteristic feature of molecules made understandable through the new quantum mechanics.

One might think that Delbrück would then continue to refer to genes as kinds of molecules, but he instead chooses a more general term, "Atomverband," translated here as "assemblage of atoms."[3] The reason he offers for the more general choice might look somewhat odd to later readers, accustomed as we are to concrete, visual models of molecules; Delbrück wishes to distinguish the notion of a molecule as a relatively rigid structure from the more abstract, somewhat positivist notion common in chemistry at the time, of a substance that produces certain observable properties when collected in large enough quantity. The latter is clearly useless in genetics, where genes produce observable properties in only one or two copies, not the moles of copies found in chemistry (a point expressed with greater clarity in Schrödinger's *What Is Life?*). The choice of "assemblage of atoms" over "molecule" also has the distinct advantage of allowing the gene to be thought of as merely a part of an even larger structure or assemblage, a fact emphasized in part 4.

In the second section, Delbrück examines the different ways that an assemblage of atoms might undergo change, and he focuses on rearrangements of the constituent atoms into different stable configurations, as these are the only kinds of change that affect the chemical character of the assemblage (which carries the implicit assumption that genetic information is stored chemically). He stresses that these rearrangements can only occur in jumps (*sprungweise*), once enough energy has been supplied to overcome the potential barrier between two stable configurations. In making these arguments, he tacitly draws on the quantum-theoretical work of Walter Heitler, Fritz London, and Lise Meitner on covalent bonds. What is curious is that he does not mention or cite any of that work; the omission is especially glaring given that Heitler was his doctoral adviser, and that he had been working under Meitner at the Kaiser Wilhelm Institute for three years by the time the 3MP was published.[4] Since this discreteness is a cardinal feature of the quantum mechanics he uses to explain the peculiar stability of genes, one might have thought he would take greater care in explaining it; however, even the term "quantum theory" appears only once in the whole part.

What follows is a discussion of transition probabilities, which are governed by activation energy, or the energy needed to surmount a potential barrier between two stable configurations. Delbrück shows that small changes in the activation energy lead to very large changes in the speeds of the associated chemical reactions. If mutations are chemical reactions that take place spontaneously on the order of between 1 and about 30,000 years, then they should have an activation energy of about 1.5 eV. Since ionizations triggered by secondary electrons, themselves dislodged by radiation, contain about 30 eV, there is plenty of energy in each ionization to initiate a mutation. The key point is that the energy per ionization is largely independent of the energy of the incident electron, which explains the independence of mutation rate on radiation wavelength (as well as the supposed linearity of the dose dependence); the secondary electron absorbs all of the energy of the incident radiation and then disperses it in 30 eV increments.

In comparing his model of mutation to experiment, Delbrück uses a quantity he refers to as the van't Hoff factor. The purpose is to use the weak temperature dependence of chemical reaction speed to construct a more convenient measure of activation energies. The van't Hoff factor that the 3MP uses is the ratio of reaction speed (i.e., mutation rate) at room temperature to that at 10°C above room temperature. The result is a quantity that still depends on the activation energy but varies less dramatically with it, thus reducing the precision required in measurement.

Delbrück finds that his model of mutation, as a transition between stable configurations of an assemblage of atoms, compares very well with the experiments discussed in the first two parts of the 3MP. Namely, mutations should have van't Hoff factors (i.e., activation energies) noticeably larger than those of developmental processes, spontaneously occurring mutations should be producible artificially through irradiation, and the rate of triggered mutations should be proportional to the total energy, or dose.

Part 4, the Three-Man Section: Theory of Gene Mutation and Structure

The final part of the 3MP, attributed to all three authors, contains general reflections on the preceding parts. The authors attempt to draw out the consequences of Delbrück's model (and the experimental data supporting it) for the theory of the gene itself, although they do not actually go any further than Delbrück does in part 3. They also reprise the warning given in part 1 concerning the lack of warrant in their conclusions for a strong genic determinism in developmental processes.

After emphasizing that mutation is an elementary process (even if the second or third step in some physical process, such as ionization following irradiation), the authors argue that, if mutation is a rearrangement of an assemblage of atoms, then a gene *is* that assemblage, that is, one in which a rearrangement is possible. This is, of course, a clear implication of Delbrück's reasoning in part 3, even if he does not stress it there. They add that the assemblage must be "largely autonomous" (267 [237]) both functionally and in relation to other genes, but they wish to leave unresolved the question of whether genes are parts of a larger functional structure, or whether they are like Morgan's beads on a string. These ideas are somewhat in tension with one another—that genes might be autonomous but also part of something larger—but the authors do not point this out, much less explore it further.

In the concluding section of part 4, the authors review possible future experiments that could test their model and lead to refinements. These include use of monochromatic ultraviolet light (where the mutation-triggering mechanism appears to be different), more careful investigation of specific mutation pathways (especially those with particularly high and low spontaneous mutation rates), and modeling mutations with specially tailored photochemical reactions.

The last three paragraphs are worth a careful look; they reiterate and expound the warning made in part 1 against situating genes as master controllers of developmental processes. The authors begin by describing the point of view they wish to critique, that some biologists wish to dissolve the cell into the "ultimate units of life," that is, the genes (270 [240]); the rhetorical construction is such that the casual reader might mistake this opposing viewpoint for the authors' own. But the stance of the 3MP is clear: "Our ideas about the gene challenge this picture" (ibid.). The authors agree that genes are "physical-chemical units" but insist that such genes cannot be responsible for generating all morphogenic substances used in developmental processes. Instead, they offer a weaker formulation: the genome collectively forms merely a foundation for only those developmental processes conditioned by heredity. As discussed in the introduction to this volume, any attribution of genic determinism to the 3MP is simply not supported by the text.

The paper closes as it opens, with an empiricist plea grounded in the authors' physics-centric methodology. They recommend that researchers in fields associated with genetics use only "conceptions of the gene that are constructed on the relevant and . . . singularly decisive data of mutation research" (271 [241]). The success of this proposal in shifting the practice of modern genetic research is the enduring legacy of the 3MP.

Notes

1. Unfortunately, this empirical fact, which is the foundation for much of the theoretical work in the 3MP, would not stand the test of time. The mutation effect of radiation on gametes depends on the age of the cells, and for gametes irradiated at a given age, the dose-response effect is nonlinear. However, when the effect is averaged over gametes of all ages (up to the five days it takes *Drosophila* sperm cells to mature), the dependence turns out to be roughly linear, and this is what Timoféeff-Ressovsky observed. See M. Perutz, "Erwin Schrödinger's *What Is Life?* and Molecular Biology," in *Schrödinger: Centenary Celebration of a Polymath*, edited by C. W. Kilmister (Cambridge: University Press, 1987), 234–51.

2. The term "target theory" is the conventional English translation of *Treffertheorie*, despite the fact that the German *Treffer* refers to the striking of the target, not the target itself. Consequently, *Treffer* is sometimes translated as "hit" and *Treffertheorie* as "hit theory." I have opted to maintain "target theory" and to translate *Treffer* as "target-strike," which has the advantage of retaining explicit reference to the notion of a target. See Beyler chapter, this volume, note 8, for a historical discussion of these terms.

3. Other translations considered were "atomic structure" or "structure of atoms"; however, both of these terms carry some ambiguity between the structure inside the atoms and the structure of which they are only parts.

4. See the introduction to this volume for more biographical detail.

On the Nature of Gene Mutation and Gene Structure

N. V. Timoféeff-Ressovsky, K. G. Zimmer, and M. Delbrück

Presented by A. Kühn at the meeting of April 12, 1935

Contents

Foreword

In the following paper, we attempt to construct a general picture of the nature of the gene and of mutation, on the basis of experimental investigations of the mutation processes in *Drosophila* and a physical analysis of the results of those experiments. Compared to previous hypotheses about the nature of the gene and of gene mutation, we believe we have taken a step forward, in that our ideas are constructed only from the experimental results of mutation research, and therefore from an area of research that concerns events directly involving the genes themselves. Most importantly, this has enabled us to avoid methodologically questionable conclusions from adjoining fields, such as phenogenetics* and developmental physiology, concerning the problem of the structure and mutation of genes.

Thus, we hope to arrive at a theory of the mutation process and gene structure that will be experimentally well-founded and that will have experimentally verifiable implications. Of course, we are far from regarding our ideas as conclusive; rather, we see their value lying in the extending of earlier approaches through the use of the concepts of physics.

This paper represents a collaboration between genetics and physics. It originated from lectures and discussions in a small, private circle of representatives from genetics, biochemistry, physical chemistry, and physics. We thank Dr. K. Wohl for special instruction on questions of the kinetics of chemical reactions. Some of the experiments were accomplished with aid from the Emergency Association for German Science within the

---

* The term "phenogenetics" was used to refer to the study of heredity via phenotypes, the observable characteristics of organisms.—Trans.

framework of the Working Group on Hereditary Damage through Radiation Effects.

Part 1: Some Facts from Mutation Research

N. V. TIMOFÉEFF-RESSOVSKY[1]

I. Introduction

The original version of the gene concept in the experimental theory of inheritance is a purely formal one. In keeping with the sense of G. Mendel's classic experiments and the vast body of experimental work done in genetics since their rediscovery, the gene can be defined as an "assorting unit," or as the "unknown" [etwas] that forms the basis for those characteristics that assort according to Mendel's laws. Accordingly, we can (1) reveal the existence of genes from crossing experiments, and (2) concern ourselves with only those genes that exist in at least two different forms (alleles) in the data we have at our disposal, or, in other words, for which mutations are known. Consequently, the gene is conceptually bound on one hand to the segregating traits, and on the other hand to mutation.

The first attempts to construct a more concrete notion of what genes and mutations are were tied to examinations of normal and mutant hereditary characteristics. The well-known "Presence-Absence"–hypothesis of W. Bateson was constructed in this way; here, the dominant (usually "normal") allele constitutes a physical "unknown" [etwas] that is lost in recessive mutations. Bateson thereby assumed the point of view (which also partly corresponds to the facts) that one can often portray the characteristics of recessive mutations as the absence of certain properties of normal alleles. Many modern theories of the gene take a similar path, inferring the nature of the gene and gene mutation from the phenomena of gene expression; extreme "morphological" (Serebrovsky 1929) as well as "physiological" versions (Goldschmidt 1928) of such theories can be distinguished.

The theories of the localization and linear arrangement of genes in chromosomes, which have developed rapidly since the prewar period and today are more or less conclusively proved, have substantially advanced the theory of the gene. We now know that the genome is a spatially constant

---

1. Genetics Division of the Kaiser Wilhelm Institute for Brain Research, Berlin-Buch.

and determinately ordered material system in which the constituent elementary parts, the genes, occupy completely determinate locations.

With the accumulation of different kinds of genetic data, it becomes ever more evident that only a small and uncertain amount can be learned about the nature of the gene and gene mutation through examination of the phenomena of gene expression. For (1) we still know almost nothing at all about the effects of genes; (2) that which we do know applies not to the isolated effects of individual genes, but rather to the total developmental system, itself modified through individual gene mutations, where we can discern as the ultimate entities not the genes but rather whole cells (and their modes of operation); and (3) the "projection of the phenomena of gene expression into the gene" is logically prohibited, because, after all, we know that the structure and development of the individual (where the effects of the gene are revealed) are fundamentally different from the structure and variation of the individual genes. On the other hand, our knowledge of the mutation process as such has grown rapidly in recent times, so that today we can quantitatively detect and analyze it in at least a few genetically well-investigated test objects. And, as was pointed out years ago by the Morgan school, and above all by H. J. Muller (1920, 1922, 1923, 1927[b], 1929a), we can in this way best hope to learn the full and exact details of the nature of the gene and gene mutation.

II. Analysis of the Mutation Process

In this section, the most important results from mutation research will be presented—in particular, the results of mutation-triggering experiments—which we can use to construct a theory of gene mutation and gene structure. *Drosophila* research will be considered here almost exclusively, since (1) *Drosophila* is our test object, and (2) *Drosophila* is the most convenient object for exact and quantitative mutation research.

1. SPONTANEOUS MUTATION. In the extensive body of fruit-fly data explored by geneticists over the course of 25 years of *Drosophila* research, many mutations have been observed to arise spontaneously. These spontaneous mutations of *Drosophila* already exhibit an array of general properties of the mutation process.

Above all, it appears that fundamentally different types of hereditary variations can occur. Without going into this question any further, the following classification of hereditary variations can be given:

A. Plasmatical hereditary changes.
   1. Changes to some organelles of the cytoplasm (e.g., chloroplast inheritance, Correns [1909]).*
   2. Adaptation of the cytoplasm to a new genotype (e.g., plasmatic accommodation in some types of hybrids, Michaelis [1933]).
   3. Enduring modifications (Jollos [1913]).
B. Genotypic hereditary changes (mutations).
   1. Gene mutations (mutations in the strict sense).
   2. Chromosomal mutations.
      a. Breaks and fragmentation.
      b. Removal of chromosomal sections (deficiency and deletion).
      c. Inversions.
      d. Duplications and simple translocations.
      e. Mutual translocations.
      f. Chromosomal fusions.
   3. Karyotype mutations.
      a. Heteroploids.
      b. Polyploids.
C. Combinations.

The most important types of genotypic hereditary changes are shown schematically in figure 1.

Henceforth, only gene mutations shall interest us. We have already obtained some information about these from the spontaneous mutation processes. It appears that gene mutation always develops heterozygotically; i.e., only one of the two alleles present in diploid organisms changes when a mutation occurs. Furthermore, it appears that dominant and recessive characteristics, including the most varied morphological traits and physiological characteristics, can each develop both very strong and totally weak deviations from the norm or initial state through mutations; as for the biological value (or viability) of the mutant forms, the whole spectrum exists, from (admittedly relatively infrequent) mutations with increased viability, to those with greatly reduced viability, and up to the (particularly frequent) lethal factors, which are fatal to homozygotes. Due to repeated

* In the original, Correns, Michaelis, and Jollos are cited, but no papers by any of these authors are included in the list of references. However, a nearly identical list, with full citations included, appears in Timoféeff-Ressovsky 1934c. We have included those citations above.— Trans.

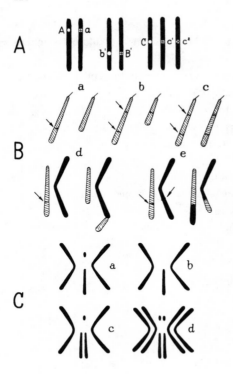

FIGURE I. Different types of mutations. A, gene mutation: the dominant, "normal" (or initial) allele A mutates to a; the recessive allele b' mutates to B'; the dominant allele C produces c¹ or c² (multiple alleles) through mutation. B, chromosomal mutations: a, removal of a chromosomal section; b, deletion; c, inversion; d, simple translocation; e, mutual translocation. C, genomic mutation: a, normal, haploid chromosomes; b–c, heteroploidy (b, haplosomy; c, trisomy); d, polyploidy.

mutation, many genes consist of groups of multiple alleles; however, the exact same mutations can occur repeatedly, and the number of elements in a group of multiple alleles is limited. Different genes display different mutation rates. Finally, back mutations are also occasionally observed, from which it must be concluded that at least part of the mutation process is reversible.

Consequently, the spontaneous mutation process of *Drosophila* has already given us a full and multifaceted qualitative picture of mutation. But one cannot infer analytically a great deal from this, for two reasons: (1) the spontaneous mutation rates are so low that an exact, quantitative analysis is for the most part infeasible in practice, and (2) we do not know the triggering causes of spontaneous mutations, and knowledge of not only the reaction, but also of the triggering stimulus is required for a fruitful analysis. Consequently, the results of experimentally triggered mutations should be much more promising. Special preference will thereby be given to radiation genetics: first, because X-ray and radium irradiation exerts the strongest, and hence most reliably reproducible, effect on the mutation rate; second, because the irradiation is light and can be exactly dosed; and third (and especially important), because short-wave radiation penetrates reliably to the genes in a well-definable form and permits an experimental, biophysical analysis of its direct mode of operation, in contrast to many other exactly dosable stimuli (e.g., temperature or chemicals), which penetrate to the genes indirectly and in a form we generally know very little about.

2. QUALITATIVE PICTURE OF THE RADIATION-INDUCED MUTATION PROCESS. Since the first successful radiation-genetic *Drosophila* experiments by H. J. Muller (1927b, 1928b, 1928c), a very extensive body of radiation-genetic data on *Drosophila* has been amassed, so that today we have at our command a full and multifaceted overview of the radiation-induced mutation process. Apart from the general ease of working with the objects themselves, the work was also facilitated by the fact that we have in the case of *Drosophila melanogaster* a group of hybridization methods particularly convenient for the detection of mutations. The two most common, the Mullerian "ClB" hybridization method and the "attached X" method of L. [Lillian Vaughn] Morgan [1922], are shown schematically in figure 2.

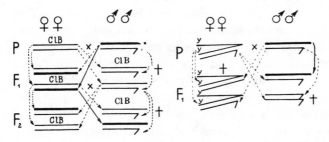

FIGURE 2. Diagrams of the "ClB" (left) and the "attached X" (right) crossing methods. The "ClB"-♀♀ have in one of their X chromosomes a crossing-over inhibitor (C), a recessive lethal factor (l), and the dominant gene Bar-eye (B); the half of their male offspring that contain the ClB chromosome do not survive; in F_2 of the crossing of irradiated P ♂♂ with ClB ♀♀, the surviving half of the ♂♂ get the irradiated X chromosome; the F_2 cultures, which have an irradiated X chromosome with a newly generated lethal factor, produce no males at all. The "attached-X" ♀♀ contains two X chromosomes attached to each other (with the yellow allele) and an extra Y chromosome; the F_1 ♂♂ get their X chromosome from the father; if irradiated ♂♂ are crossed with "attached-X" ♀♀, then all observable sex-linked mutations on F_1 ♂♂, the recessive ones included, must appear. The irradiated X chromosomes are indicated by thick bars.

The general, qualitative character of the radiation-induced mutation process can be briefly portrayed as follows.

All kinds of radiation, from ultraviolet rays to the hardest gamma rays, are capable of greatly increasing the mutation rate. In the following, however, we will limit ourselves to those results achieved through X-ray or gamma-ray irradiation, because the radiation-genetic experiments with ultraviolet rays provide somewhat meager and unreliable results, due to certain technical difficulties (strong absorption in tissue), and above all because it can already be maintained a priori that the details of the physical effect of ultraviolet rays on genes must be different from those of short-wave radiation.

The genetic effect of radiation is a very general one, insofar as mutations are in fact triggered at high rates through irradiation in all subjects and in all tissues in which the production of mutations can be ascertained see the general presentation in Muller 1930a, 1934a; Stubbe 1934; Timoféeff-Ressovsky 1931b, 1934c). All types of mutations characterized above as known through the spontaneous mutation process and portrayed in figure 1 are observed at increased rates in radiation-genetic experiments. Also, all types of gene mutations are produced through irradiation. There is thus, qualitatively, a parallelism between the spontaneous and radiation-induced mutation processes. This parallelism goes even further, in that in both cases the same types of mutations occur most infrequently ("macro" ones [große], those strongly deviating from the norm, and especially the dominant mutations; mutations with increased viability) and most frequently ("micro" [kleine] mutations; mutations with reduced viability and with lethal factors). The character of the spontaneous mutation process typical for each subject generally persists after irradiation. We therefore have no reason to presume the existence of any principal difference between "spontaneous mutations" and "radiation mutations."

It has been possible to establish that back mutations can also be generated through irradiation. Extensive experimental results related to this are presented in table 1; they show that back mutations of recessive, mutant alleles to normal, dominant initial alleles can be induced, to some extent repeatedly, on different genes of the X and the III chromosome of *Drosophila melanogaster* through X-ray irradiation. In figure 3, pairs of alleles are shown in which forward and back mutations were produced directly from one to the other in my experiments, i.e., cases in which steps in the mutation process could be triggered in two opposite directions through

TABLE I. **Back mutations of recessive, mutant alleles of *Drosophila melanogaster* induced through X-ray irradiation (4,800 r and 3,900 r, respectively). The P ♂ were irradiated. (From Timoféeff-Ressovsky 1932b; Johnston and Winchester 1934.)**

Alleles and their locations		Timoféeff-Ressovsky		Johnston and Winchester	
		Number of gametes	Number of back mutations	Number of gametes	Number of back mutations
X chromosome	0 y	21,897	0	69,923	1
	0 + sc	17,676	3	101,042	5
	1.5 w	29,233	2	—	—
	1.5 wᵉ	23,472	2	—	—
	1.5 wᵃ	—	—	69,302	0
	5.5 ec	17,676	0	57,323	0
	14 cv	16,460	2	—	—
	20 ct	12,914	0	57,323	1
	33 v	29,384	2	61,119	1
	36 m	—	—	39,923	2
	44 g	12,914	0	57,323	4
	56 f	34,811	8	130,421	15
	62 car	—	—	69,302	1
III chromosome	0 ru	27,155	0	—	—
	26 h	27,155	1	—	—
	42 th	5,681	0	—	—
	44 st	27,155	1	—	—
	48 pᵖ	21,474	4	—	—
	50 cu	5,681	0	—	—
	59 ss	21,474	0	—	—
	62 sr	5,681	0	—	—
	71 eˢ	27,155	1	—	—
	101 ca	5,681	0	—	—
			26	713,001	36
Total		390,709	0.0066%		0.0051%
Control		152,352	0	700,000	0

$$W \rightleftharpoons w^e \qquad w^e \rightleftharpoons W$$
$$\overset{\text{I}}{} \qquad \overset{\text{II}}{}$$
$$F \rightleftharpoons f \qquad f \rightleftharpoons F$$
$$\overset{\text{III}}{} \qquad \overset{\text{IV}}{}$$
$$p \rightleftharpoons P$$
$$\overset{\text{V}}{}$$

FIGURE 3. Allele pairs of *Drosophila melanogaster*, on which mutations in opposite directions were produced directly from one to the other through X-ray irradiation. I, eosin was produced from normal, and through further irradiation, normal was again produced from this eosin. II, normal was produced from spontaneously originating eosin, and eosin was again produced from this normal. III, forked was produced from normal, and from this forked, normal was produced. IV, normal was produced from a spontaneously originating forked, and forked was again produced through further irradiation from this normal. V, pink was produced from normal, as was the back mutation from this pink to normal, and from this normal, pink was again produced.

the same radiation stimulus. Within groups of multiple alleles, different steps in the mutation process can be triggered in different directions
through irradiation, as presented in figure 4 for the white-eye-allele series
of *Drosophila melanogaster*, which is certainly the most well investigated
so far in this respect. From the experimental results mentioned in this
section, it follows that the mutation-triggering radiation effect, and the
mutation process itself, cannot be purely destructive, because it involves
reversible changes.

3. QUANTITATIVE ANALYSIS OF THE MUTATION PROCESS.   In the previous
section, a general, qualitative picture of the radiation-induced mutation
process was given; we now want to turn to experiments that have the aim
of analyzing the genetic effects of radiation.

A necessary prerequisite for performing such experiments is the possibility of having access to reliably detectable, well-defined mutation rates.
We have special hybridization methods for *Drosophila melanogaster*, as
presented in figure 2, that allow us to control easily the mutability of an
entirely specific part of the genome—the X chromosome. Because, however, as we have noted, mutations are phenotypically distributed along a
very broad spectrum, from good alternative traits to barely perceptible

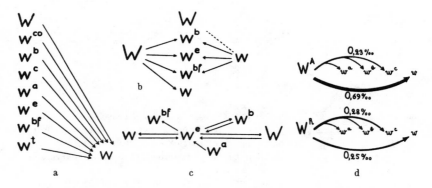

FIGURE 4. X-ray–induced mutability of the multiple W[hite-eye]–allele series of *Drosophila
melanogaster*. a, the white-eye allele could be produced from all other heretofore-examined
alleles. b, the normal allele (W) has produced mutations to blood (w^b), eosin (w^e), and buff
(w^{bf}), which are also produced as back mutations from white (w); normal also mutates directly
to white, but the theoretically conceivable mutation w→W is never observed. c, different
mutations to and from eosin. d, differences in the mutability of two different "normal" alleles
(W^A and W^R) of the white series, to white (w) and to intermediate alleles (w^a, w^b, w^c).

characteristics, we must work with a segment of the total mutation process that is easily and reliably registerable. We can take the lethal and "large" morphological mutations of the X chromosomes to be such. Admittedly, one might suppose a priori that the mutability of the X chromosome might falsely reflect the average total mutation rate of *Drosophila*; specific experiments by R. Berg (1934) and by Schapiro and Serebrovskaja (1934) have shown, as have our own unpublished experimental results, that the mutation rates of different chromosomes of *Drosophila melanogaster* are proportional to the lengths of their genetically active parts; or, put another way, that sections of the total genome of different length have the same average rates of mutation. Consequently, we can use the above-characterized mutation rate of the X chromosome, which is objectively and exactly ascertainable with the help of the "ClB" hybridization method, as an exact characterization of the mutation of *Drosophila melanogaster*.

   a. Direct Effects of Radiation and Independence from Biological Material. It is possible that irradiation does not affect genes directly, but rather influences the mutation rate in some indirect ways. This hypothesis

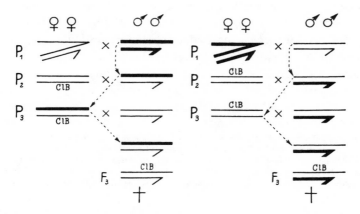

FIGURE 5. Left, scheme for crossings to ascertain the mutability of X chromosomes that were free of mutations immediately after irradiation; P_1 ♂♂ were irradiated and crossed with "attached-X" ♀♀; mutation-free F_1 ♂♂ were crossed with ClB ♀♀; and in F_3 the mutation rate of previously irradiated X chromosomes (the P ♂♂) is established. Right, scheme for crossings to ascertain the mutability of nonirradiated X chromosomes in irradiated eggs; P_1-"attached X" ♀♀ were irradiated and crossed with nonirradiated ♂♂; the F_1 ♂♂ (P_2 ♂♂) contain nonirradiated X chromosomes that were crossed into irradiated eggs; these ♂♂ were crossed with ClB ♀♀ (P_2), and in F_3 the mutation rate of nonirradiated X chromosomes situated in irradiated egg plasma was established. The irradiated chromosomes are indicated by thick bars.

can be tested experimentally, according to the hybridization scheme in figure 5, by (1) ascertaining the mutability of chromosomes that are free of mutations immediately after irradiation, and (2) testing for mutations of nonirradiated chromosomes in eggs that have been irradiated.

Hybridizations of the first type were obtained by Muller (1930a), Timoféeff-Ressovsky (1930c, 1931a), and Grüneberg (1931); hybridizations of the second type were obtained by Timoféeff-Ressovsky (1931a). All of these experiments have yielded negative results: neither an aftereffect of the irradiation nor an effect on nonirradiated chromosomes through irradiated plasma could be established. The experimental results of Timoféeff-Ressovsky are summarized in table 2.

It has been shown (Timoféeff-Ressovsky 1931a) that the origin of mutation in *Drosophila*, much as in plants (Stadler 1930), is not bound to the dividing of chromosomes and/or genes, but rather can also occur in the completely "inactive" state of the chromosome.

Mutations can be generated in both sexes through X-ray and radium exposure, in a variety of stages of development and in a variety of tissues. The question of whether equal [doses of] irradiation trigger the same rate of mutation in all cases unfortunately cannot yet be answered precisely and unambiguously, due to a series of technical difficulties. But in the opinion of most authors who have addressed these questions (Moore 1934; Neuhaus 1934a; Schapiro and Neuhaus 1933; Sidorov 1931; Timoféeff-Ressovsky 1930d, 1931a), the differences sometimes observed in the rates of mutation (in mature and immature sperm, male or female) are

TABLE 2. **Rates of sex-linked mutations in *Drosophila melanogaster*: (1) in nonirradiated control cultures; (2) in sperm containing X chromosomes irradiated a generation prior, but which were free of mutations immediately after irradiation (P_2 on the left of figure 5); (3) in sperm containing nonirradiated X chromosomes, but which were crossed into irradiated egg cells (P_2 on the right of figure 5); (4) in sperm containing X chromosomes that were irradiated directly with 3,000 r of X-rays. The P ♂ were crossed with CIB ♀.**

| | Number of | | Percentage of sex-linked |
Type of treatment	Cultures	Mutations	mutations
1. Nonirradiated control	3,708	7	0.19 ± 9.07
2. X chromosomes remaining free of mutations after irradiation	1,431	3	0.21 ± 0.12
3. Nonirradiated X chromosomes in irradiated cells	1,212	2	0.16 ± 0.12
4. Directly irradiated X chromosomes (3,000-r dose)	2,239	198	8.84 ± 0.59

caused by the effect of germinal selection (or other circumstances influencing the detection of the observed mutations) and not by some new, physiologically derived instability of the genes themselves.

The probability of obtaining mutations through irradiation is the same in any given individual (Serebrovsky et al. 1928); thus, there are no individuals especially predisposed to mutating. Whether the genotypic setting (i.e., breed) in which a certain gene is situated can influence its radiation-induced mutability is difficult to determine. So far, the only case examined in detail (the mutability of the normal alleles of the white-eye series) has shown that the genotypic environment does not influence the mutability of any particular gene, and that the difference that had existed in this case was explained by the two breeds' containing different alleles of the corresponding gene (Timoféeff-Ressovsky 1932a, 1932b, 1933b). A test of the radiation-induced mutation rates in the eyes of purebred males of *Drosophila simulans* and of hybrid males of *D. melanogaster* × *D. simulans* (thus, in the X chromosome of *Drosophila simulans*, first in purebred *simulans* and then in the hybrid genotype) produced only a slight and statistically uncertain difference (Belgovskii 1934). It is worth remarking that differences in mutations of related species and breeds can be due to different kinds of "masking" of a part of the mutations, not to differences in the actual mutation rates of the genes (Timoféeff-Ressovsky 1931b, 1934c), and therefore are to be assessed with caution.

b. Mutation Rate and Irradiation Dose. The first experiments by H. J. Muller (1928b, 1928c) showed already that the induced mutation rate is directly proportional to the applied dose. A series of special experiments, performed by various authors (Demerec 1933[b]; Hanson and Heys 1929, 1932; Oliver 1930, 1932; Schechtmann 1930; Timoféeff-Ressovsky 1931b, 1934a, 1934b, 1934c), were aimed at a precise determination of the relation between the mutation rate and the radiation dose. All of these experiments have produced the same result, that the induced mutation rates are directly and linearly proportional to the radiation dose. The results of my experiments are summarized in table 3, and the proportionality curve is presented graphically in figure 6. The experiments of the aforementioned authors cover a very broad range of doses, from around 350 r to around 9,000 r. The deviation of the highest mutation rates below the expected linear proportionality is explained by the occurrence of the saturation phenomenon; when the "ClB" hybridization methods are used, the appearance of two lethal factors (or more) in the same chromosome is usually not distinguished from those cases in which only one lethal

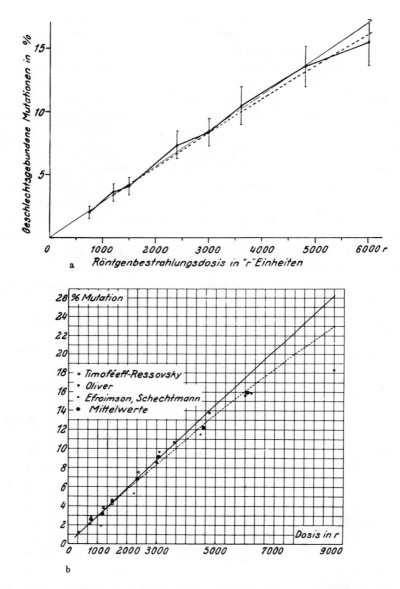

FIGURE 6. Proportionality curves for the sex-linked mutation rate of *Drosophila melanogaster* as a function of the X-ray irradiation dose. Top, experiments by Timoféeff-Ressovsky. Bottom, compilation of experiments by Efroimson, Oliver, and Timoféeff-Ressovsky. The straight line corresponds to direct linear proportionality, the dashed line to the corresponding saturation curve; the vertical lines on the top curve give the margins of error for particular experimental results.

TABLE 3. **Proportionality between the rates of sex-linked mutations and the doses of X-ray irradiation in *Drosophila melanogaster*. (50 kV, 1 mm Al[uminum filter])**

X-ray dose	Number of cultures	Number of mutations	Percentage of mutations
Control	3,058	4	0.13 ± 0.07
750 r	988	21	2.12 ± 0.46
1,200 r	718	27	3.76 ± 0.71
1,500 r	803	34	4.23 ± 0.71
2,400 r	518	39	7.53 ± 1.16
3,000 r	619	53	8.56 ± 1.12
3,600 r	430	46	10.69 ± 0.49
4,800 r	392	54	13.77 ± 1.74
6,000 r	416	65	15.62 ± 1.78

factor arises. Therefore, we must expect empirically a better agreement of experimental results with the saturation curves of linear proportionality, which is in fact just the case (Timoféeff-Ressovsky 1934a; Zimmer 1934).

Because the radiation was dosed in r units in the aforementioned experiments, it can thus be established that, within the broad range of doses tested, *the induced mutation rates are directly and linearly proportional to the ionization rates of the radiation.*

Finally, it can still be noted that the same pattern is true not only for the total mutation rate of the X chromosome (which consists of about 90% lethal factors), but also apparently, as some as yet unfinished experiments suggest, for the rates of "visible" sex-linked mutations and some individual genes.

c. Mutation Rate and Radiation Wavelength. Successful and precise mutation-triggering experiments were carried out on *Drosophila* with different kinds of radiation, from very soft X-rays to the hard gamma rays of radium (Efroimson 1931; Hanson and Heys 1929; Hanson, Heys, and Stanton 1931; Schechtmann 1930; Timoféeff-Ressovsky 1931b, 1934a, 1934b, 1934c). Experiments by Hanson, Heys, and Stanton (1931), by Schechtmann (1930), and by Timoféeff-Ressovsky (1934a) have shown that no dependence on wavelength can be established for the mutation rate of *Drosophila* within the wavelength range of X-rays (from so-called border rays up to hard X-rays). In contrast, data on radium doses (in r) in experiments by Hanson and Heys (1932) suggested that equivalent doses of radium affect the mutation rates more weakly than X-rays. Because the metering of gamma rays in r was technically difficult, this data had to be carefully inspected. Reexaminations (Pickhan 1934) have shown that

TABLE 4. **Effect of equivalent doses of (1) soft and hard X-rays and (2) somewhat soft X-rays and hard gamma rays on the rate of sex-linked mutations of *Drosophila melanogaster*. P ♂ were irradiated and crossed with CIB ♀.**

Dose	Type of radiation	Number of		Percentage of mutations
		Cultures	Mutations	
3,750 r	X-rays, 25 kV 0.5 mm Al	486	54	11.11 ± 1.42
3,750 r	X-rays, 160 kV 0.25 mm copper + 3 mm Al	516	53	10.27 ± 1.33
4,500 r	X-rays, 50 kV 1 mm Al Gamma rays from Ra[dium]	591	72	12.19 ± 1.29
4,500 r		508	59	11.61 ± 1.35
Nonirradiated control		1,827	2	0.11 ± 0.10

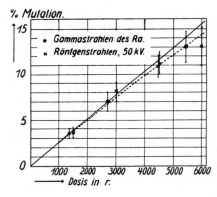

FIGURE 7. Proportionality of the mutation rates to the irradiation dose in experiments with X-rays and with hard gamma rays. Notation is identical to the top curve of figure 6.

equivalent doses of X-rays and gamma rays, measured exactly in r, trigger the same mutation rates. The experiments of Timoféeff-Ressovsky with various X-rays and the experiments of A. Pickhan with X-rays and gamma rays are presented in table 4. It thus appears that the same doses (in r) of all X-rays and gamma rays produce the same mutation-triggering effect. Moreover, experiments by A. Pickhan with a variety of staggered doses of X-rays and gamma rays (figure 7) have shown that the mutation rates, triggered by these different forms of radiation, produce the same, directly linear proportionality curves (to the dose in r).

It thus turns out that, *over the entire range of X-rays and gamma rays, the mutation rate of* Drosophila *is independent of wavelength and is solely a function of the dose.*

d. Mutation Rate and Time Factor. Many radiation-biological reactions show a certain dependence on whether the irradiation dose is administered concentrated or diluted, whole (all at once) or divided up (and distributed over a longer span of time). One thereby speaks of the effect of the "time factor," or the invalidity of the so-called $I*t$ = const. law. Because the determination of whether the time factor exerts an influence is of great analytical significance, special mutation-triggering experiments, in which the temporal distribution of the doses was varied, were performed on *Drosophila* by several authors.

Patterson (1931) irradiated *Drosophila* males with an X-ray dose of around 1,250 r, which he first applied continuously over 10 minutes, and then divided into 8 parts and distributed over time spans ranging from 4 hours to 8 days; the same mutation rate was produced in all cases. Hanson and Heys (1932) applied the same doses of radium rays in concentrated form (over 0.5 and 1 hour) and in diluted form (over 75 and 150 hours) and found that the concentration of the dose had no influence on the mutation rate. In experiments by Timoféeff-Ressovsky (1934a) and Timoféeff-Ressovsky and Zimmer (1935), similar X-ray doses were administered concentrated (300 r/min, 240 r/min), diluted (10 r/min, 1 r/min), divided (6–10 portions, distributed over 6–10 days), and diluted-divided (see table 5); the percentage of mutations triggered in these experiments was again proportional only to the size of the dose, and was not affected by its temporal distribution, despite the total time of application varying by a ratio of up to 1:1,440. In experiments by Pickhan (1934), the concentration of

TABLE 5. **Rates of sex-linked mutations in *Drosophila melanogaster*, triggered with equivalent doses of X-rays, concentrated, diluted, and divided. P ♂ were irradiated and crossed with CIB ♀. Irradiation: 1–3: 3,600 r (50 kV, 1 mm Al) in 15 min (240 r/min), in 6 hours (10 r/min), and every 5 min over 6 days (120 r/min); 4–5: 3,000 r (50 kV, 1 mm Al) in 10 min (300 r/min), and every 5 hours over 10 days (1 r/min). Differences in concentration: 1–2 = 1:24; 4–5 = 1:300. Differences in time: 1–3 = 1:576; 4–5 = 1:1,440.**

X-ray dose	Type of treatment	Number of Cultures	Number of Mutations	Percentage of sex-linked mutation
3,600 r	1. Concentrated in 15 min	493	54	10.95 ± 1.41
3,600 r	2. Diluted, over 6 hours	521	60	11.51 ± 1.39
3,600 r	3. Divided, every 5 min for 6 days	423	47	11.11 ± 1.52
3,000 r	4. Concentrated, in 10 min	531	43	8.09 ± 1.18
3,000 r	5. Diluted-divided, every 5 hours for 10 days	573	52	9.07 ± 1.18
Nonirradiated control		1,827	2	0.11 ± 0.10

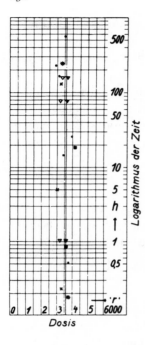

FIGURE 8. Independence of the mutation rate from the time factor in irradiation experiments with *Drosophila melanogaster*. From irradiation experiments by various authors, calculated irradiation doses (in r[oentgens]) that trigger a sex-linked mutation rate of 10 shown as a function of the time of irradiation (these in logarithmic scale); the doses range within the statistical margins of error around the ideal value, extracted from the proportionality curve, of 3,600 r (▼: Hanson and Heys 1932, strong dose; ▽: Hanson and Heys 1932, weak dose; ●: Patterson 1931; ■: Pickhan 1934; X: Timoféeff-Ressovsky 1934a; *: Timoféeff-Ressovsky and Zimmer 1935).

X-rays was varied by a ratio of 1:19 (around 70.5 r/min and around 3.7 r/min), without the mutation rate being affected. The results of all time-factor experiments are presented graphically in figure 8.

All *Drosophila* experiments have thus shown that *the radiation-induced mutation rate is independent of time factors and is proportional only to the total amount of radiation*. This leads to some important *conclusions*. First, it is a further proof that the influence of applied radiation on the gene is direct and proportional to the dose. Second, this finding shows that no minimal or "subthreshold" irradiation dose is to be expected, and that the proportionality curve may be extrapolated in the same form all the way down to the zero-point. And third, it must be concluded that, in contrast to many other biological reactions, the mutation process is a nonrestitutable one (*nichtrestituierbar*), in that the gene transitions from one stable state to another equally stable state.*

---

* The use of "nichtrestituierbar" here is unusual. Timoféeff-Ressovsky appears to be pointing to the fact that a mutation cannot *immediately* reverse itself, since the final state has a similar stability to the initial state.—Trans.

e. Combined Effect of Irradiation with Other Stimuli.   Using plant seeds, Stadler (1930) was able to show that permeation of the seeds with heavy-metal salts, which alone had no mutation-triggering effect, did increase the effect of subsequent irradiation, which likely must be due to stronger absorption of the radiation in the permeated tissue. This was also recently confirmed for *Drosophila melanogaster* (Medvedev 1933); flies raised on food supplemented with 1 percent $Pb(CH_3COO)_2$ produced more mutations after irradiation with X-rays than untreated flies had produced after the same irradiation with X-rays.

Besides treatment with heavy-metal salts, the effect of temperature on *Drosophila* during irradiation was tested. Muller (1930a) irradiated flies with the same X-ray dose at 8°C and at 34°C, without observing any statistically meaningful difference. In our experiments (Timoféeff-Ressovsky 1934a), *Drosophila melanogaster* males were irradiated with 3,000 r of X-rays, at 10°C and at 35°C; the resulting mutation rates showed no statistically significant difference (table 6).

The experiments described in this section thus show that irradiation increases the mutation rate through a direct impact on the gene (and not through physiological intermediaries [*Umwegen*]). The triggered mutation rate is directly proportional to the applied dose and is not dependent on the wavelength, the temporal distribution of the applied irradiation dose, or the temperature during irradiation. Figure 9 displays graphically the mutation rates per 1,000 r from all of our experiments with different X-ray doses, different temporal distributions of the dose, different wavelengths of the applied radiation, and different temperatures during irradiation. All mutation rates per 1,000 r are distributed randomly around the total average rate per 1,000 r of all experiments, without producing a statistically significant deviation. This figure shows especially clearly that the radiation-induced mutation rate of *Drosophila* exhibits a direct, linear proportionality to the dose, measured in r, and is independent of all other

TABLE 6. **Rates of sex-linked mutations in *Drosophila melanogaster* after irradiation with equal doses of X-rays at various temperatures during irradiation (type of radiation: 50 kV, 1 mm aluminum filter).**

Conditions of irradiation	Number of cultures	Number of mutations	Percentage of mutation
Control	1,827	2	0.11 ± 0.10
Ca. 3,000 r, at 10°C during irradiation	401	37	9.22 ± 1.44
Ca. 3,000 r, at 35°C during irradiation	368	30	8.15 ± 1.45

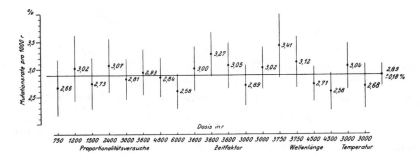

FIGURE 9. Mutation rates (of the X chromosome) per 1,000 r [units] from various irradiation experiments with *Drosophila melanogaster*. From left to right, doses from 750 r to 6,000 r of the same X-rays (table 3); time-factor experiments with concentrated, diluted, divided, and diluted-divided X-ray doses (table 5); wavelength experiments with equal doses of weaker and stronger X-rays and equal doses of X-rays and gamma rays (table 4); irradiation with the same X-ray dose at lower and higher temperatures (table 6). The horizontal line corresponding to the rate of 2.89 percent represents the total mean rate for all experiments of sex-linked mutations per 1,000 r; the vertical lines indicate the margins of error for individual rates.

tested factors. Permeation with heavy-metal salts of the tissues to be irradiated has proven to be the only effectual secondary stimulus, one that is physically self-explanatory and stands in agreement with other experimental results.

4. DEPENDENCE OF THE SPONTANEOUS MUTATION RATE ON TIME AND TEMPERATURE. Two questions regarding spontaneous mutability must now be clarified: the dependence of the mutation rate on time and on temperature.

The question of the relationship between the mutation rate and time actually resolves into two separate questions. First, it must be clarified whether the mutation process as such is time-independent, i.e., whether the probability of the gene mutating is independent of its "age" (the time it has spent unmutated). We would thus need to determine whether, if we regularly eliminated mutations as they occur, the unmutated remainder of the genes would eventually exhibit a continually increasing mutation rate, or whether the mutation rate would stay constant. To be sure, it is hardly possible to test this question experimentally, but it can be resolved by means of some reflections of sufficient clarity. The aforementioned experiment—the regular elimination of mutating genes—is continually realized both freely in nature and in our breeds. Were the "lifespan" of the alleles limited and the mutation process time-dependent, the natural mutation rate should be very high (given the relatively advanced age of current species), as well as increase regularly in older breeds (observed over a long time). We have no evidence at all that this actually takes place, and we can therefore assume that the mutation process as such is time-independent.

The second question that concerns the relationships between the mutation rate and time is the following: Should the mutation rate be defined as a percentage of mutations per unit time or per biological unit, e.g., generation? In practice, the mutation rate of *Drosophila* is established per generation, albeit on males of a certain age under fairly constant conditions, and thus per absolute unit of time as well. One could answer these questions by comparing the percentage of mutations in an inactive cell phase over different spans of time. In *Drosophila*, the mature sperm that are not reabsorbed or ejaculated and in which no germinal selection takes place are suitable for this (Harris 1929; Timoféeff-Ressovsky 1931a). Our experiments on the spontaneous rates of sex-linked lethal factors in the mature sperm of newly hatched and of 20-day-old males are presented in table 7. These experiments indicate that the mutation rate should be defined as a percentage of the mutations per unit time, despite there being too little data to yield statistically irreproachable results. The same model follows from everything that we have previously learned about the mutation process.

The question of the relationship between the spontaneous mutation rate and temperature was addressed in the first work by H. J. Muller on the determination of the measurable rate of mutation of *Drosophila* (Muller and Altenberg 1919). In a second work, Muller (1928a) included further data on the spontaneous mutation rate at various temperatures. In both works, it was possible to show that mutability is accelerated by an increase

TABLE 7. **Spontaneous rate of sex-linked lethal factors in the mature sperm of *Drosophila mela-nogaster* males, both young (immediately after the hatching of the males) and aged (15–20 days after hatching).**

		Number of		Percentage of lethal factors
Type of experiment	Age of the sperm	Culture	Lethal factors	
"CIB" experiments, run at ca. 24°C	1. P-♂♂ paired immediately after hatching	6,831	7	0.102 ± 0.038
	2. P-♂♂ paired 15–20 days after hatching	5,957	14	0.234 ± 0.062
Difference 1–2				0.132 ± 0.072

TABLE 8. **Studies on the dependence of the spontaneous rate of sex-linked lethal factors on temperature in *Drosophila melanogaster* ♂ (within normal temperature limits).**

		Number of		
Type of experiment	Treatment	F_1–F_2 cultures	Sex-linked lethal factors	Percentage of lethal factors
"CIB" experiments on	1. P-♂♂ at 14°C	6,871	6	0.087 ± 0.035
sex-linked lethal factors	2. P-♂♂ at 24°C	3,708	7	0.188 ± 0.071
	3. P-♂♂ at 28°C	6,158	20	0.325 ± 0.072

Difference 1–3: 0.24% ± 0.08%.
Development time: 14°C—22 days; 24°C—14 days; 28°C—14 days; 28°C—11.5 days.
Corrected mutation rates: 14°C—0.056%; 24°C—0.188%; 28°C—0.396%.
$t°Q_{10} = 2.6$; corrected, $t°Q_{10} = 5.1$.

in temperature. Our temperature experiments, consolidated in table 8, also led to approximately the same result. The spontaneous mutation rate is also increased by an increase in temperature. The temperature quotient for 10°C ($t°Q_{10}$) amounts to around 2.5; if one imposes a correction for accelerated development at higher temperatures, allowing for the previously established relationship between the mutation rate and time, one obtains

$$t°Q_{10} = \text{ca. } 5.$$

From this it follows, as Muller has suggested, that the mutation rate obeys van't Hoff's rule.

Thus, the experiments and considerations presented in this section have shown that the *spontaneous mutation rate should be defined as a percentage of the mutations per unit time, and, thus defined, it is time-independent but temperature-dependent*; in the latter case, it follows van't Hoff's rule.

5. RATES OF INDIVIDUAL GENE MUTATIONS. All the experiments mentioned above were concerned with a predefined portion of the total mutation rate of *Drosophila melanogaster*, thus with the sum of the mutations of many different genes. With the help of radiation-genetic methods, the mutability of individual genes, even the rates of specific, individual steps in the mutation process, can be quantitatively determined. To do this requires the examination of a very large amount of data, and we have at this time only a few individual pieces of data at our disposal.

The X-ray-induced mutability of the white-eye locus of *Drosophila melanogaster* is summarized in table 9 (Timoféeff-Ressovsky 1932b, 1933b). The experimental results show that, within this group of multiple alleles, the various steps of the mutation process occur with dissimilar frequencies. It can be concluded that the *frequency of mutation is determined by the structure of the relevant allele.*

The frequencies of *forward and back mutations* of some genes of *Drosophila melanogaster* triggered through X-ray irradiation are presented in table 10 (Patterson and Muller 1930; Timoféeff-Ressovsky 1932b, 1933a, 1933b, and unpublished). The few allele pairs examined so far show that we should expect the whole spectrum of cases, from those in which only direct mutation occurs, through those in which forward and back

TABLE 9. **Comparison of the different mutation rates within the white-eye allele group of** *Drosophila melanogaster*, **produced through equal X-ray irradiation (ca. 5,000 r).**

Mutations	Number of Gametes	Number of Mutations	Mutation rates in hundredths of a percent	Difference of the mutation rates
All direct mutations	136,000	63	0.463 ± 0.061	0.432 ± 0.062
All back mutations	190,000	6	0.031 ± 0.013	
$W \leftrightarrow w^x$	48,500	37	0.763 ± 0.125	0.708 ± 0.128
$w \leftrightarrow w^x$	54,000	3	0.055 ± 0.032	
$W \leftrightarrow w$	48,500	25	0.515 ± 0.102	0.264 ± 0.106
$w^x \leftrightarrow w$	87,500	22	0.251 ± 0.056	
$w^e \leftrightarrow w^{-e}$	72,000	22	0.306 ± 0.064	0.266 ± 0.069
$w^e \leftrightarrow w^{+e}$	72,000	3	0.040 ± 0.024	
$W \leftrightarrow w^x$	48,500	37	0.763 ± 0.125	0.393 ± 0.136
$w^{e-co} \leftrightarrow w^x$	73,000	27	0.370 ± 0.071	0.297 ± 0.078
$w^{-bf} \leftrightarrow w^x$	68,500	5	0.073 ± 0.032	

TABLE 10. **Comparison of the mutation rates in two opposite directions of five different pairs of alleles in** *Drosophila melanogaster* **under the influence of X-ray irradiation (4,800–6,000 r). Experiments by Timoféeff-Ressovsky and H. Muller and J. Patterson.**

Gene mutations	Direct mutations Gametes	Direct mutations Mutations	Back mutations Gametes	Back mutations Mutations
$W \leftrightarrow w$	69,500	28	54,000	0
$W \leftrightarrow w^e$	69,500	9	72,000	3
$F \leftrightarrow f$	43,000	11	44,000	15
$P \leftrightarrow p^p$	52,000	1	58,000	9
$B \leftrightarrow b$	69,500	0	9,000	8

TABLE 11. **Comparison of the spontaneous and X-ray–induced (4,800–6,000 r) mutation rates of *W* (normal) to *w* (white) and of *bb*^lx^ (lethal allele of bobbed) to *Bb* (normal) in *Drosophila melanogaster.***

Gene mutation	Spontaneous-mutation rate	X-ray–induced mutation rate
W → w	1 : ca. 300,000	1 : ca. 2,500
	(3 : 1,000,000)	(28 : 69,500)
bb^lx^ → Bb	1 : ca. 12,000	1 : ca. 1,800
	(3 : 35,000)	(10 : 18,000)

mutations occur with the same frequency, up to those in which back mutations occur much more frequently than "direct" mutations (of a normal to mutant allele), which appear only rarely or even just once.

Finally, table 11 shows the results of experiments in which the spontaneous and X-ray-induced rates of two different mutations are compared (Timoféeff-Ressovsky unpublished). The spontaneous mutation rates of the two genes are very different. One mutation, from normal to white (W→ w), like most mutations of normal *Drosophila melanogaster*, occurs so rarely that its rate can be estimated only approximately as 1 in about 300,000. The other mutation, from bobbed^lx^ to normal (bb^lx^→Bb), spontaneously occurs relatively frequently: approximately 1 in 12,000. These two mutations, occurring spontaneously with such different frequencies, were generated through irradiation with equal amounts of X-rays at rates of the same order of magnitude (1 in about 2,500, and 1 in about 1,800, respectively).

In connection with the experiments in table 11, we must briefly note the results obtained for the so-called *"mutable" or "unstable" genes* ("frequently mutating genes"; see reviews in Demerec 1928, 1929a; Stubbe 1933). "Mutable" alleles of the miniature gene of *Drosophila virilis* were examined most accurately and in the most detail by M. Demerec. The "mutable," mutant alleles differ dramatically and very strongly from all normal alleles in having extraordinarily high mutation rates, consisting primarily of back mutations to the stable, normal allele. The normal alleles have mutation rates of an order of magnitude between 0.001 percent and 0.0001 percent; by contrast, the mutable alleles can mutate at over 1 percent. The mutability of the "mutable" alleles would likely not stand out so dramatically if we were able to compare it not only with the mutability of normal alleles, but also with that of the mutant alleles. The latter is known in only a few cases, but, as shown by the spontaneous mutability of the mutant bb^lx^ allele presented in table 11, we can find a variety of gradations between "stable" and "unstable" genes among the various mutant alleles. That is, we must assume that the alleles that constitute the normal type are profoundly stable, because over time the unstable ones must be eliminated through natural selection and replaced by more stable ones, presuming they do not produce traits with especially high positive selection value; but even in the latter case they would likely be gradually replaced by stable genes that bring about equivalent characteristics. We do find, in fact, not only the "mutable" alleles but the more unstable ones in general only among the mutations of *Drosophila* and not in the "wild type."

In a series of very beautiful experiments, M. Demerec (1929a, 1929b, 1929c, 1932a, 1932b, 1934) has carefully examined the mutability of these unstable alleles. Two of his findings in particular interest us here. In one investigation (1932a), Demerec was able to establish that the mutability of the "mutable" mini-3-gamma allele is not visibly influenced by temperature; in each case, an increase in temperature did not increase the mutability of this allele more strongly than the overall development of the flies. And in recently published experiments, Demerec (1934) has shown that the mutation rate of the same "mutable" allele of *Drosophila virilis* likewise remains almost unaffected by X-ray irradiation; the rate of other mutations is increased many times over by such irradiation, but only a very small, statistically insignificant increase in mutations on the "mutable" miniallele can be observed following irradiation. This fact is especially interesting in conjunction with the experiments of table 11 and shows again that the effect of radiation on spontaneous mutability need not be proportional.

The experiments mentioned in this section have thus shown that the structure of an allele determines its mutability, but that genes that are especially unstable spontaneously appear to show no correspondingly high radiation-induced mutation rate. Furthermore, it was possible to establish that mutability in two opposite directions can proceed differently for different alleles; the full spectrum exists, from cases where opposing directions in the mutation process have the same probability, out to both extremes, where only one direction of mutation is realized.

III. Concluding Remarks

I. SUMMARY OF THE MOST IMPORTANT EXPERIMENTAL RESULTS. The following conclusions about the mutation process in *Drosophila melanogaster* can be drawn from the observations and experiments described above.

a. A great variety of mutations appear spontaneously, but the mutation rate is small and has a confirmed value of around 0.1%–0.2% for the sum total of all the lethal and "good" visible mutations of the X chromosome.

b. Spontaneous mutability is time-independent; i.e., an unmutated gene's disposition to mutate does not increase with time, but rather remains constant.

c. The spontaneous mutation rate should be defined as a percentage of the mutations per unit time (table 7).

d. The spontaneous mutation rate is temperature dependent and follows the van't Hoff rule with a temperature quotient of around 5 at 10°C (table 8).

e. Different genes, as well as different alleles of the same gene, show differentially high mutation rates, from which we should conclude that mutability depends on the structure of the allele. The mutation rates found in mutant alleles are higher than in normal alleles, and sometimes especially unstable, "mutable" alleles appear.

f. The mutation rate is strongly increased by irradiation with X-rays and radium. However, all types of mutations thereby produced also originate spontaneously; special "radiation mutation types" are not observed. There is thus an extensive parallelism between spontaneous and radiation-induced mutation.

g. Irradiation produces mutations in both sexes and in a variety of tissues by acting directly on the genes (and not through physiological intermediaries; table 2). The action of radiation in the gene is not purely destructive, since, in many cases, forward and back mutations can be produced directly from one another through irradiation (table 1; figures 3, 4).

h. The mutation rates triggered through irradiation are directly and linearly proportional to the applied doses (table 3, figure 6).

i. Neither the median-effective dose nor the form of the proportionality curve of mutability depends on the wavelength of the applied radiation (in the range of very soft X-rays up to the gamma rays of radium). The mutation rate is thus wavelength-independent (table 4, figure 7).

k. [sic] The temporal distribution of the irradiation dose (whether administered concentrated or diluted, whole or divided, or diluted-divided) has no influence on the percentage of mutations triggered. The mutation rate is thus independent of time factors and depends only on the total amount of radiation applied (table 5, figure 8).

l. The mutation-triggering effect of radiation is independent of the temperature during irradiation (table 6). On the other hand, the effectiveness of the irradiation dose is increased by permeating the tissue to be irradiated with heavy-metal salts (which on their own have no mutation-triggering effect).

m. The radiation-induced rate for individual genes is apparently influenced by the structure of the corresponding allele, as is the spontaneous rate. With radiation-induced forward and back mutations of specific pairs of alleles, the full spectrum is observed, from those cases in which the probabilities of the two directions are equal, to those in which a mutation can only be realized in one direction (table 10).

n. Genes that are especially unstable spontaneously need not have a correspondingly high radiation-induced mutation rate (table 11).

o. The impact of both temperature increases and irradiation on the mutation rate of "mutable" alleles is minimal.

2. THE QUESTION OF THE THEORY OF GENE MUTATION AND GENE STRUC-TURE.   In this first, genetic part of our paper, we have presented and summarized those facts from mutation research that should form the foundation for the theory of gene mutation and gene structure to be developed in the concluding section (part 4). But first, some physical questions must be subjected to a systematic analysis, because, as we mentioned at the outset, the biophysical, radiation-genetic experiments show the most promise of illuminating the nature of the mutation process.

First of all, we must clarify what is essential for the mutation-effect of irradiation; that is, from the standpoint of the target theory, which prevails in physics today, what is to be considered a "target-strike" in the triggering of a mutation. The second part of this paper is dedicated to this question.

We must then develop a physical model that conforms to the facts of mutation research and to the definition of the mutation-triggering target-strike. This model must be tested at every level of detail against the results of mutation research, by having its implications compared to experimental results. The third part of this paper is dedicated to this task.

Thus will we finally arrive at a both physically and genetically well-founded model of the general nature of gene mutation, from which we will then be able to draw certain conclusions about the nature of the gene. This will form the subject matter of the fourth part of this paper.

Part 2: The Target Theory and Its Relation to the Triggering of Mutations

K. G. ZIMMER[2]

1. Introduction

Many of the investigations described in part 1 concerned the generation of gene mutations through radiation because radiation-genetic experiments are particularly well suited for shedding light on the actual mechanism of gene mutation. Still, for the interpretation of the mutation process, a detailed analysis of the effects of radiation is required alongside the genetic investigation. This should be realized within the framework of the most successful theory to date of the biological effects of radiation, the so-called target theory.

2. Definition of the Target-Strike

Through a large number of related investigations, it has become clear that we cannot determine generally what is ultimately essential for the biological effect; that is, whether the target-strike event should be seen as

a. the absorption of a single quantum of radiation (Holweck and Lacassagne [1934], Wyckoff [1930a, 1930b]),
b. the transit of an excited electron through a biological unit, a so-called sensitive volume or target zone (Glocker [1932], Mayneord [1934]), or
c. the generation of an ion pair and, accordingly, an excitation in a target zone (Dessauer [1923], Crowther [1926]).

The definition of the target zone (sensitive volume) varies with each alternative definition of the target-strike. Especially important for our purposes is that, depending on which definition of the target-strike is applied, we arrive at different conclusions concerning the dependence of the median-effective dose ($D_{1/2}$) on the wavelength λ of the incident radiation.[3] As Glocker showed for these cases:

---

2. Radiation Unit of the Cecilienhaus, Berlin-Charlottenburg. [Ed. Cecilienhaus was a women's clinic that also conducted radiation research.]

3. The median-effective dose (Halbwertsdosis) is that needed to produce the reaction in question in half of the irradiated subjects.

(0)
$$\textbf{a. } D_{1/2} = \text{const.}\frac{1}{\lambda};$$

(0')
$$\textbf{b. } D_{1/2} = \text{const.}\frac{1}{\lambda}\frac{a}{a+R},$$

where R = actual range of action* of the secondary electrons in tissue, and a = median free path of the secondary electrons in the target zone;

(0")
$$\textbf{c. } D_{1/2} = \text{const.}^4$$

Nevertheless, we should attempt to deduce here a picture of the process of the genetic effects of radiation; most important, we should do so independently of the viewpoint provided by the experimental results disclosed in the first part. Only in the concluding remarks will the derived models be compared with the theoretical interpretation of other radiation reactions.

3. The Target-Strike in the Mutation Process

The relationship between the irradiation dose and the resulting triggered mutation rate (figure 6) can be shown by the equation (Zimmer [1934])

(1)
$$x = a(1 - e^{-kD}), \text{ where}$$

x = number of mutated genes,

a = number of irradiated genes,

D = radiation dose,

k = constant of the rate,

e = basis of the natural logarithm.

On the other hand, according to Blau and Altenburger [1923], the following equation holds, in general and independently of both the type of target

---

* *Reichweite* has been translated as "range of action" following Schrödinger's usage in *What Is Life?*, chapter 5.—Trans.

4. This holds only if the target zone is small compared to the average distance between two ionizations along the path of a secondary electron (Dessauer 1933).

zone and the definition of the target-strike, for the impact (damage) of x target zones out of a number a of equal sensitivity, where D is the dose:

(2) $$x = a\left[1 - e^{-kD}\left(1 + kD + \frac{(kD)^2}{2!} + \frac{(kD)^3}{3!} + \ldots + \frac{(kD)^{n-1}}{(n-1)!}\right)\right];$$

n = number of impacts needed for transforming activity. From this it follows that one target-strike is sufficient, because equation (2) reduces to the experimentally determined equation (1) for the impact number $n = 1$.

It follows from experiments concerning the relationship between radiation dose and triggered-mutation rate that *a single target-strike is sufficient for the triggering of a gene mutation.* Yet one still cannot draw conclusions regarding the nature of the target-strike from these experiments. That is possible only through an investigation of the importance of the wavelength of the radiation used.

The irradiation experiments with X-rays (50 kV$_s$, 1 mm aluminum filter, 0.55 Å$_{eff}$) and gamma rays (0.5 mm platinum-filter-equivalent, 0.015 Å$_{eff}$) showed that equal doses of the two kinds of radiation, measured in the same r unit, produced the same mutation rates. All values also fall on one and the same curve (figure 7), so that neither the form of the curve nor the median effective dose depends on the wavelength. This finding makes possible a determination between the three possible cases mentioned above (compare formulas 0, 0', and 0").

a. Because the number of quanta in a given dose, measured in r, drops with the wavelength, the target-strike in this reaction cannot consist of the absorption of a quantum; otherwise, the mutation rate triggered by the same dose would have to be much lower in the region of the gamma rays.

b. In considering secondary electrons, we have to be a little more careful. Mayneord [1934] has recently suggested that significant deviations from formula (0') can arise if the irradiated subject is so small that the extensive range of action of the strong electrons cannot be fully utilized. This shows up in the Glocker theory in the following way: Given

N_0 = number of the electrons generated by the unit dose per cubic centimeter,

N = number of electron paths that pass through the target zone for each unit dose,

r = radius of the target zone (assumed to be spherical),

v = volume of the target zone,

a = mean free path of the electrons in the target zone ($a = 4/3\ r$),

N_i = number of electrons produced in the interior of the target zone, and

N_a = number of electrons in the target zone supplied by the surrounding spherical shell with the radius $\rho = \sqrt{R^2 + r^2}$.

Then,

(3)
$$N = N_i + N_a = N_0 \frac{R+a}{a} v.$$

The resulting median-effective dose[5] is inversely proportional to N (compare to [0']):

(4)
$$D_{1/2} = \text{const.} \frac{1}{\lambda} \frac{a}{R+a},$$

while

(5)
$$N_0 = \text{const.} \lambda.$$

However, ignoring most of the electrons supplied by the target zone's surroundings gives

(6)
$$N = N_1 = N_0 \frac{4}{3} r^3 \pi.$$

Thus, the percentage of damage depends only on the number of electrons produced per unit dose, and not on their range.

The median-effective dose then follows from (6) and (5):

(7)
$$D_{1/2} = \text{const.} \frac{1}{\lambda},$$

i.e., the effect becomes proportional to the number of incident quanta, as in case a.

---

5. Here we will deal only with the case in which the target-strike number (number of target-strikes needed to have an effect) does not change with the wavelength. This restriction is permitted because the experiments with both wavelengths produced the same "damage curve."

In the radiation-genetic experiments considered here, the range of action of the strong electrons was not fully utilized, due to the flies' small size; however, because the flies were irradiated in a capsule whose walls were approximately as thick as the flies' bodies themselves, approaching the maximal range of the strong electrons in tissue, the energy loss was compensated for by additional electrons (for the basics of this procedure, see Friedrich and Zimmer 1934). However, equation (4) still may not be used straight away here in the discussion of the influence of the wavelength for a second reason; due to the preponderance of the Compton processes in the area of the total radiation, equation (5) is not satisfied. Without knowledge of the exact relationship between N_0 and λ, only an estimation is possible; because of the energy balance, we can set N_0 proportional to $\frac{1}{\sqrt{R}}$, where R is understood to be the average range of the electrons, as was done by Mayneord [1934] in taking into account the proportion and true range of action of the photo- and Compton electrons:

$$(8) \qquad\qquad D_{1/2} = \text{const.}\sqrt{R}\,\frac{a}{R+a}.$$

The data necessary for the analysis are collected in table 12 for a pair of wavelengths closest to those used in the experiments. We thus put the radius of the target zone at $a \approx 1\mu$. This leads to the values collected in table 13.

It has thus been shown that if one were to assume a target-strike to be the transit of a secondary electron, the median-effective dose in the region of gamma rays should be approximately 8 times smaller [than in the region of X-rays]. Because experiment showed no evidence of a dependence of the median-effective dose on wavelength, the transit of a sec-

TABLE 12. **Some physical characteristics of Roentgen and gamma-ray wavelength ranges used in the experiments in table 4.**

Characteristics	X-ray	Gamma ray
Wavelength λ in Å	0.360	0.021
Range of action (R_0) of the photoelectrons in μm	23.2	2,790
Integral range of action of the photoelectrons $R_0 * \lambda$ (relative value)	1.67	11.7
Integral range of action of the Compton electrons $R_0 * \lambda * \zeta$ (relative value)	0.150	6.40
Median integral range of action (relative value)	1.51	6.40
Mean range of action R_0 in μm	21.0	1,525

TABLE 13. **The median-effective dose of X-rays and gamma rays in the mutation experiment with** *Drosophila*, **calculated on the basis of equation (8), on the assumption that a target-strike is the transit of a secondary electron.**

Radiation	Wavelength in Å	Median-effective dose D (relative)
X-rays	0.360	0.210
Gamma rays	0.021	0.026

ondary electron can be eliminated from the group of possible target-strike events.

c. Conversely, the observed wavelength-independence of mutation triggering through radiation is entirely consistent with the assumption that a target-strike is the formation of an ion pair. Indeed, because the dose is measured by definition by the number of generated ion pairs, all effects that can be stimulated by the generation of an ion pair must proceed independently of the wavelength and in parallel with the dose.[6]

Our investigations of the target-strike in mutation-triggering experiments can be summarized as follows: *a target-strike suffices for the triggering of mutations by X-rays and gamma rays, and this strike consists of the generation of an ion pair or an excitation.* This conclusion permits us, in conjunction with the considerations presented below in the third part of this paper, to sketch an atomic physics model of gene mutation.

4. Concluding Remarks

Glocker [1932] has shown in detailed discussions and calculations that the results of many of the available radiation-biological investigations can be best explained if one takes a target-strike to be the transit of a secondary electron triggered by an incident quantum. Because this is not the case for the process of mutation triggering in *Drosophila*, it seems reasonable to surmise that the difference is based in the nature of the biological reaction, and that *genetic reactions obey different laws concerning the action of radiation than do nongenetic reactions.* For reasons explicated below in part 3, it should be assumed that gene mutation consists of the transformation of a single molecule, thus proving to be more a chemical than biological

---

6. With the qualification made in note 4.

reaction. However, chemical reactions are wavelength independent, as calculations and experiments by Glocker [1932], Risse [1931], and Berthold, as well as Bouwers* show; put differently, they proceed in parallel with the dose, measured by ionization of the air, because the kinetic energy of the photoelectrons and Compton electrons alone is determinate for both processes. Hence, the results conveyed here concerning gene mutation do not conflict with the fact that many nongenetic radiation reactions can be best explained if one views a target-strike as the transit of a secondary electron.

In conclusion, it should be noted that we have not considered the possibility of explaining the effects of radiation on a colloid-chemical basis.

Part 3: Atomic-Physics Model of Gene Mutation

M. DELBRÜCK[7]

1. Introduction

The question of whether it is appropriate to introduce speculations drawn from atomic physics into the conceptual system of genetics is perceived differently by researchers in the two fields. For this reason, we want to begin by briefly indicating the pros and cons.

As is generally known, genetics is largely a logically self-contained, rigorous science. It is quantitative, but without making use of a physical measurement system. We should make clear what this independence from chemistry and physics depends upon, for chemistry does not in any way exhibit such an independence from physics. On the contrary, chemistry was transformed from a descriptive science into a quantitative one primarily through the establishment of the principle of mass conservation and the introduction of scales, that is, through the adoption of a physical measurement system. The development of electrochemistry proceeded analogously, in that Faraday's law of equivalence enabled the measurement of electric charge to be connected to a measurement of weight, thereby laying the foundation for quantitative analysis. This development

---

* There is no entry for Berthold and Bouwers in the original paper, and it is not clear from contemporary works which paper is being cited here.—Trans.

7. Radioactive Physics Unit of the Kaiser Wilhelm Institute for Chemistry, Berlin-Dahlem.

led to the theory of the atom, the common root of physics and chemistry, which today form a unity in virtue of a unified, sharply defined, and extremely narrowly circumscribed notion of observation. This unity is made manifest in the absolute measurement system that extends through all of their branches. On the other hand, we recognize as the foundation of this common measurement system the existence of rigid measuring rods and mechanically regular clocks. These are themselves possible only on the basis of atoms that are stable and that have properties persisting in time. We can characterize this development in the following manner: Physics and chemistry, as quantitative sciences, rest upon the existence of stable atoms. Yet they [these sciences] have advanced toward these atoms only over the course of a century of development. They had to manage for a long time with only a very general account, situating the stability of matter in the existence of rigid bodies and in the constancy and specificity of their chemical nature.

Genetics differs from these sciences in that it has a natural unit for quantitative, numerical analysis: the individual organism. This circumstance makes genetics independent of a physical measurement system. Chemistry arrives at its natural unit primarily by way of the molecule concept, which for its part is made possible primarily by the law of multiple proportions, one of the first major achievements of quantitative chemistry. Accordingly, the concept of observation in genetics is completely different in nature, and is expanded tremendously. Whereas in physics all measurements must in principle lead back to measurements of position and time, the fundamental concept in genetics—the differentiation of character traits—is unlikely to be expressed easily in absolute units of measurement in even a single case, much less in general. Even with such character traits as average length or duration of development, the absolute magnitude of the length or duration will usually be without significance, because it will still depend on surrounding conditions.

On the basis of such considerations, one may assert that genetics is autonomous and that it should not be muddled with physical-chemical notions. In particular, one could think this because, in the places in biology where the use of physical and chemical concepts has so far been met with real success, no convergence with the phenomena of genetics can be seen. Rather, while genetics always deals provisionally with the isolation of processes that do indeed have, physically-chemically, an unambiguous character, they appear as only parts of processes when viewed biologically, and their relation to the whole life process remains problematic, unless their coordination is viewed as arising on the basis of some heuristic scheme in which the life process is postulated in principle as physical-chemical machinery.

But in the course of development of genetics itself, there has been an expansion in the sphere of concepts. First, the linking of genetics to cytological research proved that the gene, which was originally simply a symbolic representation of the differentiating unit, can be localized spatially and tracked in its movements. The sophisticated analysis of *Drosophila* has led to estimates of the gene's size, which are on the order of the largest distinctively structured molecules known to us. Based on this result, many researchers see the gene as nothing more than a special type of molecule whose structure is simply not yet known in detail.

One must keep in mind that there is an essential difference in the chemical definition of molecules. In chemistry, we speak of a specific kind of molecule when we have before us a substance that behaves uniformly in response to chemical stimuli. In genetics, on the other hand, there is by definition only one single instance of the "gene-molecule" in each organism, in an environment that chemically could not be more heterogeneous; we establish its identity with a gene of another individual only on the basis of similar developmental effects. Thus, we could not speak of a uniform chemical reaction even in a thought experiment, unless we imagined every corresponding gene being isolated from a large number of genetically similar organisms and the behavior of these isolated and collected genes being chemically analyzed. Unless we want to draw some specific, directly testable claims from the presumption that it is possible in principle, such a thought experiment is merely a game [*Spielerei*]. Thus understood, it seems more practical to express the underlying thought somewhat differently.

Besides the cytologically revealed minuteness of the genes, the real reason for identifying them with molecules concerns their stability, the fact that, as judged by the results of hybridization experiments, they retain their identity unchanged in the face of all surrounding influences. It seems reasonable to link this stability directly to the stability of molecules. We want, therefore, when we speak of molecules, to think not so much of uniform behavior as very generally of a well-defined assemblage of atoms, according to which we assume that the identity of two genes lies in the fact that the same atoms are stably arranged in them in the same fixed way. The stability of the configuration must be especially high compared to the chemical excitations occurring naturally in living cells; genes can take part in general metabolism only catalytically. We will thus leave unresolved the question of whether the individual gene is a polymeric form that arises through the repetition of identical structures of atoms, or whether it exhibits no such periodicity. This version of the molecular hypothesis, which is suggested by general considerations, is, as we shall see, capable of the widest range of application and test through experiment.

Before we put this to the test, we want to emphasize that the fundamental property of the gene—its identical self-replication during mitosis (where this property is itself free from mutation)—is certainly not a feature of the model of the gene by itself, but rather a mutual capacity of the gene and the surrounding matter. The compatibility of our model with this fact cannot be tested as long as this interaction is not included in the extended model (Muller 1929a).

2. Model of Mutation

Because on the one hand, as noted, we cannot yet use chemical means to determine directly a gene's atomic composition, and because on the other hand we know almost nothing about the chemical mode of action of the genes as catalysts of development, we must tackle the problem from a more primitive angle. We must first investigate the *type and limit of the stability* of the genes and see whether there is agreement with what we know from atomic theory about well-defined assemblages of atoms.

We will first review the *kinds of change that can occur in an assemblage of atoms*, as well as in detail the conditions of their occurrence, and then turn to a comparison with mutations.

In our gene model, we assume that the atoms in the assemblage have determinate central locations and that the electrons are in determinate states. With these stipulations, we find that changes to the model can occur only in jumps. They must thus consist of a sequence of elementary processes. We begin with these.

An assemblage of atoms is capable of the following modifications through elementary processes:

a. *Changes in the states of vibration*: Changes in which the central location of the atoms remains the same occur with tremendous frequency at normal temperatures; thus, no molecule is stable in this respect. But the chemical character does not change with the state of vibration. From the outset, then, the state of vibration may not be incorporated into the definition of the assemblage of atoms.

b. *Changes of the electron states through excitations of one or more electrons*: In general, a change of this kind requires an energy that is large compared to the energy of thermal motion. If it is brought about through external interference, by light or electron impact, then either it will be completely reversed through an emission of radiation or radiation-free transition to the electron's ground state[8] (a kind of regeneration of the electron configuration), or it will lead to the following process.

c. *Rearrangement of the atoms into another equilibrium state:*[9]

α. *Through fluctuations in the thermal energy*. A rearrangement of this kind can occur when, through a stochastic fluctuation of the thermal energy, a mode of vibration develops in the assemblage of atoms with an amplitude so great that the limit of stability is overcome and the atoms no longer revert to their original positions. This stability limit must necessarily lie at energies considerably greater than the mean energy of thermal motion per degree of freedom. In principle, every stability limit can be overcome in this way. The probability of this happening can be determined as follows. We designate the energy needed for overcoming the stability limit, the so-called activation energy, by U, and the mean energy of thermal motion per degree of freedom, which is proportional to the absolute temperature, T, by kT. Then the probability, W, that at any moment a degree of freedom has an energy U at temperature T, decreases exponentially with the ratio of U to kT:

(1) $$W = Ze^{\frac{-U}{kT}}.$$

For our problem, however, we must ask not about the probability that a degree of freedom has an energy U, but rather this: how long does it take on average for a degree of freedom, starting with an energy smaller than U, to obtain an energy greater than U? This duration measures the mean lifetime of the molecule with respect to the type of instability under consideration. The reciprocal of this time, the transition frequency, measures the rate of the corresponding reaction. According to a well-known theo-

---

8. In this case, the energy is converted into vibrational energy, without a chemical change taking place. The energy blends into the general energy of thermal motion.

9. For this section, compare, for example, Bonhoeffer and Harteck 1933 and Eggert 1929.

rem of the probability calculus, this duration is independent of preceding fluctuations, so that molecules that remain inactivated up to a certain point in time do not thereby acquire a greater probability of activation in the future. The rate of the reaction is therefore time-independent. In order to find its magnitude, we must know how frequently a degree of freedom changes its energy at all. We can take this frequency crudely to be the frequency of atomic vibration; the expression for the transition frequency then takes the form (1), where Z is now to be understood as the mean vibrational frequency of the atom in the molecule, a quantity that for the most part does not depend on the temperature and in general is on the order of magnitude of 10^{14} per second. In order for W, the transition frequency, to be of a measurable order of magnitude, U must be, as we already emphasized, considerably greater than kT. For greater clarity, we provide a small table showing the correlation of the reaction rate (and, accordingly, its reciprocal, the half-life) with the ratio of U to kT. Moreover, in the fourth column of table 14, we list the respective absolute value U at room temperature,[10] and in the fifth column, we list the ratio of the reaction rate at two temperatures that differ from each other by 10 degrees [Celsius]. This ratio, whose weak temperature dependence is ordinarily identified as the van't Hoff law, can be calculated from expression (1):

(2)
$$\frac{W_{T+10}}{W_T} = e^{\frac{+10U}{kT^2}}.$$

TABLE 14. **Correlation between the rate of reaction and the ratio of activation energy to mean energy of thermal motion per degree of freedom U/kT, the absolute values of U at room temperature (U in eV), and the temperature quotient for 10°C.**

$\frac{U}{kT}$	W in sec^{-1}	$\frac{1}{W}$	U in eV	$\frac{W_T + 10}{W_T}$
10	$4.5 * 10^9$	$2 * 10^{-10}$ sec	0.3	1.4
20	$2.1 * 10^5$	$5 * 10^{-6}$ sec	0.6	1.9
30	9.3	0.1 sec	0.9	2.7
40	$4.2 * 10^{-4}$	33 min	1.2	3.8
50	$1.9 * 10^{-8}$	16 months	1.5	5.3
60	$8.7 * 10^{-13}$	30,000 years	1.8	7.4

10. For a unit of energy, we have used the so-called electron volt, which is the energy that an electron gains by accelerating through a potential difference of 1 volt. The chemical binding energies are in general about a volt in size, as are the electron-activation energies. The thermal energy at room temperature is around 0.03 eV.

The most important feature of the correlations presented in the table is that very slight changes in the activation energy produce enormous changes in the rate of reaction. For example, a change in the half-life from one second to more than one year is associated with an increase in the activation energy from only 0.9 to 1.5 eV (around 70%). Since the known molecular activation energies range throughout and beyond these limits, one can from the outset expect reaction rates of every order of magnitude. In some cases, the reaction rates must be so small that we can no longer detect them with the usual measurement methods of chemistry, although these reactions may be hardly different energetically or mechanically from the easily detectable ones. The fourth column in the table shows that the van't Hoff factor increases with increasing activation energy, but only slowly. For chemical reactions with measurable reaction rates, this value lies between 2 and 5, in accordance with a well-known chemist's rule of thumb. The same has been found for the change of the developmental rates of organisms, and has been interpreted to mean that the reaction sequences in an organism must all conform to the slowest-proceeding reaction, and that these follow the van't Hoff rule.

β. *Delivery of energy to an electron from outside.* Apart from fluctuations in the thermal energy, the activation energy can also be supplied by external radiation, electron impact, or energy-producing chemical reactions. We wish to exclude this last case from our gene model, because we assume that the gene is not readily disposed to reacting chemically. In the first two cases, an electron is brought into an excited state or dislodged through ionization, as mentioned in b. The excited electron need not be situated exactly in the location of the rearrangement. The excitation first causes a sudden, violent change in the forces that had previously kept neighboring atoms in equilibrium. The neighboring atoms thus begin to vibrate violently, and this vibrational energy then rapidly propagates further to other neighboring atoms, causing the energy per degree of freedom to trail off as the added energy is shared by more and more atoms. In this way, the added energy, originally localized in a certain place, dissipates and is lost into general thermal motion, *but only if* this process does not directly affect a location that is capable of some rearrangement, given the requisite activation energy.

A particular rearrangement can therefore arise in basically two ways: either through random accumulation of thermal energy, or through dissipation of excitation energy from an electron.

3. Test of the Model

After this qualitative outline of reactions in which molecular rearrangements occur, we want now to see how far we can recover the phenomena
of mutations discussed above.

We have presented genes as well-defined molecules that do not generally
change over the course of the development of individuals or of a population.
This *stability* must have come about in some way through the conditions under which life evolves, where natural selection has surely played a decisive
role as the controlling factor in the selection of especially stable formations.
At the same time, we must expect that selection has driven this stability only
so far as to exclude changes that emerge with appreciable frequency. There
must remain, then, some rearrangements whose frequency is low relative to
lifespan. We detect these in wild strains as mutations. The corresponding
reaction rate, that is, the mutation rate, is around a few orders of magnitude
smaller than that of [embryological] development. *Accordingly, the van't
Hoff factor must be noticeably larger than that of development* (compare
table 14), in agreement with experiment (table 8). It is especially satisfying
that this deviation of the van't Hoff factor from its usual value is explained
by our model without any auxiliary assumption.

In a wild strain, a gene that has experienced a rearrangement at one
location will sometimes be capable of further rearrangement at the same
location. The conditions of natural selection do not place an upper bound
on how frequently this can happen. With artificially selected mutants, we
should therefore expect occasionally to have genes that mutate with ab-
normal frequency, in agreement with experiment ([see] p. 244 [214]). In
our model, then, frequently mutating genes do not need to be intrinsically
different from the stable ones of the wild strains; rather, their occurrence
is enabled by the fact that the conditions of artificial selection are different
from those of natural selection.

The frequent mutations have a rate of reaction that is around a few orders of magnitude larger than that of the normal ones. Their rate of reaction is already comparable to that of [embryological] development. Their
van't Hoff factor should therefore not be noticeably different from that of
development, in agreement with experiment (see pp. 244–5 [214]).

According to our hypothesis, a *particular mutation is a particular rearrangement of a particular molecule.* It must be possible, therefore, to produce
a mutation artificially through some small ionization or excitation. If we

irradiate living matter with X-rays, thereby producing ionizations at random locations, then the probability of producing a specific mutation must be proportional to and depend exclusively on the amount of ionization per cubic centimeter, the ionization density, in agreement with experiment (table 3 and figure 6).

According to our model, a mutation that spontaneously occurs more frequently than another by several orders of magnitude should have an activation energy that is scarcely different from that of the other. After irradiation with X-rays, therefore, both must occur with similar frequency, in agreement with experiment (table 11 and pp. 244–45 [214–15]).

A mutation consists of a rearrangement of a stable assemblage of atoms, according to our model, and this rearrangement results from an *elementary process*. The latter refinement is compelled by the frequently observed similarity between forward and back mutation. Were further secondary processes coupled to the first elementary process, a simple reversibility would be very difficult to understand. Of course, the possibility remains that there are different types of gene mutation, consisting of either elementary or complex processes. If we were to reason by analogy to photochemical experiments, we would actually consider the latter case to be the more frequent, because it is the rule and not the exception in photochemistry that secondary reactions are linked to the primary photochemical process. But we must take into consideration that the manner in which chemical reactions are controlled in the living cell, such that they always occur only at entirely determinate and specific locations, is still completely unknown to us. Comparisons with the kinetics of ordinary photochemistry can therefore be made only with this caveat.

We can make very precise statements about the *mechanism of the effect of X-rays*, because we understand in complete detail the mechanism of the absorption of X-rays. The breakdown of the energy of X-rays proceeds in increments. First, light quanta give their energy mostly or entirely to a fast secondary electron. The secondary electron disperses its energy in many small portions, through ionizations or excitations of atoms (figure 10). Along the way, it loses an energy of about 30 eV per ionization. This energy is small compared to the overall energy of the secondary electron, but still around 1,000 times greater than the energy of thermal motion per degree of freedom, and around 20 times greater than the activation energy required for our mutation process. Thus the distance traveled by the secondary electron between two ionizations is still very large compared to the dimensions of atoms, somewhere around 100 and 1,000 atomic diameters. The individual ionizations therefore represent completely separate events. They can produce a collective effect only through very complicated intermediate processes, as appears to be the case with radiological-physiological effects. In direct contrast to the experiments of radiation physiology, we have found, as analyzed in the second part, that in radiation genetics the mutation-triggering effect depends exclusively on the dose, i.e., the energy absorbed per unit volume. The only conclusion permitted by this basic fact is that a mutation is triggered through a single ionization or excitation, in agreement with our model (table 3, figure 6, and p. 253 [222]).

We will now offer a comparison of the fundamental principles of photochemistry and X-ray chemistry. In photochemistry, the amount of matter affected is determined by the number of absorbed light quanta, in accordance with Einstein's law of equivalence. In X-ray chemistry, that is, with chemical processes produced by X-rays, the amount of matter primarily affected is determined not by the number of absorbed light quanta, but rather by the absorbed energy per unit volume, as has been confirmed experimentally in many simple cases (Günther 1934). The simple explanation for this is that the absorption of a light quantum does not give rise to an elementary chemical process, but rather to subsequent ionizations and, accordingly, excitations, whose number is proportional to the absorbed energy, since the energy dispensed per ionization fluctuates very little on average.

We are not concerned here at all with the amount of affected matter, but rather only with a single, specific elementary process, namely the rearrangement of an assemblage of atoms. The rearrangement can, as mentioned earlier, result from an ionization or excitation; indeed, we must not expect that an entirely specific ionization is necessary for this, but rather that any bundle of energy introduced in the neighborhood can achieve the activation through dissipation. The ionization must take place close enough to the location of the rearrangement to keep the energy of the dissipation process from degenerating below the amount of 1.5 volts per degree of freedom (the activation energy). We know very little about the details of these dissipation processes. We can therefore make no certain statement about the absolute value of the dose needed for a near-certain probability of the creation of any one mutation. Even so, according to what we have just said, we can expect that the dose in question, expressed as the number of ionizations per unit volume, is smaller than the number of atoms per unit volume by a factor on the order of magnitude of around 10 or 100.

For comparison with experiment, we consider one of the mutations most frequently obtained after X-ray irradiation, namely the mutations of normal to eosin [eye color] (table 10; W→w^e), which occur after one dose of 6,000 r on average around once per 7,000 gametes. Given that the mutation rate is proportional to the dose, we can calculate that one dose of $6,000 \times 7,000 = 42,000,000$ r should produce this mutation with a probability comparable to 1. On the other hand, the unit dose (r) in a cubic centimeter of normal air produces around 2×10^9 ion pairs, and in one cubic centimeter of water or organic substance about 1,000 times that, or around 2×10^{12} ion pairs; 42×10^6 r thus produces around 1×10^{20} ion pairs with an energy of 30 eV each. Given that around 1×10^{23} atoms are contained in a cubic centimeter, at least a thousandth of the atoms are ionized in this case. We must take the fact that the reaction occurs at this dose with a probability comparable to 1 to mean that the dissipation of the energy does not occur with maximal swiftness. To account for this, different versions of the model can be developed, although we have not done so in this qualitative overview.

4. Concluding Remarks

The comparison of our model of the gene with the experimental results of mutation research has, as we have shown above, resulted in qualitative agreement in multiple respects. The view of gene mutation as an elementary process in the sense of quantum theory, in particular as a specific change in a complicated assemblage of atoms, can be considered secure. This accounts for the general parallels between the spontaneous and radiation-induced mutation processes, as well as for many individual characteristics of those processes. In the following section, we will use this model as the basis for a general discussion of the nature of the mutation process and the structure of the gene.

Part 4: Theory of Gene Mutation and Gene Structure

N. V. TIMOFÉEFF-RESSOVSKY, K. G. ZIMMER, AND M. DELBRÜCK

1. Discussion of the Mutation Process

On the basis of the experiments and considerations discussed above, we come to the following picture of the mutation process.

A mutation is produced by an external infusion of energy or a fluctuation of thermal energy (which is inevitable due to the statistical-
kinetic nature of heat), and it consists of a rearrangement of atoms into
a new equilibrium configuration within a larger assemblage of atoms. An
assemblage of atoms is defined as a structure in which specific atoms are
arranged stably in determinate locations.

"Spontaneous" mutations are produced through random thermal fluctuations; the probability of crossing the threshold for the reaction to occur
depends on the structure of the corresponding assemblage of atoms (the
corresponding allele), and this accounts for the variety in the spontaneous rates of different individual genes. Many attempts have been made to
derive the spontaneous mutation rate from the effect of "natural" ionizing
radiation; all calculations (Efroimson 1931; Muller and Mott-Smith 1930;
Timoféeff-Ressovsky 1931b) have shown that the amount of this radiation
is far too small to generate the spontaneous mutation rate. Due to the considerations above, resorting to natural radiation or other sundry mutation-
triggering sources is unnecessary.

In radiation-genetic experiments, an additional energy is introduced by
radiation quanta. Analysis of the results of experiments with irradiation
at different doses, different wavelengths, and different temporal distributions confirms the model's claim that an ionization or atomic excitation constitutes a mutation-triggering "target-strike." Figure 10 shows a
schematic portrayal of the secondary and tertiary processes following the
absorption of an X-ray, where the ionization or electron excitation is portrayed as a tertiary process.

1. Photoelektron.

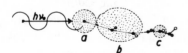

2. Comptonelektron.

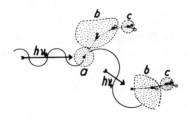

FIGURE 10. Diagram of the secondary and tertiary
processes subsequent to the transit of an X-ray or
gamma ray: a. absorption of the radiation quantum; b. transit and absorption of the secondary
electron; c. absorption of the tertiary electron =
ionization unit. White circles = ionization; struck-
through = excitation.

The idea that a mutation is an individual elementary process in the sense of quantum theory is thus suitable for giving an account of the spontaneous as well as radiation-induced mutation processes.

More specifically, we can expect that further analysis of radiation-induced mutation processes will show close parallels to photochemistry. In both cases, the primary process consists of an excitation and, accordingly, an ionization of an atom. In photochemistry, the use of monochromatic ultraviolet radiation is specially tailored for application to homogeneous chemical systems with well-defined absorption spectra; in a mutation experiment, the use of such radiation should be well suited for singling out specific groups of mutations that can be triggered by absorption of only specific wavelengths. By a rough calculation, we can readily assure ourselves that the ordinary transition probabilities of electron transitions are large enough that we can expect a measurable yield of mutations through the use of ordinary light quanta.

From photochemistry, we know that very different types of secondary processes can accompany the primary process of absorption. The primary process can result in a simple rearrangement (e.g., from maleic to fumaric acid and vice versa). Yet absorption can also lead to a dissociation of some specific bond, whereby a reactive atom or radical is released. A new group of atoms from the environment can then take the place of this atom or radical, perhaps increasing or decreasing the overall size of the molecule. Photochemical reactions with complicated secondary reactions are not generally reversible by photochemical means. Analogously, we must expect that types of mutation will occur in irradiation experiments that cannot be reversed through irradiation.

In the preceding sections, we did not discuss the *temperature experiments* involving so-called temperature shocks, or temperatures outside the normal physiological range of *Drosophila*. The results of these experiments, still somewhat contradictory, suggest that the behavior of the mutation rate differs from that found in the normal physiological temperature zone. In interpreting these phenomena, however, one must take into account what was said at the beginning of the first part (p. 227 [194]), namely that one cannot be certain whether the mutation rate can be directly manipulated through temperature, or whether the effect is indirect, for example through some energy-supplying, defensive reaction of the organism. This conjecture is strengthened by the fact that low-temperature experiments have also yielded an increase in the mutation rate. (Gottschewski 1934).

As mentioned at the beginning of the third part (p. 257 [226]), we are leaving aside the question of gene replication. Most mutations are independent of the stages in which division or replication of the genes occurs (as have been all of the experiments forming the basis for our considerations). It is not impossible, however, as M. Demerec has already pointed out (1932a), that some mutations, particularly those of the "mutable" alleles, could be linked to the replication process of the gene, in that they may involve exceptions to the fundamental property of genes—identical replication—forming so-called genetic "monsters."

2. Theory of Gene Structure

The ideas developed above apply directly to the mutation process because they were based on the results of mutation research. Yet they actually already contain our model of gene structure.

A mutation consists of a rearrangement or dissociation of a bond within an assemblage of atoms (as defined above). One can easily picture the gene as this assemblage of atoms. Accordingly, that physical-chemical unit (assemblage of atoms) in which the mutation process can occur would represent the structure of the entire gene. This picture emerges entirely naturally from the facts and considerations presented here, and satisfies the requirement from genetics that the gene be conceived of as a unit, largely autonomous in its behavior and normally incapable of further division.

On the other hand, the following objection could admittedly be made: Even if one grants that a mutation does in fact entail a change in an assemblage of atoms, one could still imagine the gene as a specific quantity of matter—a particle of matter—composed of many of the same assemblages of atoms. A mutation would then mean a change (or disintegration) of one assemblage of atoms, another mutation a change of two assemblages of atoms, etc. The following contradicts such an idea: if the gene is made of several identical units, then various ancillary hypotheses and auxiliary assumptions must be made in order to satisfy the requirements from genetics mentioned above regarding the conception of the gene (the gene as a unit). Furthermore, difficulties would arise for the explanation of back mutations: a "forward mutation" would be a consequence of the absorption of a target-strike on any of the assemblages of atoms constituting the gene; but, in order to produce a reverse mutation, not just any arbitrary

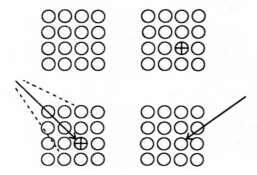

FIGURE 11. Diagram of the "forward and back mutation," assuming that a gene consists of several similar molecules. The probability of a "back mutation" is smaller than the probability of a "forward mutation" by a factor of approximately the number of molecules constituting the gene.

assemblage of atoms can be hit but rather the precise assemblage that had been modified earlier must be hit again, and this becomes more and more improbable as one assumes more and more units in the gene. This is diagrammed schematically in figure 11. Since we find similar probabilities for the forward and back mutations of individual pairs of alleles (table 10), this picture, though worth mentioning, does not do justice to the facts, at least not without auxiliary assumptions. Ultimately, a general consideration speaks against this picture: over time, initially homogeneous particles of matter (composed of the same molecules) must become heterogeneous through repeated mutating, provided one does not employ auxiliary assumptions about some automated regulating processes; we would thus return to a view similar in principle to ours, and the gene could be conceived as a structural unit or assemblage of atoms with different individual parts, perhaps showing a certain periodicity.

Consequently, *we view the gene as an assemblage of atoms within which a mutation can proceed as a rearrangement of atoms or dissociation of bonds* (triggered by thermal fluctuations or external infusion of energy), and which is largely autonomous in its operations and in relation to other genes. At this early stage, there is little point in trying to concretize these concepts any further. We also leave open for now the question of whether the individual genes are separate, individual assemblages of atoms, or whether they form largely autonomous parts of a bigger structural unit, and thus whether a chromosome is a string of beadlike genes or a physical-chemical continuum (Koltzoff 1928). This question, as well as the problem of identical gene duplication prior to cell division, is reserved for future analysis, until suitable experimental data can be obtained.

3. Consequences

The considerations developed here give rise both to practical questions for future mutation experiments and to consequences for some more general genetic and biological notions.

The radiation-genetic experiments concerning the relation of mutation rate to radiation dose, wavelength, and time dependence of the dose—i.e., those important for the definition of "target-strikes"—can be regarded as essentially completed, especially since a series of experiments performed independently by different researchers have produced results that agree very well with one another. The X-ray– and radium-induced mutation rate in *Drosophila* can continue, potentially, to be used as a known and well-defined reaction in certain radiation-biological and perhaps also radiation-physics experiments.

We can expect interesting results, as was shown earlier (p. 266 [236]), from the use of monochromatic ultraviolet light. With it, we should be able to produce specific groups of mutations independently.

We still do not know much with precision about the mutability of individual genes. We need to examine a set of individual genes in extensive radiation-genetic experiments concerning the relation of their mutation rates to the dose and the type of radiation. Results could thus be obtained that might alter or reinforce and deepen the ideas developed here. Above all, we must test how genes with high and low spontaneous mutation rates behave in irradiation experiments, and whether they follow the prediction presented on page 261 [231], which we used to reconcile the differences in the spontaneous-mutation rate in the X-ray irradiation experiment. The known data on this subject are still very meager (p. 244 [214] and table 11).

A more significant matter must also be decided: whether, as the model requires, a mutation that practically never occurs spontaneously can occur through irradiation with X-rays.

For individual genes, we need to establish more data on the van't Hoff factor, in order to test the implications of table 14, that the van't Hoff factor of genes with a higher spontaneous-mutation rate should be smaller than that of genes with a low rate.

It would be important to investigate the influence of variations in the chemical environment on the radiation-induced mutability of genes. One could thus perhaps determine whether and which type of secondary reactions could be linked to the primary process of excitation. Hopefully, we could have a direct test of the independence of the radiation-induced mutation rate from the physiological condition of the irradiated tissue; the currently available data (pp. 232–33 [200–20]) are still very poor.

Finally, it would be very instructive to compare the results of radiation genetics with specific photochemical experiments tailored as closely as possible to our model and to known data, and carried out under controlled conditions.

As for the general consequences of the ideas developed here, they can be summarized briefly as follows.

According to the conception of many biologists, the genome is a highly complicated physical-chemical structure, consisting of a series of specific, chemical pieces of matter—the individual genes. Some attempts have been made to project back theoretically, by way of the hereditarily modifiable, ontogenetic developmental sequences, from the organism to its individual genes. The genes are thus conceived as the immediate "starting points" of the chains of reactions constituting the developmental processes. On one hand, this conception requires that we assume a highly complicated structure and mode of operation for the genes, and that we deal with the gene problem from the standpoint of the requirements of developmental physiology. On the other hand, it leads to an explicit or implicit critique of cell theory; the cell, thus far proving itself so magnificently as the unit of life, dissolves into the "ultimate units of life," the genes.

Our ideas about the gene challenge this picture. Genes are physical-chemical units; perhaps the whole chromosome (to be sure the part containing genes) consists of such a unit, a large assemblage of atoms, with many individual, largely autonomous subgroups. Such genes are likely incapable of directly forming the morphogenic substances; they also can hardly be thought of as the "starting points" of developmental sequences. Nevertheless, such a genome can be thought of as the foundation for specific, heredity-conditioned morphogenesis, by providing a steady, form- and function-determining framework for the cell (Koltzoff 1928). Changes to its individual parts (gene mutation) would influence the overall functioning of the cell in specific ways and, thus, the individual development processes as well. Therefore, we need not dissolve the cell into genes, and the "starting points" of the developmental sequences are not attributed to individual genes, but rather to operations of the cell, or even to intercellular processes (which are all eventually controlled by the genome).

These are primarily speculations, which rest on still shaky ground. Yet, general ideas can sometimes influence the posing of concrete questions. And we believe that it would be advisable for related fields to work with conceptions of the gene that are constructed on the relevant and, as mentioned in the introduction, singularly decisive data of mutation research.

References in the Three-Man Paper

Prepared by James Barham

Items with asterisks are not cited in the body of the paper.—Trans.

*Alexander, J. and C. B. Bridges. 1928. "Some Physico-chemical Aspects of Life, Mutation and Evolution." In *Colloid Chemistry: Theoretical and Applied*, vol. 2, edited by J. Alexander, 9–58. New York: Chemical Catalog.

*Altenburg, E. 1928. "The Limit of Radiation Frequency Effective in Producing Mutations." *American Naturalist*, 62: 540–45.

*———. 1930. "The Effect of *Ultra*-violet Radiation on Mutation." Abstract. *Anatomical Record*, 47: 383.

*———. 1933. "The Production of Mutations by Ultra-violet Light." *Science*, 78: 587.

*———. 1934. "The Artificial Production of Mutations by Ultra-violet Light." *American Naturalist*, 68: 491–507.

Belgovskij [Belgovskii], M. L. 1934. "Hybridization Effect on the Mutability of the White Gene in *Drosophila simulans*." (In Russian, with substantial English summary.) *Doklady Akademii Nauk SSSR*, n.s., 5: 108–12.

Berg, R. L. 1934. "The Relative Mutation Frequencies in *Drosophila* Chromosomes." (In Russian, with substantial English summary.) *Doklady Akademii Nauk SSSR*, n.s., 5: 234–39.

Blau, M. and K. Altenburger. 1923. "Über einige Wirkungen von Strahlen II." *Zeitschrift für Physik*, 12: 315–29.

Bonhoeffer, K. F. and P. Harteck. 1933. *Grundlagen der Photochemie*. Dresden/Leipzig: T. Steinkopff.

*Cronheim, G., S. Götzky, and P. Günther. 1931. "Der Zerfall des Benzophenondiazids unter dem Einfluß von Röntgenstrahlen." In *Bodenstein-Festband, Zeitschrift für physikalische Chemie: Ergänzungsband*, edited by C. Drucker, G. Joos, and F. Simon, 785–91. Leipzig: Akademische Verlagsgesellschaft.

*Cronheim, G. S. and P. Günther. 1930. "Die Energieausbeute bei der Zersetzung von Chloroform durch Röntgenstrahlen und der Mechanismus dieser und ähnlicher Röntgenreaktionen." *Zeitschrift für physikalische Chemie, Abteilung B*, 9: 201–28.

Crowther, J. A. 1926. "The Action of X-Rays on *Colpidium colpoda.*" *Proceedings of the Royal Society of London, Series B*, 100: 390–404.

Demerec, M. 1928. "The Behavior of Mutable Genes." In *Verhandlungen des fünften internationalen Kongresses für Vererbungswissenschaft, Berlin, 1927.* (*Zeitschrift für induktive Abstammungs- und Vererbungslehre, Supplement Band I*). Vol. 1, edited by H. Nachtsheim, 183–93. Leipzig: Gebrüder Bornträger.

———. 1929a. "Mutable Genes in *Drosophila virilis.*" In *Proceedings of the International Congress of Plant Sciences, Ithaca, NY, Aug. 16–23, 1926.* Vol. 1, edited by B. M. Duggar, 943–46. Menasha, WI: Collegiate Press/George Banta Publishing Co.

———. 1929b. "Changes in the Rate of Mutability of the Mutable Miniature Gene of *Drosophila virilis.*" *Proceedings of the National Academy of Sciences (USA)*, 15: 870–76.

———. 1929c. "Genetic Factors Stimulating Mutability of the Miniature-Gamma Wing Character of *Drosophila virilis.*" *Proceedings of the National Academy of Sciences (USA)*, 15: 834–38.

———. 1932a. "Effect of Temperature on the Rate of Change of the Unstable Miniature-3 Gamma Gene *of Drosophila virilis.*" *Proceedings of the National Academy of Sciences (USA)*, 18: 430–34.

———. 1932b. "Rate of Instability of Miniature-3 Gamma Gene *of Drosophila virilis* in the Males in the Homozygous and in the Heterozygous Females." *Proceedings of the National Academy of Sciences (USA)*, 18: 656–58.

*———. 1933a. "What Is a Gene?" *Journal of Heredity*, 24: 369–78.

*———. 1933b. "The Effect of X-Ray Dosage on Sterility and Number of Lethals in *Drosophila melanogaster.*" *Proceedings of the National Academy of Sciences (USA)*, 19: 1015–20.

———. 1934. "Effect of X-Rays on the Rate of Change in the Unstable Miniature-3 Gene of *Drosophila virilis.*" *Proceedings of the National Academy of Sciences (USA)*, 20: 28–31.

Dessauer, F. 1923. "Über einige Wirkungen von Strahlen I." *Zeitschrift für Physik*, 12: 38–47.

———. 1933. "Quantenphysik der biologischen Röntgenstrahlenwirkungen." *Zeitschrift für Physik*, 84: 218–21.

*Dubinin, N. P. 1929. "Allelomorphentreppen bei *Drosophila melanogaster.*" *Biologisches Zentralblatt*, 49: 328–39.

*———. 1932a. "Step-allelomorphism in *Drosophila melanogaster.*" *Journal of Genetics*, 25: 163–81.

*———. 1932b. "Step-allelomorphism and the Theory of Centres of the Gene Achaete-scute." *Journal of Genetics*, 26: 37–58.

Efroimson, W. P. 1931. "Die transmutierende Wirkung der X-Strahlen und das Problem der genetischen Evolution." *Biologisches Zentralblatt*, 51: 491–506.

Eggert, J. 1929. *Lehrbuch der physikalischen Chemie in elementarer Darstellung.* Leipzig: S. Hirzel.

Friedrich, W. and K. Zimmer. 1934. "Beiträge zum Problem der Radiumdosimetrie. Teil III. Probleme der Dosismessung in der Praxis." *Strahlentherapie*, 51: 32–45.

Glocker, R. 1932. "Quantenphysik der biologischen Röntgenstrahlenwirkung." *Zeitschrift für Physik*, 77: 653–75.

*Glocker, R. and A. Reuß. 1933 "Über die Wirkungen von Röntgenstrahlen verschiedener Wellenlänge auf biologische Objekte. V." *Strahlentherapie*, 47: 28–34.

Goldschmidt, R. 1928. "The Gene." *Quarterly Review of Biology*, 3: 307–24.

Gottschewski, G. 1934. "Untersuchungen an *Drosophila melanogaster* über die Umstimmbarkeit des Phänotypus und Genotypus durch Temperatureinflüsse." *Zeitschrift für induktive Abstammungs- und Vererbungslehre*, 67: 477–528.

*Gowen, J. W. and E. H. Gay. 1933. "Gene Number, Kind, and Size in *Drosophila*." *Genetics*, 18: 1–31.

*Griffith, H. D. and K. G. Zimmer. 1935. "The Time-Intensity Factor in Relation to the Genetic Effects of Radiation." *British Journal of Radiology*, n.s., 8: 40–47.

*Grüneberg, H. 1931. "Über die zeitliche Begrenzung genetischer Röntgenwirkungen bei *Drosophila melanogaster*." *Biologisches Zentralblatt*, 51: 219–25.

Günther, P. 1934. "Reaktionsanregung durch Röntgenstrahlen und durch Ionen." In *Anwendungen der Röntgen- und Elektronenstrahlen: Ergebnisse der technischen Röntgenkunde*, vol. 4, edited by J. Eggert and E. Schiebold, 100–112. Leipzig: Akademische Verlagsgesellschaft.

*Hanson, F. B. 1928. "The Effect of X-Rays in Producing Return Gene Mutations." *Science*, 67: 562–63.

Hanson, F. B. and F. Heys. 1929. "Analysis of the Effects of the Different Rays of Radium in Producing Lethal Mutations in *Drosophila*." *American Naturalist*, 63: 201–13.

———. 1932. "Radium and Lethal Mutations in *Drosophila*: Further Evidence of the Proportionality Rule from a Study of the Effects of Equivalent Doses Differently Applied." *American Naturalist*, 66: 335–45.

*Hanson, F. B., F. Heys, and E. Stanton. 1931. "The Effect of Increasing X-Ray Voltages on the Production of Lethal Mutations in *Drosophila melanogaster*." *American Naturalist*, 65: 134–43.

Harris, B. B. 1929. "The Effects of Aging of X-Rayed Males upon Mutation Frequency in *Drosophila*." *Journal of Heredity*, 20: 299–302.

Holweck, F. and A. Lacassagne. 1934. "Le problème des quanta en radiobiologie." Abstract. In *IV. [Vierter] internationaler Radiologenkongress, Zürich, 1934. Referate*, vol. 2, 407–8. Leipzig: Georg Thieme.

Johnston, O. and A. M. Winchester. 1934. "Studies on Reverse Mutations in *Drosophila melanogaster*." *American Naturalist*, 68: 351–58.

Koltzoff [Kol'tsov], N. K. 1928. "Physikalisch-chemische Grundlage der Morphologie." *Biologisches Zentralblatt*, 48: 345–69.

*————. 1930. "On the Experimental Production of Mutations." (In Russian only.) *Zhurnal Eksperimental'noi Biologii*, 6: 237–49.

Mayneord, W. V. 1934. "The Physical Basis of the Biological Effects of High Voltage Radiations." *Proceedings of the Royal Society of London, Series A*, 146: 867–79.

Medvedev, N. N. 1933. "The Production of Mutations in *Drosophila melanogaster* by the Combined Influence of X-Rays and Salts of Heavy Metals." (In Russian, with substantial English summary.) *Doklady Akademii Nauk SSSR*, n.s., 5: 230–36.

Moore, W. G. 1934. "A Comparison of the Frequency of Visible Mutations Produced by X-Ray Treatment in Different Developmental Stages of *Drosophila*." *Genetics*, 19: 209–22.

*Morgan, L. V. 1922. "Non-Criss-Cross Inheritance in *Drosophila melanogaster*." *Biological Bulletin*, 42: 267–74.

*Morgan, T. H. 1926. *The Theory of the Gene*. New Haven, CT: Yale University Press.

*Morgan, T. H., C. B. Bridges, and A. H. Sturtevant. 1925. "The Genetics of *Drosophila*." *Bibliographia Genetica*, 2: 1–258.

Muller, H. J. 1920. "Further Changes in the White-Eye Series of *Drosophila* and Their Bearing on the Manner of Occurrence of Mutations." *Journal of Experimental Zoology*, 31: 443–73.

————. 1922. "Variation due to Change in the Individual Gene." *American Naturalist*, 56: 32–50.

————. 1923. "Mutation." In *Eugenics, Genetics, and the Family: Scientific Papers of the Second International Congress of Eugenics, Held at the American Museum of Natural History, New York, September 22–28, 1921*. Vol. 1, 106–12. Baltimore: Williams & Wilkins Co.

*————. 1927a. "Quantitative Methods in Genetic Research." *American Naturalist*, 61: 407–19.

————. 1927b. "Artificial Transmutation of the Gene." *Science*, 66: 84–87.

————. 1928a. "The Measurement of Gene Mutation Rate in *Drosophila*, Its High Variability, and Its Dependence upon Temperature." *Genetics*, 13: 279–357.

————. 1928b. "The Problem of Genic Modification." In *Verhandlungen des fünften internationalen Kongresses für Vererbungswissenschaft, Berlin, 1927*. (*Zeitschrift für induktive Abstammungs- und Vererbungslehre, Supplement Band I*), vol. 1, edited by H. Nachtsheim, 234–60. Leipzig: Gebrüder Bornträger.

————. 1928c. "The Production of Mutations by X-Rays." *Proceedings of the National Academy of Sciences (USA)*, 14: 714–26.

————. 1929a. "The Gene as the Basis of Life." In *Proceedings of the International Congress of Plant Sciences, Ithaca, NY, Aug. 16–23, 1926*, vol. 1, edited by B. M. Duggar, 897–921. Menasha, WI: Collegiate Press/George Banta Publishing Co.

*————. 1929b. "The Method of Evolution." *Scientific Monthly*, 29: 481–505.

————. 1930a. "Radiation and Genetics." *American Naturalist*, 64: 220–51.

*———. 1930b. "Types of Visible Variations Induced by X-Rays in *Drosophila*." *Journal of Genetics*, 22: 299–334.

*———. 1932a. "Further Studies on the Nature and Causes of Gene Mutations." In *Proceedings of the Sixth International Congress of Genetics, Ithaca, New York, 1932*, vol. 1, edited by D. F. Jones, 213–55. Menasha, WI: Brooklyn Botanic Garden.

*———. 1932b. "Heribert Nilsson's Evidence against the Artificial Production of Mutations." *Hereditas*, 16: 160–68.

———. 1934a. "The Effects of Roentgen Rays upon the Hereditary Material." In *The Science of Radiology*, edited by O. Glasser, 305–18. Springfield, IL: Charles C. Thomas.

*———. 1934b. "Radiation Genetics." Abstract. In *IV. [Vierter] internationaler Radiologenkongress, Zürich, 1934: Referate*, vol. 2, 100–102. Leipzig: Georg Thieme.

Muller, H. J. and E. Altenburg. 1919. "The Rate of Change of Hereditary Factors in *Drosophila*." *Proceedings of the Society for Experimental Biology and Medicine*, 17: 10–4.

*———. 1930. "The Frequency of Translocations Produced by X-Rays in *Drosophila*." *Genetics*, 15: 283–311.

Muller, H. J. and L. M. Mott-Smith. 1930. "Evidence That Natural Radioactivity Is Inadequate to Explain the Frequency of 'Natural' Mutations." *Proceedings of the National Academy of Sciences (USA)*, 16: 277–85.

*Muller, H. J. and F. Settles. 1927. "The Non-functioning of the Genes in Spermatozoa." *Zeitschrift für induktive Abstammungs- und Vererbungslehre*, 43: 285–312.

Neuhaus [Neigauz], M. E. 1934a. "Mutability of the Locus of Bobbed in *Drosophila melanogaster*." (In Russian, with brief English summary.) *Biologicheskii Zhurnal*, 3: 342–45.

*———. 1934b. "The Effect of X-Rays on the Mutational Process in Mature and Immature Sex Cells of Males *of Drosophila melanogaster*." (In Russian, with brief English summary.) *Biologicheskii Zhurnal* 3: 602–8.

*Noethling, W. and H. Stubbe. 1934. "Untersuchungen über experimentelle Auslösung von Mutationen bei *Antirrhinum majus*. V. Die Auslösung von Genmutationen nach Bestrahlung reifer männlicher Gonen mit Licht." *Zeitschrift für induktive Abstammungs- und Vererbungslehre*, 67: 152–72.

Oliver, C. P. 1930. "The Effect of Varying the Duration of X-Ray Treatment upon the Frequency of Mutation." *Science*, 71: 44–46.

———. 1932. "An Analysis of the Effect of Varying the Duration of X-Ray Treatment upon the Frequency of Mutations." *Zeitschrift für induktive Abstammungs- und Vererbungslehre*, 61: 447–88.

*Patterson, J. T. 1929. "X-Rays and Somatic Mutations." *Journal of Heredity*, 20: 261–67.

————. 1931. "Continuous versus Interrupted Irradiation and the Rate of Mutation in *Drosophila.*" *Biological Bulletin*, 61: 133–38.

*————. 1932. "Lethal Mutations and Deficiencies Produced in the X-Chromosome of *Drosophila* by X-Radiation." *American Naturalist*, 66: 193–206.

Patterson, J. T. and H. J. Muller 1930. "Are Progressive Mutations Produced by X-Rays?" *Genetics*, 15: 495–577.

*Patterson, J. T., W. Stone, S. Bedichek, and M. Suche. 1934. "The Production of Translocations in *Drosophila.*" *American Naturalist*, 68: 359–69.

Pickhan, A. 1934. "Vergleich der mutationsauslösenden Wirkung von gleichen Dosen Röntgen- und Gammastrahlen." Abstract. In *IV. [Vierter] internationaler Radiologenkongress, Zürich, 1934: Referate*, vol. 2, 102–3. Leipzig: Georg Thieme.

*Promptov, A. N. 1932. "The Effect of Short Ultra-violet Rays on the Appearance of Hereditary Variations in *Drosophila melanogaster.*" *Journal of Genetics*, 26: 59–74.

*Rajewsky, B. 1930. "Physikalische Darstellung des Schädigungsvorganges und ihre experimentelle Prüfung." In *Zehn Jahre Forschung auf dem physikalisch-medizinischen Grenzgebiet: Bericht des Instituts für physikalische Grundlagen der Medizin an der Universität Frankfurt am Main*, edited by F. Dessauer, 202–35. Leipzig: Georg Thieme.

*————. 1934. "Theorie der Strahlenwirkung und ihre Bedeutung für die Strahlentherapie." In *Wissenschaftliche Woche zu Frankfurt am Main, 2.–9. September 1934. Band II: Carcinom*, edited by W. Kolle, 75–90. Leipzig: Georg Thieme.

Risse, O. 1931. "Die physikalischen Grundlagen der Photochemie (Licht und Röntgenstrahlen)." *Ergebnisse der medizinischen Strahlenforschung*, 5: 71–128.

*Schapiro (Shapiro), N. I. 1931. "Einfluß des Alters der Geschlechtszellen auf die Entstehung von Translokationen bei *Drosophila melanogaster.*" (In Russian only.) *Zhurnal Eksperimental'noi Biologii*, 7: 340–48.

Schapiro (Shapiro), N. I. and M. E. Neuhaus. 1933. "Versuch einer vergleichenden Analyse des Mutationsprozesses bei Männchen und Weibchen von *Drosophila melanogaster.*" (In Russian only.) *Biologicheskii Zhurnal*, 2: 425–46.

Schapiro (Shapiro), N. I. and R. I. Serebrovskaja. 1934. "Relative Mutability of the X and Second Chromosomes *of Drosophila melanogaster.*" (In Russian, with substantial English summary.) *Doklady Akademii Nauk SSSR*, n.s., 5: 228–33.

Schechtmann, J. (Ia. L. Shekhtman). 1930. "Der Mutationseffekt und die quantitative Gesetzmäßigkeit der Röntgenstrahlenwirkung." (In Russian, with brief German summary.) *Zhurnal Eksperimental'noi Biologii*, 6: 271–79.

*Schreiber, H. 1934. "Der photobiologische Wirkungsmechanismus der Röntgenstrahlen." *Die Naturwissenschaften*, 22: 536–40.

Serebrovsky (Serebrovskii), A. S. 1929. "A General Scheme for the Origin of Mutations." *American Naturalist*, 63: 374–78.

Serebrovsky, A. S., N. P. Dubinin, I. I. Agol, V. N. Slepkov, and B. E. Al'tshuler.

1928. "The Origin of Mutations through the Influence of X-Rays in *Drosophila melanogaster.*" (In Russian.) *Zhurnal Eksperimental'noi Biologii*, 4: 161–80.

Sidorov, B. N. 1931. "Zur Frage über die Wirkung der X-Strahlen auf den Mutationsprozeß in unreifen Geschlechtszellen der Männchen von *Drosophila melanogaster.*" (In Russian only.) *Zhurnal Eksperimental'noi Biologii*, 7: 349–51.

*————. 1934. "Einfluß der X-Strahlen auf die Mutationsrate verschiedener Gene im X-Chromosom von *Drosophila melanogaster.*" (In Russian, with brief German summary.) *Biologicheskii Zhurnal*, 3: 123–25.

Stadler, L. J. 1930. "Some Genetic Effects of X-Rays in Plants." *Journal of Heredity*, 21: 3–20.

*————. 1932. "On the Genetic Nature of Induced Mutations in Plants." In *Proceedings of the Sixth International Congress of Genetics, Ithaca, New York, 1932*, vol. 1, edited by D. F. Jones, 274–94. Menasha, WI: Brooklyn Botanic Garden.

*Stubbe, H. 1930–34. "Untersuchungen über experimentelle Auslösung von Mutationen bei *Antirrhinum majus*, I–IV." *Zeitschrift für induktive Abstammungs- und Vererbungslehre*, 1930, 56: 1–38, 202–32; 1932, 60: 474–513; 1934, 64: 181–204.

————. 1933. "Labile Gene." *Bibliographia Genetica*, 10: 299–355.

————. 1934. "Probleme der Mutationsforschung." In *Wissenschaftliche Woche zu Frankfurt am Main, 2.–9. September 1934. Band I: Erbbiologie*, edited by W. Kolle, 71–88. Leipzig: Georg Thieme.

*Sturtevant, A. H. 1925. "The Effects of Unequal Crossing Over at the Bar Locus in *Drosophila.*" *Genetics*, 10: 117–47.

*Timoféeff-Ressovsky, N. V. 1925. "A Reverse Genovariation in *Drosophila funebris.*" (In Russian.) *Zhurnal Eksperimental'noi Biologii*, 1: 143–4. (English translation: *Genetics*, 1927, 12: 125–27.)

*————. 1927. "Eine somatische Rückgenovariation bei Drosophila melanogaster." *Wilhelm Roux' Archiv für Entwicklungsmechanik der Organismen*, 113: 620–25.

*————. 1929a. "The Effect of X-Rays in Producing Somatic Genovariations of a Definite Locus in Different Directions in *Drosophila melanogaster.*" *American Naturalist*, 63: 118–24.

*————. 1929b. "Der Stand der Erzeugung von Genovariationen durch Röntgenbestrahlung." *Journal für Psychologie und Neurologie*, 39: 432–37.

*————. 1929b. "Rückgenovariationen und die Genovariabilität in verschiedenen Richtungen. I. Somatische Genovariationen der Gene *W*, *w^e* und *w* bei *Drosophila melanogaster* unter dem Einfluß der Röntgenbestrahlung." *Wilhelm Roux' Archiv für Entwicklungsmechanik der Organismen*, 115: 620–35.

*————. 1930a. "Rückgenovariationen und die Genovariabilität in verschiedenen Richtungen. II. Rückgenovariationen bei *Drosophila melanogaster* unter dem Einfluß der Röntgenbestrahlung." (In Russian, with brief German summary.)

Zhurnal Eksperimental'noi Biologii, 6: 3–8. (English translation: "Reverse Genovariations and Gene Mutations in Different Directions." *Journal of Heredity*, 1931, 22: 67–70.)

*———. 1930b. "Das Genovariieren in verschiedenen Richtungen bei *Drosophila melanogaster* unter dem Einfluß der Röntgenbestrahlung." *Die Naturwissenschaften*, 18: 434–37.

———. 1930c. "Does X-Ray Treatment Produce a Genetic Aftereffect?" (In Russian.) *Zhurnal Eksperimental'noi Biologii*, 6: 79–83. (English translation: *Journal of Heredity*, 1931, 22: 221–23.)

———. 1930d. "On the Question of the Functioning of the Genes in the Germ Cells." (In Russian.) *Zhurnal Eksperimental'noi Biologii*, 6: 181–87.

*———. 1930e. "Röntgenbestrahlungsversuche mit *Drosophila funebris*." *Die Naturwissenschaften*, 18: 431–34.

———. 1931a. "Einige Versuche an *Drosophila melanogaster* über die Art der Wirkung der Röntgenstrahlen auf den Mutationsprozeß." *Wilhelm Roux' Archiv für Entwicklungsmechanik der Organismen*, 124: 654–65.

———. 1931b. "Die bisherigen Ergebnisse der Strahlengenetik." *Ergebnisse der medizinischen Strahlenforschung*, 5: 129–228.

———. 1932a. "Verschiedenheit der "normalen" Allele der white-Serie aus zwei geographisch getrennten Populationen von *Drosophila melanogaster*." *Biologisches Zentralblatt*, 52: 468–76.

———. 1932b. "Mutations of the Gene in Different Directions." In *Proceedings of the Sixth International Congress of Genetics, Ithaca, New York, 1932*, vol. 1, edited by D. F. Jones, 308–30. Menasha, WI: Brooklyn Botanic Garden.

———. 1933a. "Rückmutationen und die Genmutabilität in verschiedenen Richtungen. III. Röntgenmutationen in entgegengesetzten Richtungen am forked-Locus von *Drosophila melanogaster*." *Zeitschrift für induktive Abstammungs- und Vererbungslehre*, 64: 173–75.

———. 1933b. "Rückgenmutationen und die Genmutabilität in verschiedenen Richtungen. IV. Röntgenmutationen in verschiedenen Richtungen am white-Locus von *Drosophila melanogaster*." *Zeitschrift für induktive Abstammungs- und Vererbungslehre*, 65: 278–92.

*———. 1933c. "Rückmutationen und die Genmutabilität in verschiedenen Richtungen. V. Gibt es ein wiederholtes Auftreten identischer Allele innerhalb der white-Allelenreihe von *Drosophila melanogaster*?" *Zeitschrift für induktive Abstammungs- und Vererbungslehre*, 66: 165–69.

———. 1934a. "Einige Versuche an *Drosophila melanogaster* über die Beziehungen zwischen Dosis und Art der Röntgenbestrahlung und der dadurch ausgelösten Mutationsrate." *Strahlentherapie*, 49: 463–78.

———. 1934b. "Beziehungen zwischen der Mutationsrate und der Dosis und Art der Bestrahlung." Abstract. In *IV. [Vierter] internationaler Radiologenkongress, Zürich, 1934: Referate*, vol. 2, 104–5. Leipzig: Georg Thieme.

———. 1934c. "The Experimental Production of Mutations." *Biological Reviews*, 9: 411–57.

*———. 1934d. "Auslösung von Vitalitätsmutationen durch Röntgenbestrahlung bei *Drosophila melanogaster*." *Strahlentherapie*, 51: 658–63.

*———. 1934e. "Über den Einfluß des genotypischen Milieus und der Außenbedingungen auf die Realisation des Genotyps." *Nachrichten aus der Biologie: Nachrichten von der Gesellschaft der Wissenschaften zu Göttingen. Mathematisch-Physikalische Klasse. Neue Folge. Fachgruppe VI: 1934–35*, vol. 1, 53–106. Göttingen: Vandenhoeck & Ruprecht. [Report communicated in 1934, but true publication date was 1935.]

*———. 1935. "Auslösung von Vitalitätsmutationen durch Röntgenbestrahlung bei *Drosophila melanogaster*." *Nachrichten aus der Biologie. Nachrichten von der Gesellschaft der Wissenschaften zu Göttingen. Mathematisch-Physikalische Klasse. Neue Folge. Fachgruppe VI: 1934–35*, vol. 1, 163–80. Göttingen: Vandenhoeck & Ruprecht.

Timoféeff-Ressovsky, N. V. and K. G. Zimmer. 1935. "Strahlengenetische Zeitfaktorversuche an *Drosophila melanogaster*." *Strahlentherapie*, 53: 134–38.

Wyckoff, R. W. G. 1930a. "The Killing of Certain Bacteria by X-Rays." *Journal of Experimental Medicine*, 52: 435–46.

———. 1930b. "The Killing of Colon Bacilli by X-Rays of Different Wavelengths." *Journal of Experimental Medicine*, 52: 769–80.

*Zeleny, C. 1921. "The Direction and Frequency of Mutations in the Bar-eye Series of Multiple Allelomorphs in *Drosophila*." *Journal of Experimental Zoology*, 34: 203–33.

Zimmer, K. G. 1934. "Ein Beitrag zur Frage nach der Beziehung zwischen Röntgenstrahlendosis und dadurch ausgelöster Mutationsrate." *Strahlentherapie*, 51: 179–84.

References in the Chapters

Archives, web publications, and unpublished materials appear in separate sections at the end of this reference list.

Aaserud, F. *Redirecting Science: Niels Bohr, Philanthropy, and the Rise of Nuclear Physics*. Cambridge: Cambridge University Press, 1990.

Abir-Am, P. G. "The Biotheoretical Gathering, Trans-disciplinary Authority and the Incipient Legitimation of Molecular Biology in the 1930s: New Perspective on the Historical Sociology of Science." *History of Science* 25 (1987): 1–70.

———. "The Discourse of Physical Power and Biological Knowledge in the 1930s: A Reappraisal of the Rockefeller Foundation's 'Policy' in Molecular Biology." *Social Studies of Science* 12 (1982): 341–82.

———. "The First American and French Commemorations in Molecular Biology: From Collective Memory to Comparative History." In *Commemorative Practices in Science*, edited by P. G. Abir-Am and C. A. Elliott, 324–72. Chicago: University of Chicago Press, 2000.

———. "From Multidisciplinary Collaboration to Transnational Objectivity: International Space as Constitutive of Molecular Biology, 1930–1970." In *Denationalizing Science: The Contexts of International Scientific Practice*, edited by E. Crawford, T. Shinn, and S. Sörlin, 153–86. Dordrecht: Kluwer Academic, 1992.

———. "A Historical Ethnography of a Scientific Anniversary in Molecular Biology: The First Protein X-Ray Photograph (1984, 1934)." *Social Epistemology* 6 (1992): 323–54.

———. "Molecular Biology in the Context of British, French, and American Cultures." *International Social Science Journal* 168 (2001): 187–99.

———. "The Molecular Transformation of Twentieth-Century Biology." In *Science in the Twentieth Century*, edited by J. Krige and D. Pestre, 495–524. Amsterdam: Harwood, 1997.

———. "The Politics of Macromolecules: Molecular Biologists, Biochemists, and Rhetoric." *Osiris*, 2nd ser., 7 (1992): 164–91.

———. "The Rockefeller Foundation and Refugee Biologists: European and American Careers of Leading RF Grantees from England, France, Germany, and Italy." In *The "Unacceptables": American Foundations and Refugee Scholars between the Two Wars and After*, edited by G. Gemelli, 217–40. Brussels: P.I.E.–Peter Lang, 2000.

———. "Themes, Genres and Orders of Legitimation in the Consolidation of New Scientific Disciplines: Deconstructing the Historiography of Molecular Biology." *History of Science* 23 (1985): 73–117.

Adams, M. "Sergei Chetverikov, the Kol'tsov Institute, and the Evolutionary Synthesis." In *The Evolutionary Synthesis: Perspectives on the Unification of Biology*, edited by E. Mayr and W. Provine, 242–78. Cambridge, MA: Harvard University Press, 1980.

Anderson, J. M. "Changing Concepts about the Distribution of Photosystems I and II between Grana-Appressed and Stroma-Exposed Thylakoid Membranes." In *Discoveries in Photosynthesis*, edited by J. T. Beatty, H. Gest, and J. F. Allen, 729–36. New York: Springer, 2005.

Anonymous. Review of *Introduction to Biophysics*, by Otto Stuhlman. *Quarterly Review of Biology* 18 (1943): 393–94.

Ash, M. G. *Gestalt Psychology in German Culture, 1890–1967*. Cambridge: Cambridge University Press, 1995.

Autrum, H., and F. L. Boschke. "Obituary for Ernst Lamla." *Die Naturwissenschaften* 73 (1986): 689.

Bauer, H. "Genetische Wirkungen von kurzwelligen und Korpuskularstrahlungen." In *Biophysik*, edited by B. Rajewsky and M. Schön, 1:66–88. Wiesbaden: Dieterich'sche Verlagsbuchhandlung, 1948.

Beatty, J. T., H. Gest, and J. F. Allen, eds. *Discoveries in Photosynthesis*. New York: Springer, 2005.

Beller, M. "The Birth of Bohr's Complementarity: The Context and the Dialogues." *Studies in History and Philosophy of Science* 23 (1992): 147–80.

Berg, R. L. "In Defense of Timoféeff-Ressovsky." *Quarterly Review of Biology* 65 (1990): 457–79.

Beurton, J., R. Falk, and H.-J. Rheinberger, eds. *The Concept of the Gene in Development and Evolution*. Cambridge: Cambridge University Press, 2000.

Beyerchen, A. *Scientists under Hitler*. New Haven: Yale University Press, 1977.

Beyler, R. H. " 'Consider a Cube Filled with Biological Material': Re-conceptualizing the Organic in German Biophysics, 1918–45." In *Biophysics 1918–1945: Fundamental Changes in Cellular Biology in the XXth Century*, edited by C. Galperin, S. F. Gilbert, and B. Hoppe, 39–46. Turnhout, Belgium: Brepols, 1999.

———. "Evolution als Problem für Quantenphysiker." In *Evolutionsbiologie von Darwin bis Heute*, edited by R. Brömer, U. Hossfeld, and N. Rupke, 137–60. Berlin: Verlag für Wissenschaft und Bildung, 1999.

————. "Exporting the Quantum Revolution: Pascual Jordan's Biophysical Initiatives." In *Pascual Jordan (1902–1980): Mainzer Symposium zum 100. Geburtstag*, edited by J. Ehlers et al., 69–81. Preprint no. 329. Berlin: Max-Planck-Institut für Wissenschaftsgeschichte, 2007.

————. "Targeting the Organism: The Scientific and Cultural Context of Pascual Jordan's Quantum Biology, 1932–1947." *Isis* 87 (1996): 248–73.

Bielka, H. *Geschichte der medizinisch-biologischen Institute Berlin-Buch.* 2nd ed. Berlin: Springer, 2002.

Blau, M., and K. Altenburger. "Über einige Wirkungen von Strahlen, II." *Zeitschrift für Physik* 12 (1922): 315–29.

Bleuler, E. "Die Beziehungen der neueren physikalischen Vorstellungen zur Psychologie und Biologie." *Viertel Jahrschrift der naturforschenden Gesellschaft in Zurich* 78 (1933): 152–97.

Bloch, W. *Einführung in die Relativitätstheorie.* Leipzig: Teubner, 1918.

Bohr, Niels. *Atomic Theory and the Description of Nature.* Vol. 1 of *The Philosophical Writings of Niels Bohr.* Woodbridge, CT: Ox Bow Press, 1987. (English translation originally published by Cambridge: Cambridge University Press, 1934.)

————. "Die Atomtheorie und die Prinzipien der Naturbeschreibung." *Die Naturwissenschaften* 18 (1930): 73–78.

————. *Causality and Complementarity: Supplementary Papers.* Vol. 4 of *The Philosophical Writings of Niels Bohr.* Edited by J. Faye and H. J. Folse. Woodbridge, CT: Ox Bow Press, 1998.

————. "Discussion with Einstein on Epistemological Problems in Atomic Physics." In *Albert Einstein: Philosopher Scientist*, Library of Living Philosophers, vol. 7, edited by P. Schilpp, 200–241. Evanston, IL: Library of Living Philosophers, 1949.

————. *Essays 1932–1957 on Atomic Physics and Human Knowledge.* Vol. 2 of *The Philosophical Writings of Niels Bohr.* Woodbridge, CT: Ox Bow Press, 1987. (English translation originally published in New York by Wiley & Sons, 1958.)

————. *Essays 1958–1962 on Atomic Physics and Human Knowledge.* Vol. 3 of *The Philosophical Writings of Niels Bohr.* Woodbridge, CT: Ox Bow Press, 1987. (English translation originally published in New York by Wiley & Sons, 1963.)

————. "Licht und Leben." *Die Naturwissenschaften* 21 (1933): 245–50.

————. "Light and Life." *Nature* 131 (1933): 421–23, 457–59.

————. "Light and Life Revisited." In *Essays 1958–1962 on Atomic Physics and Human Knowledge: The Philosophical Writings of Niels Bohr*, 3: 23–29. Woodbridge, CT: Ox Bow Press, 1987. (Originally published in 1962.)

————. *Niels Bohr: Collected Works.* Vol. 10. Edited by D. Favrholdt. Amsterdam: Elsevier, 1999.

————. "Wirkungsquantum und Naturbeschreibung." *Die Naturwissenschaften* 17 (1929): 483–86.

Born, M., W. Heisenberg, and P. Jordan. "Zur Quantenmechanik." *Zeitschrift für Physik* 34 (1925): 858–88; 35 (1926): 557–615.

Braun, R., and H. Holthusen. "Einfluß der Quantengröße auf die biologische Wirkung verschiedener Röntgenstrahlenqualitäten." *Strahlentherapie* 34 (1929): 707–34.

Bunde, E. "In Memoriam Richard Glocker." In *Medizinische Physik in der klinischen Routine*, edited by H. J. Schopka, 33–40. Medizinische Physik vol. 9. Heidelberg: Alfred Hüthig Verlag, 1978.

Burian, R. M., and J. Gayon. "The French School of Genetics: From Physiological and Population Genetics to Regulative Molecular Genetics." *Annual Review of Genetics* 33 (1999): 313–49.

Burian, R. M., J. Gayon, and D. Zallen. "The Singular Fate of Genetics in the History of French Biology." *Journal of the History of Biology* 21 (1988): 357–402.

Cairns, J., G. S. Stent, and J. D. Watson, eds. *Phage and the Origins of Molecular Biology*. Cold Spring Harbor, NY: Cold Spring Harbor Press, 1966.

Carlson, E. A. *The Gene: A Critical History*. Philadelphia: W. B. Saunders, 1966.

———. *Genes, Radiation, and Society: The Life and Work of H. J. Muller*. Ithaca, NY: Cornell University Press, 1981.

———. *Mendel's Legacy: The Origin of Classical Genetics*. Cold Spring Harbor, NY: Cold Spring Harbor Press, 2004.

———. "An Unacknowledged Founding of Molecular Biology: H. J. Muller's Contributions to Gene Theory, 1910–1936." *Journal of the History of Biology* 4 (1971): 149–70.

Catsch, A. "Zum 60. Geburtstag von K. G. Zimmer." *Strahlentherapie* 142 (1971): 124–25.

Catsch, A., A. Kanellis, G. Radu, and P. Welt. "Über die Auslösung von Chromosomenmutationen bei *Drosophila melanogaster* mit Röntgenstrahlen verschiedener Wellenlänge." *Die Naturwissenschaften* 32 (1944): 228.

Ceccarelli, L. *Shaping Science with Rhetoric: The Cases of Dobzhansky, Schrödinger, and Wilson*. Chicago: University of Chicago Press, 2001.

Chadarevian, S. de. *Designs for Life: Molecular Biology after World War II*. Cambridge: Cambridge University Press, 2002.

———."Reconstructing Life: Molecular Biology in Postwar Britain."*Studies in History and Philosophy of Biological and Biomedical Sciences* 33 (2002): 431–48.

Chadarevian, S. de, and B. J. Strasser. "Molecular Biology in Postwar Europe: Towards a 'Glocal' Picture." *Studies in History and Philosophy of Biological and Biomedical Sciences* 33 (2002): 361–65.

Chadwick, J., and E. Beiler. "The Collision of α-Particles with Hydrogen Nuclei." *Philosophical Magazine* 42 (1922): 923–40.

Chevalley, C. "Niels Bohr's Words and the Atlantis of Kantianism." In *Niels Bohr and Contemporary Philosophy*, edited by J. Faye and H. J. Folse, 33–55. Dordrecht: Kluwer Academic, 1994.

Clarke, A. E., and J. H. Fujimura, eds. *The Right Tools for the Job: At Work in Twentieth-Century Life Sciences*. Princeton, NJ: Princeton University Press, 1992.

Condon, E. U., and H. M. Terrill. "Quantum Phenomena in the Biological Action of X-Rays." *Journal of Cancer Research* 11 (1927): 324–33.

Creager, A. N. H. "Building Biology across the Atlantic." *Journal of the History of Biology* 36 (2003): 579–89.

———. *The Life of a Virus: Tobacco Mosaic Virus as an Experimental Model, 1930–1965*. Chicago: University of Chicago Press, 2002.

Crick, F. *Of Molecules and Men*. Seattle: University of Washington Press, 1966.

———. "Recent Research in Molecular Biology: An Introduction." *British Medical Bulletin* 21 (1965): 183–86.

Crow, J. F. "Sixty Years Ago: The 1932 International Congress of Genetics." *Genetics* 131 (1992): 764–66.

Crowther, J. A. "The Action of X-Rays on *Colpidium Colpoda*." *Proceedings of the Royal Society of London B* 100 (1926): 390–404.

———. *Ions, Electrons, and Ionizing Radiations*. 8th ed., London: E. Arnold, 1961. 1st ed., London: Longmans, Green & Co., 1919.

———. "Some Considerations Relative to the Action of X-Rays on Tissue Cells." *Proceedings of the Royal Society of London B* 96 (1924): 207–11.

Curie, M. "Radium and Radioactivity." *Century Magazine* (January 1904): 461–66.

———. "Sur l'étude des courbes de probabilité relatives à l'action des rayons X sur les bacilles." *Comptes rendus hebdomadaires des séances de l'Académie des sciences* 188 (1929): 202–4.

Dehlinger, U. "Über die Morphologie des Gens und den Mechanismus der Mutation." *Die Naturwissenschaften* 25 (1937): 138.

Deichmann, U. *Biologists under Hitler*. Translated by T. Dunlap. Cambridge, MA: Harvard University Press, 1996.

———. "A Brief Review of the Early History of Genetics and Its Relationship to Physics and Chemistry." In *Max Delbrück and Cologne: An Early Chapter of German Molecular Biology*, edited by S. Wenkel and U. Deichmann, 3–18. Hackensack, NJ: World Scientific, 2007.

———. "Emigration, Isolation and the Slow Start of Molecular Biology in Germany." *Studies in History and Philosophy of Biological and Biomedical Sciences* 33 (2002) : 449–71.

Delbrück, M. "Biophysics." In *Commemoration of the Fiftieth Anniversary of Niels Bohr's First Papers on Atomic Constitution, Held in Copenhagen on 8–15 July 1963*, 41–67. Duplicated typescript, Institute for Theoretical Physics, University of Copenhagen, 1963.

———. "Experiments with Bacterial Viruses (Bacteriophages)." *Harvey Lectures* 41 (1945–46): 161–87.

————. "Light and Life III." *Carlsberg Research Communications* 41 (1976): 299–309.

————. "Mind from Matter?" *American Scholar* 47 (1978): 339–53.

————. *Mind from Matter? An Essay on Evolutionary Epistemology*, edited by G. S. Stent, E. P. Fischer, S. Golomb, D. Presti, and H. Seiler. Oxford: Blackwell Scientific Publications, 1986.

————. "A Physicist Looks at Biology." *Transactions of the Connecticut Academy of Arts and Sciences* 38 (December 1949): 173–90. Reprinted in *Phage and the Origins of Molecular Biology*, edited by J. Cairns, G. S. Stent, and J. D. Watson, 9–22. Cold Spring Harbor, NY: Cold Spring Harbor Press, 1966.

————. "A Physicist's Renewed Look at Biology: Twenty Years Later." *Science* 168 (1970): 1312–15.

————. "Possible Existence of Multiply Charged Particles of Mass One." *Nature* 130 (1932): 626–27.

————. "Radiation and the Hereditary Mechanism." *American Naturalist* 74 (1940): 350–62.

————. "Opening Address: Virology Revisited." In *Molecular Basis of Host-Virus Interaction: International Symposium on Molecular Basis of Host-Virus Interaction*, IUB Symposium no. 78, edited by M. Chakravorty, 1–11. Princeton, NJ: Science Press, 1979.

————. "What Is Life? And What Is Truth?" *Quarterly Review of Biology* 20 (1945): 370–72.

————. "Zusatz bei der Korrektur." Note added to L. Meitner and H. Kösters, "Ueber die Streuung kurzwelliger Y-strahlen." *Zeitschrift für Physik* 84 (1933): 137–44.

Delbrück, M., and G. Gamow. "Uebergangswahrscheinlichkeiten von angeregten Kernen." *Zeitschrift für Physik* 72 (1931): 492–99.

Demerec, M. "The Effect of X-Ray Dosage on Sterility and Number of Lethals in *Drosophila melanogaster*." *Proceedings of the National Academy of Sciences* 19 (1933): 1015–20.

————. "Hereditary Effects of X-Ray Radiation." *Radiology* 30 (1938): 212–20.

————. "The Role of Genes in Evolution." *American Naturalist* 69 (1935): 125–38.

Demerec, M., and U. Fano. "Mechanism of the Origin of X-Ray-Induced Notch Deficiencies in *Drosophila melanogaster*." *Proceedings of the National Academy of Sciences* 27 (1941): 24–31.

Dempster, E. R. "Dominant vs. Recessive Lethal Mutation." *Proceedings of the National Academy of Sciences* 27 (1941): 249–50.

Dessauer, F. "Problemstellung und Theorie." In *Zehn Jahre Forschung auf dem physikalisch-medizinischen Grenzgebiet*, edited by F. Dessauer, 176–201. Leipzig: Georg Thieme, 1931.

————. "Über den Grundvorgang der biologischen Strahlenwirkung." *Strahlentherapie* 27 (1928): 364–81.

———. "Über einige Wirkungen von Strahlen, I." *Zeitschrift für Physik* 12 (1922): 38–47.

———. "Über einige Wirkungen von Strahlen, IV." *Zeitschrift für Physik* 20 (1923): 288–98.

———. "Zur Erklärung de biologische Strahlenwirkungen." *Strahlentherapie* 16 (1924): 208–21.

Dietrich, M. R. "From Gene to Genetic Hierarchy: Richard Goldschmidt and the Problem of the Gene." In *The Concept of the Gene in Development and Evolution: Historical and Epistemological Perspectives*, edited by P. J. Beurton, R. Falk, and H.-J. Rheinberger, 91–114. Cambridge: Cambridge University Press, 2000.

Dirac, P. *Die Prinzipien der Quantenmechanik*. Translated by W. Bloch. Leipzig: Hirzel, 1930.

Dobzhansky, T., and S. Wright. "Genetics of Natural Populations: V, Relations between Mutation Rate and Accumulation of Lethals in Populations of *Drosophila pseudoobscura*." *Genetics* 26 (1941): 23–51.

Dommann, M. *Durchsicht, Einsicht, Vorsicht: Eine Geschichte der Röntgenstrahlen 1896–1963*. Interferenzen: Studien der Kulturgeschichte der Technik, vol. 5. Zurich: Chronos, 2003.

Domondon, A. T. "Bringing Physics to Bear on the Phenomenon of Life: The Divergent Positions of Bohr, Delbrück and Schrödinger." *Studies in History and Philosophy of Biological and Biomedical Sciences* 37 (2006): 433–58.

Doyle, R. *On Beyond Living: Rhetorical Transformations of the Life Sciences*. Stanford, CA: Stanford University Press, 1997.

Dronamraju, K. R. "Erwin Schrödinger and the Origins of Molecular Biology." *Genetics* 153 (1999): 1071–76.

Dupré, J. *The Disorder of Things: Metaphysical Foundations of the Disunity of Science*. Cambridge, MA: Harvard University Press, 1993.

East, E. M. "The Concept of the Gene." In *Proceedings of the International Congress of Plant Sciences, Ithaca, New York, August 16–24, 1926*, vol. 1, edited by B. M. Duggar, 889–95. Menasha, WI: George Banta, 1929.

Einstein, A. *The Collected Papers of Albert Einstein*. Vol. 3, *The Swiss Years: Writings 1909–1911*, edited by M. J. Klein, A. J. Knox, J. Renn, and R. Schulmann. Princeton, NJ: Princeton University Press, 1993.

———. "Über einen die Erzeugung und Verwandlung des Lichtes betreffenden heuristischen Gesichtspunkt." *Annalen der Physik* 322 (1905): 132–44.

Einstein, A., B. Podolsky, and N. Rosen. "Can Quantum-Mechanical Description of Reality Be Considered Complete?" *Physical Review* 47 (1935): 777–80.

Emmons, C. W., and A. Hollaender. "The Action of Ultraviolet Radiation on Dermatophytes: II, Mutations Induced in Cultures of Dermatophytes by Exposure of Spores to Monochromatic Ultraviolet Radiation." *American Journal of Botany* 26 (1939): 467–75.

Everson, T. *The Gene: A Historical Perspective*. Westport, CT: Greenwood, 2007.

Fabry, C. "Fernand Holweck." *Pensée* 4 (1944): 65–69.

———. "Fernand Holweck (1890–1941)." *Cahiers de physique* 8 (1942): 1–16.

Fano, U. "On the Interpretation of Radiation Experiments in Genetics." *Quarterly Review of Biology* 17 (1942): 244–52.

———. "The Significance of the Hit Theory of Radiobiological Actions." *Journal of Applied Physics* 12 (1941): 347.

Favrholdt, D. "General Introduction: Complementarity beyond Physics." In *Niels Bohr: Collected Works*, vol. 10, edited by D. Favrholdt, xxiii–xlix. Amsterdam: Elsevier, 1999.

———. "Introduction, Part 1: Complementarity in Biology and Related Fields." In *Niels Bohr: Collected Works*, vol. 10, edited by D. Favrholdt, 3–27. Amsterdam: Elsevier, 1999.

———. *Niels Bohr's Philosophical Background*. Copenhagen: Munksgaard, 1992.

Faye, J. "The Bohr-Høffding Relationship Reconsidered." *Studies in History and Philosophy of Science* 19 (1988): 321–46.

———. *Niels Bohr: His Heritage and Legacy: An Anti-realist View of Quantum Mechanics*. Dordrecht: Kluwer Academic, 1991.

———. "Once More: Bohr-Høffding." *Danish Yearbook of Philosophy* 29 (1994): 106–13.

Faye, J., and H. J. Folse, eds. *Niels Bohr and Contemporary Philosophy*. Dordrecht: Kluwer Academic, 1994.

Fewtrell, L., and J. Bartram. *Water Quality: Guidelines, Standards, and Health*. London: IWA, 2001.

Fischer, E. P. "Max Delbrück," *Genetics* 177 (2007): 673–76.

Fischer, E. P., and C. Lipson. *Thinking about Science: Max Delbrück and the Origins of Molecular Biology*. New York: Norton, 1988.

Fleming, D. "Émigré Physicists and the Biological Revolution." *Perspectives in American History* 2 (1968): 176–213. Reprinted in *The Intellectual Migration: Europe and America, 1930–1960*, edited by D. Fleming and B. Bailyn, 152–89. Cambridge, MA: Harvard University Press, 1969.

Folse, H. J. "Complementarity and the Description of Nature in Biological Science." *Biology and Philosophy* 5 (1990): 221–24.

———. *The Philosophy of Niels Bohr: The Framework of Complementarity*. Amsterdam: Elsevier and North-Holland, 1985.

Franck, J. "Beitrag zum Problem der Kohlensäure-Assimilation." *Die Naturwissenschaften* 14 (1935): 226–29.

Franck, J., and H. Levi. "Zum Mechanismus der Sauerstoff-Aktivierung durch fluoreszenzfähig Farbstoffe." *Die Naturwissenschaften* 14 (1935): 229–30.

French, C. S. "Fifty Years of Photosynthesis." *Annual Review of Plant Physiology* 30 (1979): 1–30.

Fricke, H., and M. Demerec. "The Influence of Wave-Length on Genetic Effects of X-Rays." *Proceedings of the National Academy of Sciences* 23 (1937): 320–27.

Fruton, J. S. *Molecules and Life: Historical Essays on the Interplay of Chemistry and Biology*. New York: Wiley-Interscience, 1972.

———. *Proteins, Enzymes, Genes: The Interplay of Chemistry and Biology*. New Haven: Yale University Press, 1999.

Fryer, H. C., and J. W. Gowen. "An Analysis of Data on X-Ray-Induced Visible Gene Mutations in *Drosophila melanogaster*." *Genetics* 27 (1942): 212–27.

Fuerst, J. A. "The Role of Reductionism in the Development of Molecular Biology: Peripheral or Central?" *Social Studies of Science* 12 (1982): 241–78.

Gaffron, H. "Chemical Aspects of Photosynthesis." *Annual Review of Biochemistry* 8 (1939): 483–502.

———. "Photosynthesis." In *Encyclopaedia Britannica*, 14th edition (1949), 17: 848a–848b.

Gaffron, H., and K. Wohl. "Zur Theorie der Assimilation." *Die Naturwissenschaften* 24 (1936): 81–90, 103–7.

Galison, P. L. "Marietta Blau: Between Nazis and Nuclei." *Physics Today* 50, no. 11 (1997): 42–48.

Gates, R. R. "The Species Concept in the Light of Cytology and Genetics." *American Naturalist* 72 (1938): 340–49.

Gaudillière, J.-P. "Paris–New York Roundtrip: Transatlantic Crossings and the Reconstruction of the Biological Sciences in Post-War France." *Studies in History and Philosophy of Biological and Biomedical Sciences* 33 (2002): 389–417.

Gaudillière, J.-P., and H.-J. Rheinberger, eds. *From Molecular Genetics to Genomics: The Mapping Cultures of Twentieth-Century Genetics*. London: Routledge, 2004.

———. "Life Stories." *Studies in History and Philosophy of Biological and Biomedical Sciences* 35 (2004): 753–64.

Gausemeier, B. *Natürliche Ordnungen und politische Allianzen: Biologische und biochemische Forschung an Kaiser-Wilhelm-Instituten, 1933–1945*. Göttingen: Wallstein, 2005.

Giles, N. "The Effect of Fast Neutrons on the Chromosomes of *Tradescantia*." *Proceedings of the National Academy of Sciences* 26 (1940): 567–75.

Gillett, C. and B. Loewer, eds. *Physicalism and Its Discontents*. Cambridge: Cambridge University Press, 2001.

Gimbel, J. *Science, Technology, and Reparations: Exploitation and Plunder in Postwar Germany*. Stanford, CA: Stanford University Press, 1990.

Glass, B. "Two Prize Nobelists." *Quarterly Review of Biology* 64 (1989): 323–28.

Glocker, R. "Das Grundgesetz der physikalischen Wirkung von Röntgenstrahlen verschiedener Wellenlänge und seine Beziehung zum biologischen Effekt." *Strahlentherapie* 26 (1927): 147–55.

———. "Quantenphysik der biologischen Röntgenstrahlenwirkung." *Zeitschrift für Physik* 77 (1932): 653–75.

———. "Die Wirkung der Röntgenstrahlen auf die Zelle als physikalisches Problem." *Strahlentherapie* 33 (1929): 199–205.

Glocker, R., E. Hayer, and O. Jüngling. "Über die biologische Wirkung verschiedener Röntgenstrahlenqualitäten bei Dosierung in R-Einheiten." *Strahlentherapie* 32 (1929): 1–38.

Glocker, R., H. Langendorff, and A. Reuß. "Über die Wirkungen von Röntgenstrahlen verschiedener Wellenlänge auf biologische Objekte, III." *Strahlentherapie* 46 (1933): 517–28.

Glocker, R., and A. Reuß. "Über die Wirkungen von Röntgenstrahlen verschiedener Wellenlänge auf biologische Objekte, I." *Strahlentherapie* 46 (1933): 137–60.

———. "Über die Wirkungen von Röntgenstrahlen verschiedener Wellenlänge auf biologische Objekte, V." *Strahlentherapie* 47 (1933): 28–34.

Goenner, H. "Albert Einstein and Friedrich Dessauer: Political Views and Political Practice." *Physics in Perspective* 5 (2003): 21–66.

Goodspeed, T. H., and F. M. Uber. "Radiation and Plant Cytogenetics." *Botanical Review* 5 (1939): 1–48.

Govindjee. "On the Requirement of Minimum Number of Four versus Eight Quanta of Light for the Evolution of One Molecule of Oxygen in Photosynthesis: A Historical Note." *Photosynthesis Research* 59 (1999): 249–54.

Gowen, J. W., and E. H. Gay. "Gene Number, Kind, and Size in *Drosophila*." *Genetics* 18 (1933): 1–31.

Granin, D. *The Bison: A Novel about a Scientist Who Defied Stalin*. Translated by A. W. Bouis. New York: Doubleday, 1990.

Griffith, H. D., and K. G. Zimmer. "The Time-Intensity Factor in Relation to the Genetic Effects of Radiation." *British Journal of Radiology*, n.s., 8 (1935): 40–47.

Gross, B. "Recollections." *IEEE Transactions on Electrical Insulation* E1-21 (June 1986): 245–47.

Gulick, A. "What Are the Genes? I, The Genetic and Evolutionary Picture." *Quarterly Review of Biology* 13 (1938): 1–18.

Haffer, J. "Beiträge zoologischer Systematiker und einiger Genetiker zur evolutionären Synthese in Deutschland (1937-1950)." In *Die Entstehung der synthetischen Theorie: Beiträge zur Geschichte der Evolutionsbiologie in Deutschland 1930–1950*, edited by T. Junker and E.-M. Engels, 121–50. Berlin: Verlag für Wissenschaft und Bildung, 1999.

Hagen, U., and J. T. Lett. "In Memoriam: Karl G. Zimmer." *Radiation and Environmental Biophysics* 27 (1988): 245–46.

Haldane, J. S. *Mechanism, Life and Personality: An Examination of the Mechanistic Theory of Life and Mind*. New York: E. P. Dutton, 1923.

Hanson, F. B., and F. Heys. "An Analysis of the Effects of the Different Rays of Radium in Producing Lethal Mutations in *Drosophila*." *American Naturalist* 63 (1929): 201–13.

———. "Radium and Lethal Mutations in *Drosophila*: Further Evidence of the Proportionality Rule from a Study of the Effects of Equivalent Doses Differently Applied." *American Naturalist* 66 (1932): 335–45.

Harrington, A. *Reenchanted Science: Holism in German Culture from Wilhelm II to Hitler*. Princeton, NJ: Princeton University Press, 1996.

Hartmann, M. *Gesammelte Vorträge und Aufsätze*, 2 vols. Stuttgart: Fischer, 1956.

———. *Die Kausalität in Physik und Biologie*. Berlin: De Gruyter, 1937.

———. *Philosophie der Naturwissenschaften*. Berlin: Springer, 1937.

Harwood, J. *Styles of Scientific Thought: The German Genetics Community, 1900–1933*. Chicago: University of Chicago Press, 1993.

Hayes, W. "Max Ludwig Henning Delbrück, 4 September 1906–10 March 1981." *Biographical Memoirs of Fellows of the Royal Society* 28 (1982): 58–90. Reprinted in *Biographical Memoirs of the National Academy of Sciences* 62 (1993): 67–117.

Heilbron, J. "The Earliest Missionaries of the Copenhagen Spirit." *Revue d'histoire des sciences* 38 (1985): 195–230.

———. "The Scattering of α and β Particles and Rutherford's Atom." *Archive for History of Exact Sciences* 4 (1968): 247–307.

Heim, S., C. Sachse, and M. Walker. *The Kaiser Wilhelm Society under National Socialism*. Cambridge: Cambridge University Press, 2009.

Herrlich, P. "In Memoriam: Karl Günther Zimmer (1911–1988)." *Radiation Research* 116 (1988): 178–80.

Hessenbruch, A. "Geschlechterverhältnis und rationalisierte Röntgenologie." In *Geschlechterverhältnisse in Medizin, Naturwissenschaft, und Technik*, edited by C. Meinel and M. Renneberg, 148–58. Stuttgart: Verlag für Geschichte der Naturwissenschaft und der Technik, 1996.

Høffding, H. *Begrebet Analogi*. Copenhagen: A. F. Høst & Son, 1923.

———. *Bemerkninger om Erkendelsesteoriens nuvaerende Stilling*. Copenhagen: A. F. Høst & Son, 1930.

———. "Mindetale over Christian Bohr." *Tilskueren* (March 1911): 209–12.

———. "Om Vitalisme." In *Mindre Arbejder*, 40–50. Copenhagen: Nordisk Forlag, 1899.

———. *Psykologi*. Copenhagen: P. G. Philipsens Forlag, 1882.

Hoffmann, D., and M. Walker. "Friedrich Möglich: A Scientist's Journey from Fascism to Communism." In *Science and Ideology: A Comparative History*, edited by M. Walker, 227–60. London: Routledge, 2003.

Holmes, F. L. "Seymour Benzer and the Convergence of Molecular Biology with Classical Genetics." In *From Molecular Genetics to Genomics: The Mapping Cultures of Twentieth-Century Genetics*, edited by J.-P. Gaudillière and H.-J. Rheinberger, 42–62. London: Routledge, 2004.

Holthusen, H. "Bemerkungen zu obigen Ausführungen Dessauers über meine Arbeit 'Der Grundvorgang der biologischen Strahlenwirkung.'" *Strahlentherapie* 27 (1928): 382–85.

———. "Der Grundvorgang der biologischen Strahlenwirkung." *Strahlentherapie* 25 (1927): 157–73.

———. "Über die Dessauersche Punktwärmehypothese." *Strahlentherapie* 19 (1925): 285–306.

Holweck, F. "Essai d'interprétation énergétique de l'action des rayons K [*sic*] de l'aluminium sur les microbes." *Comptes rendus hebdomadaires des séances de l'Académie des sciences* 186 (1928): 1318–19.

———. "Mesure des dimensions élémentaires des virus par la méthode d'ultra-micrométrie statistique." *Comptes rendus hebdomadaires des séances de l'Académie des sciences* 209 (1939): 380–82.

———. "Production de rayons X monochromatiques de grande longueur d'onde: Action quantique sur les microbes." *Comptes rendus hebdomadaires des séances de l'Académie des sciences* 188 (1929): 197–99.

Holweck, F., S. Luria, and E. Wollman. "Recherches sur le mode d'action des radiations sur les bactériophages." *Comptes rendus hebdomadaires des séances de l'Académie des sciences* 210 (1940): 639–42.

Homann, P. H. "Hydrogen Metabolism of Green Algae: Discovery and Early Research—A Tribute to Hans Gaffron and His Coworkers." In *Discoveries in Photosynthesis*, edited by J. T. Beatty, H. Gest, and J. F. Allen, 119–29. New York: Springer, 2005.

Hossfeld, U., and M. Walker. "Hero or Villain? Stasi Archives Shed Light on Russian Scientist." *Nature* 411 (2001): 237.

Howard, D. "Who Invented the Copenhagen Interpretation? A Study in Mythology." *Philosophy of Science* 71 (2004): 669–82.

Hoyningen-Huene, P. "Niels Bohr's Argument for the Irreducibility of Biology to Physics." In *Niels Bohr and Contemporary Philosophy*, edited by J. Faye and H. J. Folse, 231–55. Dordrecht: Kluwer Academic, 1993.

Joliet, P. "Period-Four Oscillations of the Flash-Induced Oxygen Formation in Photosynthesis." In *Discoveries in Photosynthesis*, edited by J. T. Beatty, H. Gest, and J. F. Allen, 371–78. New York: Springer, 2005.

Jordan, P. "Biologische Strahlenwirkung und Physik der Gene." *Physikalische Zeitschrift* 39 (1938): 345–66.

———. "Ergänzende Bemerkungen über Biologie und Quantenmechanik." *Erkenntnis* 5 (1935): 348–52.

———. *Physik und das Geheimnis des organischen Lebens*. Braunschweig, Germany: Friedrich Vieweg & Sohn, 1941.

———. "Die Quantenmechanik und die Grundprobleme der Biologie und Psychologie." *Die Naturwissenschaften* 20 (1932): 815–21.

———. "Quantenphysikalische Bemerkungen zur Biologie und Psychologie." *Erkenntnis* 4 (1934): 215–52.

———. "Über die Elementarprozesse der biologischen Strahlenwirkung." *Radiologica* 2 (1938): 16–35; 166–84.

———. "Über die Rolle atomphysikalischer Einzelprozesse im biologischen Geschehen." *Radiologica* 1 (1937): 21–25.

Judson, H. F. *The Eighth Day of Creation: Makers of the Revolution in Biology*. New York: Simon & Schuster, 1979.

Junker, T. *Die zweite Darwinische Revolution: Geschichte des synthetischen Darwinismus in Deutschland 1924 bis 1950.* Marburg: Bailisken-Presse, 2004.

Junker, T., and E.-M. Engels, eds. *Die Entstehung der synthetischen Theorie: Beiträge zur Geschichte der Evolutionsbiologie in Deutschland 1930–1950.* Berlin: Verlag für Wissenschaft und Bildung, 1999.

Just, T. "Review of [*Experimentelle*] *Mutationsforschung in der Vererbungslehre*, by N. V. Timoféeff-Ressovsky." *American Midland Naturalist* 18 (1937): 308.

Kant, I. *Critique of Judgment Including the First Introduction.* Translated by W. S. Pluhar. Indianapolis: Hackett, 1987.

———. *Kritik der Urteilskraft.* Berlin: Philosophische Bibliothek, 1926.

Karlsch, R. "Boris Rajewsky und das Kaiser-Wilhelm-Institut für Biophysik in der Zeit des Nationalsozialismus." In *Gemeinschaftsforschung, Bevollmächtige und der Wissenstransfer: Die Rolle der Kaiser-Wilhelm-Gesellschaft im System kriegsrelevanter Forschung des Nationalsozialismus*, edited by H. Maier, 395–452. Göttingen: Wallstein, 2007.

Kaufmann, B. P. "Reversion from Roughest to Wild Type in *Drosophila melanogaster*." *Genetics* 27 (1942): 537–49.

Kay, L. E. "Conceptual Models and Analytical Tools: The Biology of Physicist Max Delbrück." *Journal of the History of Biology* 18 (1985): 207–46.

———. "Life as Technology: Representing, Intervening, and Molecularizing." In *The Philosophy and History of Molecular Biology*, edited by S. Sarkar, 87–100. Dordrecht: Kluwer Academic, 1996.

———. *The Molecular Vision of Life: Caltech, the Rockefeller Foundation, and the Rise of the New Biology.* New York: Oxford University Press, 1993.

———. "Problematizing Basic Research in Molecular Biology." In *Private Science: Biotechnology and the Rise of the Molecular Sciences*, edited by A. Thackray, 20–38. Philadelphia: University of Pennsylvania Press, 1998.

———. "Quanta of Life: Atomic Physics and the Reincarnation of Phage." *History and Philosophy of the Life Sciences* 14 (1992): 3–21.

———. "The Secret of Life: Niels Bohr's Influence on the Biology Program of Max Delbrück." *Rivista di storia della scienza* 2 (1985): 487–510.

———. *Who Wrote the Book of Life? A History of the Genetic Code.* Stanford, CA: Stanford University Press, 2000.

Keller, E. F. *The Century of the Gene.* Cambridge, MA: Harvard University Press, 2000.

———. "Physics and the Emergence of Molecular Biology: A History of Cognitive and Political Synergy." *Journal of the History of Biology* 23 (1990): 389–409.

———. *Refiguring Life: Metaphors of Twentieth-Century Biology.* New York: Columbia University Press, 1995.

———. *Secrets of Life, Secrets of Death: Essays on Language, Gender, and Science.* New York: Routledge, 1992.

Kofink, W. "Statistische Thermodynamik unter Zugründelegung der Quanten-mechanik in Hartreescher Näherung." *Annalen der Physik*, ser. 5, 28 (1937): 264–96.

———. "Zur Diracschen Theorie des Elektrons." *Annalen der Physik*, ser. 5, 38 (1940): 428–35, 436–55, 562–82, 583–600.

Kohler, R. E. *Lords of the Fly:* Drosophila *Genetics and the Experimental Life.* Chicago: University of Chicago Press, 1994.

———. "Warren Weaver and the Rockefeller Foundation Program in Molecular Biology: A Case Study in the Management of Science." In *The Sciences in the American Context: New Perspectives*, edited by N. Reingold, 249–93. Washington, DC: Smithsonian Institution Press, 1979.

Korogodin, V., G. Polikarpov, and V. Velkov. "The Blazing Life of N. V. Timoféeff-Ressovsky." *Journal of Bioscience* 25 (2000): 125–31.

Krönig, B., and W. Friedrich. *Die physikalischen und biologischen Grundlagen der Strahlentherapie*. Berlin: Urban & Schwarzenberg, 1918. English translation: *The Principles of Physics and Biology of Radiation Therapy*. New York: Rebman, 1922.

Lacassagne, A. "Action des rayons K [*sic*] de grande longueur d'onde sur les microbes: Établissement de statistiques précises de la mortalité des bactéries irradiées." *Comptes rendus hebdomadaires des séances de l'Académie des sciences* 188 (1929): 200–202.

———. "Action des rayons X de l'aluminium sur quelques microbes." *Comptes rendus hebdomadaires des séances de l'Académie des sciences* 186 (1928): 1316–17.

Lacassagne, A., and E. Wollman. "Évaluation des dimensions des bactériophages au moyen des rayons X." *Annales de l'Institut Pasteur* 64 (1940): 4–39.

Lagemann, R. T., and W. G. Holladay. *To Quarks and Quasars: A History of Physics and Astronomy at Vanderbilt University*. Nashville, TN: Vanderbilt University Department of Physics and Astronomy, 2000.

Lakatos, I. "Falsification and the Methodology of Scientific Research Programmes." In *Criticism and the Growth of Knowledge*, edited by I. Lakatos and A. Musgrave, 91–195. Cambridge: Cambridge University Press, 1970.

Lamla, E. *Grundriss der Physik für Naturwissenschaftler, Mediziner und Pharmazeuten*. Berlin: Springer, 1925.

Langendorff, H. "Wirkungen auf Zellen und Zellkomplexe." In *Biophysik*, edited by B. Rajewsky and M. Schön, 1:51–65. Wiesbaden: Dieterich'sche Verlagsbuchhandlung, 1948.

Langendorff, H., M. Langendorff, and A. Reuß. "Über die Wirkung von Röntgenstrahlen verschiedener Wellenlänge auf biologische Objekte, II, IV." *Strahlentherapie* 46 (1933): 289–92, 655–62.

Langendorff, H., and K. Sommermeyer. "Strahlenwirkung auf Drosophilaeier, I." *Fundamenta Radiologica* 4 (1939): 196–209.

———. "Strahlenwirkung auf Drosophilaeier, II–V." *Strahlentherapie* 67 (1940): 110–29; 68 (1940): 42–52, 656–68.

Lawrence, C., and G. Weisz, eds. *Greater than the Parts: Holism in Biomedicine, 1920–1950*. New York: Oxford University Press, 1998.

Lea, D. E. *Actions of Radiations on Living Cells*. Cambridge: Cambridge University Press, 1946.

———. "Determination of the Sizes of Viruses and Genes by Radiation Methods." *Nature* 146 (1940): 137–38.

———. "A Radiation Method for Determining the Number of Genes in the X-Chromosome of *Drosophila*." *Journal of Genetics* 39 (1940): 181–87.

———. "A Theory of the Action of Radiations on Biological Materials Capable of Recovery." *British Journal of Radiology* 11 (1937): 489–97, 554–66.

Lea, D. E., and D. G. Catcheside. "The Bearing of Radiation Experiments on the Size of the Gene." *Journal of Genetics* 47 (1945): 41–50.

Lea, D. E., R. B. Haines, and C. A. Coulson. "The Action of Radiations on Bacteria: III–γ-Rays on Growing and on Non-proliferating Bacteria." *Proceedings of the Royal Society of London B* 123 (1936): 1–21.

———. "The Mechanism of the Bactericidal Action of Radioactive Radiations." *Proceedings of the Royal Society of London B* 120 (1936): 47–76.

Lea, D. E., and K. M. Smith. "The Inactivation of Plant Viruses by Radiation." *Parasitology* 32 (1940): 405–16.

Lenoir, T. *The Strategy of Life: Teleology and Mechanics in Nineteenth-Century German Biology*. Chicago: University of Chicago Press, 1982.

Lewis, G. N. "The Conservation of Photons." *Nature* 118 (1926): 874–75.

Lewis, S. *Arrowsmith*. New York: Harcourt Brace, 1945.

Loeb, J. *The Mechanistic Conception of Life: Biological Essays*. Chicago: University of Chicago Press, 1912.

Luria, S. E. "Actions des radiations sur le *Bacterium coli*." *Comptes rendus hebdomadaires des séances de l'Académie des sciences* 209 (1939): 604–6.

———. "Radiobiologie quantique." *Paris médical* 30, no. 26 (1940): 305–11.

———. *A Slot Machine, a Broken Test Tube: An Autobiography*. New York: Harper & Row, 1984.

Mackenzie, K., and H. J. Muller. "Mutation Effects of Ultra-Violet Light in *Drosophila*." *Proceedings of the Royal Society of London B* 129 (1940): 491–517.

Macrakis, Kristie. "The Survival of Basic Biological Research in National Socialist Germany." *Journal of the History of Biology* 26 (1993): 519–43.

———. *Surviving the Swastika: Scientific Research in Nazi Germany*. New York: Oxford University Press, 1993.

Manning, W. M., J. F. Stauffer, B. M. Duggar, and F. Daniels. "Quantum Efficiency of Photosynthesis in *Chlorella*." *Journal of the American Chemical Society* 60 (1938): 266–74.

Marshak, Alfred. "Relative Effects of X-Rays and Neutrons on Chromosomes in Different Parts of the 'Resting Stage.'" *Proceedings of the National Academy of Sciences* 28 (1942): 29–35.

Mayneord, W. V. "The Physical Basis of the Biological Effects of High Voltage Radiations." *Proceedings of the Royal Society of London A* 146 (1934): 867–79.

McKaughan, D. J. "The Influence of Niels Bohr on Max Delbrück: Revisiting the Hopes Inspired by 'Light and Life.'" *Isis* 96 (2005): 507–29.

Meitner, L., and M. Delbrück. *Die Aufbau der Atomkerne: Natürliche und künstliche Kernumwandlungen.* Berlin: Springer, 1935.

Melis, A., and T. Happe. "Trails of Green Alga Hydrogen Research—From Hans Gaffron to New Frontiers." In *Discoveries in Photosynthesis*, edited by J. T. Beatty, H. Gest, and J. F. Allen, 681–89. New York: Springer, 2005.

Möglich, F., R. Rompe, and N. V. Timoféeff-Ressovsky. "Bemerkungen zu physikalischen Modellvorstellungen über Energieausbreitungsmechanismen im Treffbereich bei strahlenbiologischen Vorgängen." *Die Naturwissenschaften* 30 (1942): 409–19.

———. "Über die Indeterminiertheit und die Verstärkererscheinungen in der Biologie." *Die Naturwissenschaften* 32 (1944): 316–24.

Möglich, F., R. Rompe, N. V. Timoféeff-Ressovsky, and K. G. Zimmer. "Über Energiewanderungsvorgänge und ihre Bedeutung für einige biologische Prozesse." *Protoplasma* 38 (1943): 105–26.

Molière, G. "Quantenmechanische Theorie der Röntgenstrahlinterferenzen in Kristallen, I–II." *Annalen der Physik*, ser. 5, 35 (1939): 272–313.

———. "Zur Strahlungstheorie, I." *Annalen der Physik*, ser. 5, 37 (1940): 415–20.

Molière, G., and M. Delbrück. "Statistische Quanten-mechanik und Thermodynamik." *Akademie der Wissenschaften, Berlin: Physisches-Mathematisches Klasse Abhandlungen* 1 (1936): 1–42.

Moore, W. *Schrödinger: Life and Thought.* Cambridge: Cambridge University Press, 1989.

Morange, M. *A History of Molecular Biology.* Translated by M. Cobb. Cambridge, MA: Harvard University Press, 1998.

Morgan, T. H. "The Relation of Genetics to Physiology and Medicine." *Scientific Monthly* 41 (1935), 7–8. Reprinted in *Physiology or Medicine*, 1: 315–16. Nobel Foundation. Amsterdam: Elsevier, 1967.

———. *The Theory of the Gene.* New Haven: Yale University Press, 1926.

Morgan, T. H., A. H. Sturtevant, H. J. Muller, and C. B. Bridges. *The Mechanism of Mendelian Heredity.* New York: Holt, 1915.

Moss, L. *What Genes Can't Do.* Cambridge, MA: MIT Press, 2003.

Muller, H. J. "An Analysis of the Process of Structural Change in Chromosomes of *Drosophila*." *Journal of Genetics* 40 (1940): 1–66.

———. "Artificial Transmutation of the Gene." *Science* 66 (1927): 84–87.

———. "The Gene as the Basis of Life." In *Proceedings of the International Congress of Plant Sciences, Ithaca, New York, August 16–24, 1926*, vol. 1, edited by B. M. Duggar, 897–921. Menasha, WI: George Banta, 1929. Reprinted in H. J. Muller, *Studies in Genetics: The Selected Papers of Hermann Joseph Muller*, 188–204. Bloomington: Indiana University Press, 1962.

————. "Gene Mutations Caused by Radiation." In *Symposium on Radiobiology: The Basic Aspects of Radiation Effects on Living Systems*, edited by J. J. Nickson, 296–332. New York: John Wiley & Sons, 1952.

————. "A Physicist Stands Amazed at Genetics." *Journal of Heredity* 32 (1946): 90–92.

————. "Physics in the Attack on Fundamental Problems of Genetics." *Scientific Monthly* 44 (1937): 210–14.

————. "The Problem of Genic Modification." In *Verhandlungen des V. Internationalen Kongresses für Vererbungswissenschaft, Berlin 1927*, 234–60. Berlin: Borntraeger, 1928.

————. "The Production of Mutations by X-Rays." *Proceedings of the National Academy of Sciences* 14 (1928): 714–26.

————. *Studies in Genetics: The Selected Papers of Hermann Joseph Muller*. Bloomington: Indiana University Press, 1962.

Muller, H. J., and A. A. Prokofyeva. "The Individual Gene in Relation to the Chromomere and the Chromosome." *Proceedings of the National Academy of Sciences* 21 (1935): 16–26.

Müller-Hill, B. "Heroes and Villains." *Nature* 336 (1988): 721–22.

Murdoch, D. *Niels Bohr's Philosophy of Physics*. Cambridge: Cambridge University Press, 1987.

Murphy, M. P., and L. A. J. O'Neill, eds. *What Is Life? The Next Fifty Years*. Cambridge: Cambridge University Press, 1995.

Myers, J. "Conceptual Developments in Photosynthesis, 1924–1974." *Plant Physiology* 54 (1974): 420–26.

Nagel, E. "The Eighth International Congress of Philosophy." *Journal of Philosophy* 31 (1934): 589–601.

Nakashima, Y. "Einige Versuche zum Grundvorgang der biologischen Strahlenwirkung." *Strahlentherapie* 24 (1926): 1–36.

Needham, J. *Order and Life*. Cambridge, MA: MIT Press, 1968. 1st ed., Yale University Press, 1936.

Nickelsen, K. "The Construction of a Scientific Model: Otto Warburg and the Building Block Strategy." *Studies in History and Philosophy of Biological and Biomedical Sciences* 40 (2009): 73–86.

Norton, J. "Thought Experiments in Einstein's Work." In *Thought Experiments in Science and Philosophy*, edited by T. Horowitz and G. Massey, 129–49. Savage, MD: Rowman & Littlefield, 1991.

Nye, M. J. "Historical Sources of Science-as-Social Practice: Michael Polanyi's Berlin." *Historical Studies in the Physical and Biological Sciences* 37 (2007): 411–36.

Olby, R. C. "The Molecular Revolution in Biology." In *Companion to the History of Modern Science*, edited by R. C. Olby, G. N. Cantor, and J. R. R. Christie, 503–20. New York: Routledge, 1996.

————. *The Path to the Double Helix*. New York: Dover, 1994. First published 1974.

————. "Quiet Debut for the Double Helix." *Nature* 421 (2003): 402–6.

————. "Schrödinger's Problem: What Is Life?" *Journal of the History of Biology* 4 (1971): 119–48.

Oleynikov, P. V. "German Scientists in the Soviet Atomic Project." *Non-Proliferation Review* (summer 2000): 1–30.

Packard, C. "The Biological Roentgen." *Radiology* 27 (1936): 191–95.

Pais, A. *Inward Bound: Of Matter and Forces in the Physical World.* Oxford: Oxford University Press, 1986.

Pais, A. *Subtle is the Lord: The Science and the Life of Albert Einstein.* Oxford: Oxford University Press, 1982.

Papineau, D. "The Rise of Physicalism." In *Physicalism and Its Discontents*, edited by C. Gillett and B. Loewer, 3–36. Cambridge: Cambridge University Press, 2001.

Pätau, K., and N. V. Timoféeff-Ressovsky. "Die Genauigkeit der Bestimmung spontaner und strahleninduzierter Mutationsraten nach der 'CLB'-Kreuzungsmethode bei *Drosophila melanogaster.*" *Zeitschrift für induktive Abstammungs- und Vererbungslehre* 81 (1943): 181–90.

————. "Statistische Prüfung des Unterschiedes der temperaturkoeffizienten höher und normaler Mutationsraten nebst einem Beispiel für die Planung von Temperaturversuchen." *Zeitschrift für induktive Abstammungs- und Vererbungslehre* 81 (1943): 62–71.

Paul, D. B., and C. B. Krimbas. "Nikolai V. Timoféeff-Ressovsky." *Scientific American* 266 (February 1992): 86–92.

Perutz, M. "Erwin Schrödinger's *What Is Life?* and Molecular Biology." In *Schrödinger: Centenary Celebration of a Polymath*, edited by C. W. Kilmister, 234–51. Cambridge: Cambridge University Press, 1987.

————. "Physics and the Riddle of Life." *Nature* 326 (1987): 555–58.

Plough, H. H., and G. P. Child. "Autosomal Lethal Mutation Frequencies in *Drosophila.*" *Proceedings of the National Academy of Sciences* 23 (1937): 435–40.

Pordes, F. "Zum biologischen Wirkungsmechanismus der Röntgenstrahlen." *Strahlentherapie* 19 (1925): 307–24.

Portin, P. "The Concept of the Gene: Short History and Present Status." *Quarterly Review of Biology* 68 (1993); 173–223.

Prigge, R., and H. von Schelling. "Zur Analyse der Antigenwirkung." *Die Naturwissenschaften* 30 (1942): 661.

Rabinowitch, E. *Photosynthesis.* Vol. 1. New York: Interscience, 1945.

Rajewsky, B. "Friedrich Dessauer zum Gedächtnis." In *Quantenbiologie: Einführung in einen neuen Wissenszweig*, edited by F. Dessauer, 2nd ed., xi–xix. Berlin: Springer, 1964.

————. "Grundlagen der Treffertheorie der biologischen Strahlenwirkung." In *Biophysik*, edited by B. Rajewsky and M. Schön, 1: 9–14. Wiesbaden: Dieterich'sche Verlagsbuchhandlung, 1948.

————. "The Limits of the Target Theory of the Biological Action of Radiation." *British Journal of Radiology*, n.s., 25 (1950): 550–52.

————. "Theorie der Strahlenwirkung und ihre Bedeutung für die Strahlenthera-pie." *Wissenschaftliche Woche zu Frankfurt am Main* 2 (1934): 75–90.

Rajewsky, B., and M. Schön, eds. *Biophysik*. 2 vols. Naturforschung und Medizin in Deutschland 1939–1946 (FIAT Review of German Science), vols. 21–22. Wiesbaden: Dieterich'sche Verlagsbuchhandlung, 1948.

Rajewsky, B., and N. V. Timoféeff-Ressovsky. "Höhenstrahlung und die Muta-tionsrate von *Drosophila melanogaster*." *Zeitschrift für induktive Abstam-mungs- und Vererbungslehre* 77 (1939): 488–500.

Rasmussen, N. "Instruments, Scientists, Industrialists and the Specificity of 'Influ-ence': The Case of RCA and Biological Electron Microscopy." In *The Invisible Industrialist: Manufactures and the Production of Scientific Knowledge*, edited by J.-P. Gaudillière and I. Löwy, 173–208. Basingstoke, UK: Macmillan, 1998.

————. "The Mid-Century Biophysics Bubble: Hiroshima and the Biological Rev-olution in America, Revisited." *History of Science* 35 (1997): 245–99.

————. *Picture Control: The Electron Microscope and the Transformation of Biol-ogy in America, 1940–1960*. Stanford, CA: Stanford University Press, 1997.

Rentetzi, M. *Trafficking Materials and Gendered Experimental Practices: Radium Research in Early 20th-Century Vienna*. New York: Columbia University Press, 2008. Electronic book, available at http://www.gutenberg-e.org/rentetzi/.

Rescher, N. "The Berlin School of Logical Empiricism and Its Legacy." *Erkenntnis* 64 (2006): 281–304.

Rick, C. "The Genetic Nature of X-Ray-Induced Changes in Pollen." *Proceedings of the National Academy of Sciences* 28 (1942): 518–25.

Riehl, N., N. V. Timoféeff-Ressovsky, and K. G. Zimmer. "Mechanismus der Wirkung ionisierender Strahlen auf biologische Elementareinheiten." *Die Naturwissenschaften* 29 (1941): 625–39.

Roll-Hansen, N. "The Application of Complementarity to Biology: From Niels Bohr to Max Delbrück." *Historical Studies in the Physical and Biological Sci-ences* 30 (2000): 417–42.

————. "Critical Teleology: Immanuel Kant and Claude Bernard on the Limitations of Experimental Biology." *Journal of the History of Biology* 9 (1976): 59–91.

————. "*Drosophila* Genetics: A Reductionist Research Program." *Journal of the History of Biology* 9 (1978): 159–210.

————. "E. S. Russell and J. H. Woodger: The Failure of Two Twentieth-Century Opponents of Mechanistic Biology." *Journal of the History of Biology* 17 (1984): 399–428.

Rosenberg, A. *Instrumental Biology, or The Disunity of Science*. Chicago: Univer-sity of Chicago Press, 1994.

————. *The Structure of Biological Science*. Cambridge: Cambridge University Press, 1985.

Rubin, E. "Erkendelsens Uafslutteligbed som et Grundmotiv hos Høffding" [The inconclusiveness of knowledge as a basic motive of Høffding]. In *Harald Høff-ding in Memoriam*, 1–23. Copenhagen: Gyldendal, 1932.

Rürup, R., and M. Schüring. *Schicksale und Karrieren: Gedenkbuch für die von den Nationalsozialisten aus der Kaiser-Wilhelm-Gesellschaft vertriebenen Forscherinnen und Forscher.* Berlin: Wallstein, 2008.

Russell, B. *An Outline of Philosophy.* 1927. Reprint, London: Unwin 1979.

Rutherford, E. "Collision of α-Particles with Light Atoms." *Philosophical Magazine* 37 (1919): 537–87.

Sachse, C., and M. Walker, eds. *Politics and Science in Wartime: Comparative International Perspectives on the Kaiser Wilhelm Institute. Osiris,* 2nd ser., 20 (2005).

Satzinger, H., and A. Vogt. *Elena Aleksandrovna und Nikolaj Vladimirovic Timoféeff-Ressovsky (1898–1973; 1900–1981).* Preprint no. 112. Berlin: Max-Planck-Institut für Wissenschaftsgeschichte, 1999.

Sax, K. "Chromosome Aberrations Induced by X-Rays." *Genetics* 23 (1938): 494–516.

Sax, K., and E. V. Enzmann. "The Effect of Temperature on X-Ray-Induced Chromosome Aberrations." *Proceedings of the National Academy of Sciences* 25 (1939): 397–405.

Schaffner, K. F. "The Peripherality of Reductionism in the Development of Molecular Biology." *Journal of the History of Biology* 7 (1974): 111–39.

Schirrmacher, A. "Physik und Politik in der frühen Bundesrepublik Deutschland: Max Born, Werner Heisenberg, und Pascual Jordan als politische Grenzgänger." *Berichte zur Wissenschaftsgeschichte* 30 (2007): 13–31.

Schlick, M. "Über den Begriff der Ganzheit." *Erkenntnis* 5 (1935): 52–53.

Schnarrenberger, C. "Botany at the Kaiser Wilhelm Institutes." *Englera* 7 (1987): 105–46.

Schön, M. "Energiewanderung in Molekülkomplexen und Struktureinheiten." In *Biophysik,* edited by B. Rajewsky and M. Schön, 1:25–43. Wiesbaden: Dieterich'sche Verlagsbuchhandlung, 1948.

Schrödinger, E. "Indeterminism and Free Will." *Nature* 138 (1936): 13–14.

———. "Warum sind die Atome so klein?" *Forschungen und Fortschritte* 9 (1933): 125–26.

———. *What Is Life? The Physical Aspect of the Living Cell, with Mind and Matter and Autobiographical Sketches.* With forward by R. Penrose. Cambridge: Cambridge University Press, 2000. First published 1944.

Schwartz, J. "The Differential Concept of the Gene: Past and Present." In *The Concept of the Gene in Development and Evolution,* edited by J. Beurton, R. Falk, and H.-J. Rheinberger, 26–39. Cambridge: Cambridge University Press, 2000.

———. *In Pursuit of the Gene: From Darwin to DNA.* Cambridge, MA: Harvard University Press, 2008.

Schwerin, A. von. *Experimentalisierung des Menschen: Der Genetiker Hans Nachtsheim und die vergleichende Erbpathologie, 1920–1945.* Göttingen: Wallstein, 2004.

Segrè, G. *Faust in Copenhagen: A Struggle for the Soul of Physics.* New York: Viking, 2007.

Setlow, R. P., and E. C. Pollard. *Molecular Biophysics*. Reading, MA: Addison-Wesley, 1962.

Solzhenitsyn, A. I. *The Gulag Archipelago, 1918–1956*. Translated by T. P. Whitney. New York: Harper & Row, 1974.

Sommermeyer, K. "Bemerkung zur Theorie des strahlenbiologischen Sättigungseffektes." *Die Naturwissenschaften* 30 (1942): 104–5.

———. *Quantenphysik der Strahlenwirkung in Biologie und Medizin*. Probleme der Bioklimatologie, vol. 2. Leipzig: Geest & Portig, 1952.

———. "Über die Treffertheorie und ihre Anwendung in der Theorie der Genmutation." *Die Naturwissenschaften* 26 (1938): 154.

Sommermeyer, K., and U. Dehlinger. "Beiträge zur Diskussion eines Gen-Modells." *Physikalische Zeitschrift* 40 (1939): 67–70.

Stadler, L. J. "The Gene." *Science* 120 (1954): 811–19.

———. "Mutations in Barley Induced by X-Rays and Radium." *Science* 68 (1928): 186–87.

Stent, G. S. "Introduction: Waiting for the Paradox." In *Phage and the Origins of Molecular Biology*, edited by J. Cairns, G. S. Stent, and J. D. Watson, 3–8. Cold Spring Harbor, NY: Cold Spring Harbor Press, 1966.

———. "Looking for Other Laws of Physics." *Journal of Contemporary History* 33 (1998): 371–97.

———. *Molecular Biology of Bacterial Viruses*. San Francisco: W.H. Freeman and Co. 1963.

———. "That Was the Molecular Biology That Was." *Science* 160 (1968): 390–94.

Steuwer, R. H. *The Compton Effect: Turning Point in Physics*. New York: Science History Publications, 1975.

Strangeways, T. S. P., and H. E. H. Oakley. "The Immediate Changes Observed in Tissue Cells after Exposure to Soft X-Rays." *Proceedings of the Royal Society of London B* 95 (1923): 373–81.

Strasser, B. J. "Building Molecular Biology in Post-war Europe: Between the Atomic Age and the American Challenge." In *Max Delbrück and Cologne: An Early Chapter of German Molecular Biology*, edited by S. Wenkel and U. Deichmann, 58–65. Singapore: World Scientific, 2007.

———. *La fabrique d'une nouvelle science: La biologie moléculaire à l'âge atomique (1945–1964)*. Florence: Leo S. Olschki, 2006.

———. "Who Cares about the Double Helix?" *Nature* 422 (2003): 803–4.

Strohmeier, B., and R. Rosner. *Marietta Blau, Stars of Disintegration: Biography of a Pioneer of Particle Physics*. Edited by P. F. Dvorak. Riverside, CA: Ariadne Press, 2003.

Thomas, J. A. "Le martyre de Fernand Holweck." *Pensée* 127 (1949): 21–28.

Thomson, J. J. "On the Number of Corpuscles in an Atom." *Philosophical Magazine* 11 (1906): 769–81.

Timoféeff-Ressovsky, N. V. "Eine biophysikalische Analyse des Mutationsvorganges." *Nova Acta Leopoldina*, n.s., 9 (1940): 209–40.

————. "The Experimental Production of Mutations." *Biological Reviews* 9 (1934): 411–57.

————. *Experimentelle Mutationsforschung in der Vererbungslehre: Beeinflussung der Erbanlagen durch Strahlung und andere Faktoren.* Wissenschaftliche Forschungsberichte: Naturwissenschaftliche Reihe, vol. 42, edited by R. E. Liesegang. Dresden: Theodor Steinkopff, 1937.

————. "Le mécanisme des mutations et la structure du gène." In *Réunion internationale de physique-chimie-biologie: Congrès du Palais de la découverte, Paris, Octobre 1937.* Actualités scientifiques et industrielles, no. 725; ser. Biologie, no. 8, part 2, 83–104. Paris: Hermann & Cie, 1938.

————. *Le mécanisme des mutations et la structure du gène.* Actualités scientifiques et industrielles, no. 812, ser. Génétique, edited by B. Ephrussi, no. 4. Paris: Hermann & Cie, 1939.

Timoféeff-Ressovsky, N. V., and M. Delbrück. "Strahlengenetische Versuche über sichtbare Mutation und die Mutabilität einzelner Gene bei *Drosophila melanogaster.*" *Zeitschrift für induktive Abstammungs- und Vererbungslehre* 71 (1936): 322–34.

Timoféeff-Ressovsky, N. V., and K. G. Zimmer. *Biophysik.* Vol. 1, *Das Trefferprinzip in der Biologie.* Leipzig: S. Hirzel, 1947. (No further volumes.)

————. "Neutronbestrahlungsversuche zur Mutationsauslösung an *Drosophila melanogaster.*" *Die Naturwissenschaften* 26 (1938): 362–65.

————. "Strahlengenetische Zeitfaktorversuche an *Drosophila melanogaster.*" *Strahlentherapie* 53 (1935): 134–38.

————. "Über Zeitproportionalität und Temperaturabhängigkeit der spontanen Mutationsrate von *Drosophila.*" *Zeitschrift für induktive Abstammungs- und Vererbungslehre* 79 (1941): 530–37.

Timoféeff-Ressovsky, N. V., K. G. Zimmer, and M. Delbrück. "Über die Natur der Genmutation und der Genstruktur." *Nachrichten von der Gesellschaft der Wissenschaften zu Göttingen, mathematisch-physikalische Klasse, Fachgruppe VI: Biologie* 1 (1935): 189–245.

Timoféeff-Ressovsky, N. V., K. G. Zimmer, and F. A. Heyn. "Auslösung von Mutationen an *Drosophila melanogaster* durch Schnelle Li+D-Neutronen." *Die Naturwissenschaften* 26 (1938): 625–26.

Van Niel, C. B. "On the Morphology and Physiology of the Purple and Green Sulphur Bacteria." *Archiv für Mikrobiologie* 3 (1931): 1–112.

Warburg, E. "Über die Anwendung der Quantenhypothese auf die Photochemie." *Die Naturwissenschaften* 30 (1917): 67.

Warburg, O., and E. Negelein. "Über den Einfluß der Wellenlänge auf den Energieumsatz bei der Kohlensäureassimilation." *Zeitschrift für physikalischen Chemie* 102 (1922): 191–218.

————. "Über den Energieumsatz bei der Kohlensäureassimilation." *Zeitschrift für physikalischen Chemie* 102 (1922): 235–66.

Watson, J. D. "The Properties of X-Ray-Inactivated Bacteriophage: I, Inactivation by Direct Effect." *Journal of Bacteriology* 60 (1950): 697–718.

———. "The Properties of X-Ray-Inactivated Bacteriophage: II, Inactivation by Indirect Effects." *Journal of Bacteriology* 63 (1952): 473–85.

Weaver, W. "Molecular Biology: Origins of the Term." *Science* 170 (1970): 582.

Welt, E. *Berlin Wild: A Novel.* New York: Viking, 1986.

Wenkel, S., and U. Deichmann, eds. *Max Delbrück and Cologne: An Early Chapter of German Molecular Biology.* Hackensack, NJ: World Scientific, 2007.

Wilson, C. T. R. "On an Expansion Apparatus for Making Visible the Tracks of Ionizing Particles in Gases and Results Obtained from Its Use." *Proceedings of the Royal Society of London A* 87 (1912): 277–92.

Wise, M. N. "Pascual Jordan: Quantum Mechanics, Psychology, National Socialism." In *Science, Technology, and National Socialism*, edited by M. Renneberg and M. Walker, 224–54. Cambridge: Cambridge University Press, 1994.

Wisniak, J. "Kurt Wohl: His Life and Work." *Educación Química* 14 (2003): 36–46.

Wohl, K. "The Mechanism of Photosynthesis in Green Plants." *New Phytologist* 39 (1940): 33–64.

Wollman, E., F. Holweck, and S. Luria. "Effect of Radiations on Bacteriophage C_{16}." *Nature* 145 (1940): 935–36.

Wollman, E., and A. Lacassagne. "Évaluation de la taille relative des bactériophages par leur radiosensibilité." *Comptes rendus des séances de la Société de biologie et de ses filiales* 131 (1939): 959–61.

Wolters, G. "Wrongful Life: Logico-Empiricist Philosophy of Biology." In *Experience, Reality, and Scientific Explanation*, edited by M. C. Galavotti and A. Pagnini, 187–208. Dordrecht: Kluwer Academic, 1999.

Wright, S. "Genes as Physiological Agents: General Considerations." *American Naturalist* 79 (1945): 289–303.

Wurmbach, P. "Miassovo See–Berlin–Cold Spring Harbor." In *Wissenschaften in Berlin*, edited by T. Buddensieg et al., 3:90–93. Berlin: Mann, 1987.

Wyckoff, R. W. G. "The Killing of Certain Bacteria by X-Rays." *Journal of Experimental Medicine* 52 (1930): 435–46.

———. "The Killing of Colon Bacilli by Ultraviolet Light." *Journal of General Physiology* 15 (1932): 351–61.

———. "The Killing of Colon Bacilli by X-Rays of Different Wavelengths." *Journal of Experimental Medicine* 52 (1930): 769–80.

Yoxen, E. J. "Giving Life a New Meaning: The Rise of the Molecular Biology Establishment." In *Scientific Establishment an Hierarchies*, edited by N. H. Elias, H. H. Martins, and R. R. Whitley, 123–43. Dordrecht: Reidel, 1982.

———. "Where Does Schrödinger's 'What Is Life?' Belong in the History of Molecular Biology?" *History of Science* 17 (1979): 17–52.

Zallen, D. T. "The 'Light' Organism for the Job." *Journal of the History of Biology* 26 (1993): 269–79.

————. "Redrawing the Boundaries of Molecular Biology: The Case of Photosynthesis." In *The Philosophy and History of Molecular Biology*, edited by S. Sarkar, 45–65. Dordrecht: Kluwer Academic, 1996.

Zilsel, E. "P. Jordans Versuch, den Vitalismus Quantenmechanisch zu Retten." *Erkenntnis* 5 (1935): 56–63.

Zimmer, K. G. "Ein Beitrag zur Frage nach der Beziehung zwischen Röntgenstrahlendosis und dadurch ausgelöster Mutationsrate." *Strahlentherapie* 51 (1934): 179–84.

————. "Dosimetrische und strahlenbiologische Versuche mit schnellen Neutronen, I." *Strahlentherapie* 63 (1938): 517–27.

————. "N. W. Timoféeff-Ressovsky, 1900–1981." *Mutation Research* 106 (1982): 191–93.

————. *Radiumdosimetrie: Verfahren und bisherige Ergebnisse*. Leipzig: Georg Thieme Verlag, 1936.

————. *Strahlungen: Wesen, Erzeugung und Mechanismus der biologischen Wirkung*. Leipzig: Georg Thieme Verlag, 1937.

————. "The Target Theory." In *Phage and the Origins of Molecular Biology*, edited by J. Cairns, G. S. Stent, and J. D. Watson, 33–42. Cold Spring Harbor, NY: Cold Spring Harbor Press, 1966.

————. "That Was the Basic Radiobiology That Was: A Selected Bibliography and Some Comments." *Advances in Radiation Biology* 9 (1981): 411–67.

————. "Zur Berücksichtigung der 'biologischen Variabilität' bei der Treffertheorie der biologischen Strahlenwirkung." *Biologisches Zentralblatt* 61 (1941): 208–20.

————. "Zur treffertheoretischen Analyse der Antigenwirkung." *Die Naturwissenschaften* 30 (1942): 452–53.

Zimmer, K. G., and N. V. Timoféeff-Ressovsky. "Dosimetrische und strahlenbiologische Versuche mit schnellen Neutronen, II." *Strahlentherapie* 63 (1938): 528–36.

————. "Über einige physikalische Vorgänge bei der Auslösung von Genmutationen durch Strahlung." *Zeitschrift für induktive Abstammungs- und Vererbungslehre* 80 (1942): 353–72.

Archival Sources

Niels Bohr Library. Center for History of Physics, American Institute of Physics, College Park, MD.

Niels Bohr Microfilm Manuscripts. Niels Bohr Library and Archives, Center for History of Physics, American Institute of Physics, College Park, MD.

Niels Bohr Papers. Institute for Theoretical Physics, Copenhagen.

Max Delbrück Papers. California Institute of Technology.

Dessauer Papers. Kommission für Zeitgeschichte, Bonn.

James Franck Papers. Regenstein Library, University of Chicago.
Institut für Stadtgeschichte, Frankfurt am Main. Magistratsakten.
H. J. Muller Papers. Lilly Library, Indiana University.
Richard Prigge Nachlaß. Archivzentrum, Universitätsbibliothek Frankfurt.

Electronic Sources

Brigandt, I., and A. Love. "Reductionism in Biology." In *The Stanford Encyclopedia of Philosophy (Fall 2008 Edition)*, edited by E. N. Zalta. http://plato.stanford.edu/entries/reduction-biology/.
Harding, Carolyn. Interview with Max Delbrück, July 14–September 11, 1978. California Institute of Technology online Delbrück archives, http://oralhistories.library.caltech.edu/16/01/OH_Delbruck_M.pdf.
"Karl Günter Zimmer." Wikipedia article. http://en.wikipedia.org/wiki/Karl_Zimmer. Accessed April 8, 2008.
Lagemann, R. T., and W. G. Holladay. "Max Delbrück at Vanderbilt." In *To Quarks and Quasars: A History of Physics and Astronomy at Vanderbilt University*. Nashville, TN: Vanderbilt University Department of Physics and Astronomy, 2000. http://www.vanderbilt.edu/delbruck/documents/Lagemann_Delbruck_Chapter.pdf.
Pohlit, W. "Friedrich Dessauer 1881–1963." Web page. 1999. http://www.physik.uni-frankfurt.de/paf/paf84.html.
Rentetzi, M. *Trafficking Materials and Gendered Experimental Practices: Radium Research in Early 20th-Century Vienna*. New York: Columbia University Press, 2007. Electronic book, available at http://www.gutenberg-e.org/rentetzi/.

Unpublished Materials

Beyler, R. "From Positivism to Organicism: Pascual Jordan's Interpretations of Modern Physics in Cultural Context." PhD dissertation, Harvard University, 1994.
Selya, R. E. "Salvador Luria's Unfinished Experiment: The Public Life of a Biologist in a Cold War Democracy." PhD Dissertation, Department of the History of Science, Harvard University, 2002.
Yoxen, E. J. "The Social Impact of Molecular Biology." PhD dissertation, Department of History and Philosophy of Science, Cambridge University, 1978.

Contributors

RICHARD H. BEYLER is professor of history at Portland State University, where he teaches courses in history of science and European intellectual history and coordinates the department's graduate studies program. After receiving his doctorate in history of science from Harvard, he was a postdoctoral fellow at the Max Planck Institute for History of Science in Berlin and at the German Historical Institute in Washington before joining the PSU faculty in 1996. His current research focuses on the political relations of German science after World War II and on biophysics in the first half of the twentieth century.

BRANDON FOGEL is collegiate assistant professor in the Humanities Collegiate Division and a Harper Fellow in the Society of Fellows at the University of Chicago. He studied physics and English as an undergraduate at the University of Pennsylvania and received his doctorate in history and philosophy of science from the University of Notre Dame. His research specializes in history and philosophy of physics; particular focuses include Bell's theorem and Weyl and Einstein studies.

DANIEL J. MCKAUGHAN is assistant professor in the Philosophy Department at Boston College. He received his PhD in the history and philosophy of science from the University of Notre Dame. Prior to joining the faculty at Boston College, he taught at the Chesapeake Biological Laboratory for the University of Maryland Center for Environmental Sciences. His research specializes in philosophy of science, empiricism, and epistemology, and he has published articles on American pragmatism, science and values, and the historical foundations of molecular biology.

NILS ROLL-HANSEN is professor emeritus of history and philosophy of science, University of Oslo. His main research interest has been history of genetics, including its applied and political aspects. *The Lysenko Effect: The Politics of Science* (2005) investigates how certain science policy doctrines undermined the rationality and autonomy of science. *Eugenics and the Welfare State* (1995, 2005), co-edited with Gunnar Broberg, investigates eugenics and sterilization policy in Scandinavia. Present research focuses on the genotype theory of Wilhelm Johannsen in early-twentieth-century genetics and evolution.

PHILLIP R. SLOAN is professor emeritus in the Program of Liberal Studies and in the Graduate Program in History and Philosophy of Science at the University of Notre Dame. His research specializes in the history and philosophy of life science, and includes studies on Buffon, Darwin, Richard Owen, and the history of classification in biology. He is also the editor of *Controlling Our Destinies: Historical, Philosophical, Ethical, and Theological Perspectives on the Human Genome Project* (2000). He is currently completing a book on the conception of life in contemporary biophysics.

WILLIAM C. SUMMERS is professor of history of medicine, of molecular biophysics and biochemistry, and of therapeutic radiology, and he is in the Program of History of Science and Medicine at Yale University. His formal education at the University of Wisconsin included mathematics, molecular biology, and medicine, and he received the MD and PhD in 1967. His research has included molecular biology and genetics of cancer and viruses, history of medicine and science, and the relations between science and the humanities. He has taught and published on topics ranging from quantum mechanics to viral genetics to the biology of gender and sexuality, as well as East Asian studies. He is currently completing books on the geopolitics of plague in Manchuria and on the history of the American Phage Group.

Index